DATE DUE			

SECOND EDITION

Biochemistry and Physiology of Protozoa

Volume 1

CONTRIBUTORS

Philip B. Applewhite

Lesley J. Borowitzka

A. D. Brown

Annette W. Coleman

John D. Dodge

Elisabeth Gantt

T. W. Goodwin

S. H. Hutner

Hideo Iwasaki

Dag Klaveness

M. Levandowsky

Eystein Paasche

Graham W. Pettigrew

Edward J. Schantz

Frederick L. Schuster

Beatrice M. Sweeney

SECOND EDITION

Biochemistry and Physiology of Protozoa

Volume 1

Edited by

M. LEVANDOWSKY

S. H. HUTNER
Haskins Laboratories of Pace University
New York, New York

Consulting Editor

LUIGI PROVASOLI
Department of Biology
Yale University
New Haven, Connecticut

1979

ACADEMIC PRESS

A Subsidiary of Harcourt Brace Jovanovich, Publishers

New York   London   Toronto   Sidney   San Francisco

ACADEMIC PRESS, INC.
111 Fifth Avenue, New York, New York 10003

United Kingdom Edition published by
ACADEMIC PRESS, INC. (LONDON) LTD.
24/28 Oval Road, London NW1   7DX

Library of Congress Cataloging in Publication Data
Main entry under title:

Biochemistry and physiology of protozoa.

    First ed. published in 1951−64, entered under A. Lwoff.
    Includes bibliographies.
    1. Protozoa−−Physiology. I. Levandowsky, Michael.
II. Hutner, Seymour Herbert, Date   III. Lwoff,
Andre, Date   ed.   Biochemistry and physiology of
protozoa.
QL369.2.L87   1979   593'.1'041   78−20045
ISBN 0−12−444601−9 (vol. 1)

Contents

9 The Bioluminescence of Dinoflagellates
Beatrice M. Sweeney

10 Sexuality in Colonial Green Flagellates
Annette W. Coleman

11 Learning in Protozoa
Philip B. Applewhite

12 Physiological Ecology of Red Tide Flagellates
Hideo Iwasaki

Appendix to Chapter 12: On a Class of Mathematical Models for *Gymnodinium breve* Red Tides
M. Levandowsky

13 Poisonous Dinoflagellates
Edward J. Schantz

List of Contributors

Numbers in parentheses indicate the pages on which the authors' contributions begin.

Philip B. Applewhite (341), Biology Department, Yale University, New Haven, Connecticut 06520

Lesley J. Borowitzka (139), Roche Research Institute of Marine Pharmacology, Dee Why 2099, Australia

A. D. Brown (139), Department of Biology, University of Wollongong, Wollongong, N.S.W. 2500, Australia

Annette W. Coleman (307), Division of Biology and Medicine, Brown University, Providence, Rhode Island 02912

John D. Dodge (7), Department of Botany, Royal Holloway College, University of London, Egham Hill, Egham, Surrey, England

Elisabeth Gantt (121), Radiation Biology University, Smithsonian Institution, Rockville, Maryland 20852

T. W. Goodwin (91), Department of Biochemistry, University of Liverpool, Liverpool L69 3BX, England

S. H. Hutner (1), Haskins Laboratories of Pace University, New York, New York 10038

Hideo Iwasaki (357), Faculty of Fisheries, Mie University, Tsu, Japan

Dag Klaveness (191), Department of Marine Biology and Limnology, University of Oslo, Oslo, Norway

M. Levandowsky (1, 394), Haskins Laboratories of Pace University, New York, New York 10038

Eystein Paasche (191), Department of Marine Biology and Limnology, University of Oslo, Oslo, Norway

Graham W. Pettigrew* (59), School of Biological Sciences, Sussex University, Falmer, Brighton BN1 9QG, United Kingdom

* Present address: Veterinary Unit, Department of Biochemistry, Royal (Dick) School of Veterinary Studies, Summerhall, Edinburgh, EH9 1QH, United Kingdom.

Edward J. Schantz (403), Department of Food Microbiology and Toxicology, University of Wisconsin, Madison, Wisconsin 53706

Frederick L. Schuster (215), Department of Biology, Brooklyn College, CUNY, Brooklyn, New York 11210

Beatrice M. Sweeney (287), Department of Biological Sciences, University of California, Santa Barbara, California 93106

Preface to the Second Edition

This inaugurates, some 15 years after its predecessor, a multivolume second edition of "Biochemistry and Physiology of Protozoa." In this sense, in retrospect the preceding volumes (three in all) constitute a first edition, but as the intervals between the new volumes will be measured in months and a year or two rather than decades, and the new volumes have been planned as a whole, "second edition" seems fitting, and emphasizes that protozoology has vastly expanded in recent years and, by most evidence, will continue expanding.

The causes of this expansion are easily detected. That the gulf separating prokaryotes and eukaryotes seems evolutionarily the widest among extant organisms is unchallenged. Kluyverian unity of biochemistry remains firmly established, but is perceived in a perspective at once deeper and more practical. How easy it was to find drugs against prokaryotes, how cursedly hard and expensive to find them against the eukaryotic parasites—protozoa, fungi, and helminths!

The World Health Organization designates malaria, leishmaniasis, and trypanosomiasis as three of the six infectious diseases posing the most important global challenges. In the developed countries recognition of the grudging pace of progress with chronic diseases and aging is widening the demand for eukaryotic "models" which will be easier to handle than conventional laboratory animals. The more conspicuously animal protozoa are increasingly meeting this hunger for expeditious approaches to eukaryotic fundamentals; they will not be neglected in this new series nor will the pathogens (taking into account that several other volumes on parasitic protozoa have recently been published or are listed in press).

Advances in identifying molecular kinships have attracted biochemists (and other gentry not formally protozoologists) into the enterprise of building abutments for bridges between eukaryotes and prokaryotes. Fittingly, therefore, this edition leads off with an overview of phytoflagellate phylogeny.

The increase in knowledge of metabolic pathways and descriptive biochemistry permits more penetrating analyses of the protozoan equivalents of endocrinology, neurology, especially as manifest in behavior. I therefore am delighted to welcome as senior editor for this edition my colleague Dr. Michael Levandowsky. In doing so I follow the precedent set up by the founder of this enterprise, Andre Lwoff, when he invited me to serve as senior editor with him for Volume II of what was, in retrospect, a three-volume first edition, with long intervals between Volumes II and III. The pace has quickened; the old verities need new kinds of substantiations.

S. H. Hutner

Preface to Volume 1

The flux of information has swelled enormously since the first edition of this treatise (Lwoff, 1951; Hutner and Lwoff, 1955; Hutner, 1964) and the later comprehensive survey of Kidder (1967); the time seems ripe again for sifting and synthesis.

Science does not progress predictably; by definition, it advances through trial and error, with long quiet spells punctuated by bursts of activity. Protozoological metaphors come to mind, such as *Amoeba* sending out exploratory pseudopodia, or the biased random-walk progression of a swimming *Paramecium*. Therefore we have not tried to be comprehensive or grimly systematic in this collection, but have followed the story line as we see it, choosing topics of great activity or promise.

Obsessed with the human side of science, we are author-minded; we tried to choose interesting authors and to stay out of their way as much as possible, entering the fray from time to time to referee jurisdictional squabbles, or to rescue the English language.

We dedicate this multivolume second edition to Andre Lwoff, master craftsman, savant, artist, philosopher, and progenitor of this enterprise. We fancy that the very diversity of topics dealt with in these volumes is in keeping with the multifariousness yet scholarly depth of his work. What else is there to say about a man whose protozoological interests extend from marine astomatous ciliates, photosynthetic and colorless flagellates, to trypanosomes? Blame him, not us, for the diversity of topics here!

No great talent as generalist or seer is demanded to conclude that these new volumes reflect not merely the growth of protozoology, but recognition that biological man does not live by *E. coli* alone; that with that groundwork well laid, the time is ripe to confront the realities of our eukaryotic identity. And this too is in keeping with Lwoffian precedents, for he went from protozoology to bacteriology and virology. We dare conclude that, with due respect to the Master, the time is ripe for turnabout, and what would be more strategic at this stage in the development of biology, than to exploit the seeming simplicities of protozoa (perhaps often misleading, but that is for the future to decide), before pursuing knowledge of the metaphytan and metazoan cell, and the devices leading to multicellularity and to the evolution of consciousness.

We must thank D. P. Petrylak and C. C. Wang for helpful discussions of certain chapters, and the editors at Academic Press for general comments.

References

Hutner, S. H., ed. (1964). "Biochemistry and Physiology of Protozoa," 1st ed., Vol. III. Academic Press, New York.

Hutner, S. H., and Lwoff, A., eds. (1955). "Biochemistry and Physiology of Protozoa," 1st ed., Vol. II. Academic Press, New York.

Kidder, C., ed. (1967). "Chemical Zoology I: Protozoa." Academic Press, New York.

Lwoff, A., ed. (1951). "Biochemistry and Physiology of Protozoa," 1st ed., Vol. I. Academic Press, New York.

M. Levandowsky
S. H. Hutner

Contents of Other Volumes

* In preparation.

* In preparation.

Introduction

1

M. LEVANDOWSKY AND S. H. HUTNER

Our bias appears at the very beginning, in the title of this series: Who are the Protozoa? Many modern biologists prefer the category Protista and one can certainly criticize a classification that lumps extremely diverse organisms—ciliates, dinoflagellates, euglenids—while leaving several obviously monophyletic groups split between the protozoa and the algae. Why include *Dunaliella* and *Volvox,* but not *Scenedesmus, Chlorella,* or *Ulva*? Why discuss *Ochromonas* while ignoring the obviously related diatoms? Our reasons for retaining the old-fashioned classification reflect the physiologist's view of evolution, though they may not satisfy some with a passion-imbued sense of the phylogenetic fitness of things.

Protozoa, in this view, are unicellular eukaryotes trying to be animals. Now, what are animals? We think immediately of nutrition, and one of us went so far as to define animals by phagotrophy (Hutner, 1961). But that definition is much too narrow to justify the contents of these volumes, and we seek now a more fundamental criterion.

To most biologists animality implies behavior, motility. Motility in turn implies navigation, which requires sensory equipment and a decision-making system—a *transducer* between sensory input and behavioral output. Animals detect and respond to transient stimuli, constantly choosing from a behavioral repertoire. We therefore define them by their sensory and decision-making abilities—their latent intellects, in fanciful language.

Volvox and *Dunaliella,* phytoflagellates playing prominent roles in this volume, are plants nutritionally, with virtually no heterotrophic tendencies. They do behave though, responding rapidly and precisely to light and other stimuli. Evidence just surfacing suggests that these organisms, like other protozoa, have fundamental biophysical homologies with metazoan neuromuscular cells. Iwasa (personal communication; Sakaguchi *et al.,* 1977) finds that the photolocomotory response of *Volvox* is potassium-dependent; experiments with specific ionophores strongly suggest in-

1

BIOCHEMISTRY AND PHYSIOLOGY OF PROTOZOA
SECOND EDITION, VOL. 1

volvement of transient membrane changes—action potentials—in the transduction from light signal to swimming response.

The last statement, however, depends to some extent on speculative extrapolations from known behavioral physiology of better-studied though phylogenetically remote protozoa, especially ciliates (Naitoh and Eckert, 1974; also to be discussed in later volumes). That is, we have used our definition to generate a prediction, demonstrating thereby its usefulness to the physiologist.

At a less abstract level, the definition of animality may ultimately be reducible, in part at least, to ionic gating mechanisms in the cell membrane, reflecting a general latent eukaryotic capacity that has developed many times independently into a behavior-controlling principle. Prokaryotic harbingers probably exist: Smelcman and Adler (1976) reported indirect evidence of membrane potential changes in bacteria responding to chemical cues, and Ordal (1977) found that such behavior was calcium-dependent.

Nutritionally, parallel evolutionary trends can be seen in groups that are extremely far apart phylogenetically: chloromonads, chrysomonads, dinoflagellates, euglenids. In each, one finds a series of forms ranging by degrees from self-sufficient photoautotrophs to obligate heterotrophs devoid of chlorophyll. In habitats with large amounts of organic acids and other reduced carbon sources (e.g., sewage lagoons), one finds the heterotrophs *Polytoma* and *Polytomella,* and facultatively heterotrophic relatives such as *Chlamydomonas.* These belong to the acetate flagellates, a polyphyletic group of organisms with a certain biochemical unity that fascinated early students of protozoan nutrition (Lwoff, 1943; Pringsheim, 1963; Hutner and Provasoli, 1955; to be reviewed in a later volume). *Euglena* is an acetate flagellate also, with the additional capacity to incorporate protein pinocytotically (Kivic and Vesk, 1974). This ability in turn becomes a true phagotrophy in chlorophyll-less relatives such as *Peranema* and (more distant) *Bodo.* Via *Bodo,* a probable evolutionary link between euglenids and kinetoplastids, an even greater stage of nutritional dependency is reached in the blood and intracellular parasites (*Trypanosoma, Leishmania,* etc.)

These trends appear in all the eukaryotic groups, blurring phyletic distinctions. Carrying this view to a perhaps ridiculous extreme, one might say that even higher plants have animal-like tendencies which crop up from time to time, as evidenced in recent reports of uptake of macromolecules and particles by root hairs (e.g., Nishizawa and Mori, 1977; Risner, 1978).

To us, then, the protozoa are a fairly logical group, though phylogenetically helter-skelter. They share a complex of physiological traits and tendencies, forming a predictive category in a physiological taxonomy. We shall no doubt return in later volumes to this interesting opposition of phylogeny and physiology, which seems related to fundamental questions raised many years ago by the first editor of this series (Lwoff, 1943; see also Provasoli, 1938).

In this first volume, we have indeed essayed a phyletic reconnaissance from several very different vantage points. Dodge scrutinizes the phytoflagellates with a botanist's eye, and starts us off with a phylogenetic shrub based largely but not entirely on ultrastructure. At the molecular level, Pettigrew summarizes current work on cytochrome primary structure. The curious lack of phylogenetic meaning in these molecules, except at the highest taxonomic levels—proteins functioning on the very cutting edge of natural selection one would think—seems a suitably baffling conundrum with which to start. Goodwin next provides a symphony of formulas to summarize present knowledge of sterol and carotenoid distributions in the various groups—a subject needing review because of the recent appearance of many new techniques.

Gantt's review of phycobiliproteins in cryptomonads deals with another phylogenetic puzzle. Found elsewhere only in blue-green bacteria and in red algae, these odd molecules appear uncorrelated with other biochemical or ultrastructural features in their phyletic distribution.

From puzzles we turn to a striking case of physiologic adaptability, Brown and Borowitzka's chapter on halotolerance in *Dunaliella*. These astonishing phytoflagellates, which thrive in habitats ranging from weakly brackish waters to saturated brines, combine a virtually impermeable cell membrane with metabolic virtuosity in regulating glycerol levels, to produce a tightly controlled unicellular *milieu interieur*. But here too we have an interesting evolutionary conundrum: Why is *Dunaliella* so alone among phytoflagellates in this capacity? Why haven't the euglenids, chrysomonads, or dinoflagellates evolved the ability to colonize brines?

Klaveness and Paasche's description of coccolith formation gives a glimpse of the complexity of intracellular morphogenesis—a hint of ontogenetic mechanisms. Schuster's review of ameboflagellates is in the same vein. The dramatic morphogenetic response of these cells to changing external conditions will be much studied in the near future as the medical importance of pathogenic strains becomes well known; epidemiologic evidence now suggests that undetected outbreaks of amebic meningitis may not be rare at all. Phylogenetically, we venture to predict, on the rather skimpy evidence available, that a natural grouping of highly versatile tiny amebas will emerge, which will include not only ameboflagellates but also the ontogenetically virtuosic cellular slime molds (to which we return in a later volume).

Sweeney deals with a dramatic phenomenon, restricted to a single group of phytoflagellates: bioluminescence. Very little is known of either the biochemical basis or the adaptive value of this odd trait, despite extensive studies in several laboratories. However, Sweeney and her colleagues have opened up new avenues for future research. Another striking phenomenon often involving dinoflagellates is "red tides"—massive blooms of such dense populations that the water is discolored. Toxins produced in these sometimes devastate fisheries, causing severe economic and public health problems. Iwasaki discusses red tides from a physiologist's viewpoint, uncovering some interesting problems of biochemical ecology. Schantz describes the most severe and best known of the toxins, saxitoxin, produced by species of *Gonyaulax*. This is a neurotoxin that blocks sodium channels; chemically and pharmacologically it is remarkably similar to tetrodotoxin, produced by several vertebrates and a species of octopus—another puzzle. The evolutionary question here: What is the function of these toxins, and why do some, but not all red tide species produce them?

A curious suggestion was made recently. Saxitoxin and the gonyautoxins have purinelike structures, and Mickelson and Yentsch (1979) report evidence of a role in nucleic acid metabolism. Perhaps, as with some of the polyamines, membrane effects may prove a coincidental property of molecules whose primary metabolic roles are elsewhere, as cofactors. Are toxic red tides, then, simply biochemical accidents?

The beginnings of a multicellular organization are explored in Coleman's chapter on volvocids. Appropriately, from our point of view, the emphasis is on mating signals and hormonal control systems.

Indeed, if we define protozoa by their animality, in what way (other than their unicellularity) are they different from metazoa? Are there any broad distinctions at a biochemical level? One might perhaps seek these in the hormonal and neuronal tissues—peculiarly metazoan coordinating systems. However, the metazoan intra- and intercellular messenger molecules—neurotransmitters, hormones, cyclic nucleotides, polyamines— are also found in the protozoa, though in most cases their functional role there is unknown. Behavioral responses to such molecules are observed, and specific receptors appear to be present in at least some cases (see, e.g., Levandowsky and Hauser, 1978). Prostaglandins may be present in slime molds, where they are known to stimulate cAMP secretion (Coe and Kuo, 1976).

A possible biochemical distinction might be the hydroxamic acids, powerful and specific metal-binding compounds secreted by many aerobic prokaryotes and fungi. These appear to be absent from normal metazoan tissues; some hydroxamates are easily reduced to highly carcinogenic amines (Neilands, 1974). Whether they are produced by protozoa is not yet known, though, Van Baalen (1978) reports evidence of their presence in a diatom lipid extract.

In general, though, there seems to be more similarity than dissimilarity between metazoa and protozoa at the biochemical level. Even the peptide hormones may be shared. Csaba and Lantos (1973) have reported behavioral responses in *Tetrahymena* to low levels of insulin and glucagon, and a number of insulin responses have been reported in the alga *Acetabularia* (Bourgeois *et al.*, 1978).

But, given the rich behavioral repertoire and well-developed sensory abilities of many protozoa (to be explored in another volume), a biophysical question arises: Can they think? Applewhite recounts a gloomy tale of unsuccessful attempts to demonstrate learning in protozoa. Simple habituation has been demonstrated, and is not controversial, but attempts to reproduce experiments in which associative learning was claimed were unsuccessful. Is this a fundamental distinction? Does associative learning require more than one cell?

Some of these questions may be partly answered in following volumes, but only to be replaced by new ones.

REFERENCES

Bourgeois, P., Beckelandt, P., Henry, E.-E., and Legros, F. (1978). *In* "Progress in Acetabularia Research" (C. L. F. Woodcock, ed.), pp. 271–286. Academic Press, New York.

Coe, E. L., and Kuo, W. J. (1976). *Fed. Proc., Fed. Am. Soc. Exp. Biol.* **35,** 1367.

Csaba, G., and Lantos, T. (1973). *Acta Protozool.* **13,** 409–413.

Hutner, S. H. (1961). *Symp. Soc. Gen. Microbiol.* **11,** 1–18.

Hutner, S. H., and Provasoli, L. (1955). *In* "Biochemistry and Physiology of Protozoa" (S. H. Hutner and A. Lwoff, eds.), 1st ed., Vol. II, pp. 17–44. Academic Press, New York.

Kivic, P. A., and Vesk, M. (1974). *Arch. Microbiol.* **96,** 155–159.

Levandowsky, M., and Hauser, D. C. R. (1978). *Int. Rev. Cytol.* **53,** 145–210.

Lwoff, A. (1943). "L'Evolution Physiologique." Hermann, Paris.

Mickelson, C., and Yentsch, C. M. (1979). *In* "Toxic Dinoflagellate Blooms" (D. L. Taylor and H. H. Seliger, eds.), p. 495. Elsevier/North Holland, New York.

Naitoh, Y., and Eckert, R. (1974). *In* "Cilia and Flagella" (M. A. Sleigh, ed.), pp. 305–345. Academic Press, New York.

Neilands, J. B., ed. (1974). "Microbial Iron Metabolism." Academic Press, New York.

Nishizawa, N., and Mori, S. (1977). *Plant Cell Physiol.* **18,** 767–782.

Ordal, G. (1977). *Nature (London)* **270,** 66–67.

Pringsheim, E. (1963). "Farblose Algen." Fischer, Stuttgart.

Provasoli, L. (1938). *Boll. Zool. Agrar. Bachic.* **9,** 3–124.

Risner, R. J. (1978). *In Vitro* **14,** 334. (Abstr.)

Sakaguchi, H., Saito, S., and Iwasa, K. (1977). *J. Phycol.* **13,** Suppl., p. 59.

Smelcman, S., and Adler, J. (1976). *Proc. Natl. Acad. Sci. U.S.A.* **73,** 4387–4388.

Van Baalen, C. (1979). *J. Gen. Microbiol.* (in press).

The Phytoflagellates: Fine Structure and Phylogeny

2

JOHN D. DODGE

I. INTRODUCTION

To the protozoologist phytoflagellates form a fairly distinctive group of organisms which possess flagella and normally have plantlike nutrition as they contain chloroplasts. To the botanist or phycologist, however, the picture is not so straightforward, for although phytoflagellates constitute the main part of several algal classes (Cryptophyceae, Dinophyceae, Euglenophyceae, Chloromonadophyceae, Chrysophyceae, and Haptophyceae), they form only a portion of others (Chlorophyceae and Prasinophyceae). In addition most of the groups which are basically flagellate

7

BIOCHEMISTRY AND PHYSIOLOGY OF PROTOZOA
SECOND EDITION, VOL. 1

also contain some members which are nonmotile (e.g., coccoid) or filamentous, although flagellate stages may be involved in reproduction.
There is also the problem of the algal classes, Phaeophyceae, Xanthophyceae, Bacillariophyceae, Eustigmatophyceae, and the larger part
of the Chlorophyceae, where flagellates are restricted to zoospore stages
or gametes. To botanists these classes provide a very useful basic cross-
reference to the similar level of organization in the flagellate groups. In
this chapter the emphasis will be placed on the eight classes which consist
mainly of flagellates, but mention will be made of the others when necessary for phylogenetic discussion.

Botanical nomenclature will normally be utilized but, to assist readers

Table I Classification of the Phytoflagellate Groups

Phylum or division	Class	Orders containing flagellates	Protozoological classification: Order[b]
	Botanical classification[a]		
Chlorophyta	Chlorophyceae (in part)	Volvocales	Volvocida
	Prasinophyceae	Pedinomonadales	
		Pterospermatales	
		Pyramimonadales	
		Prasinocladiales	
Haptophyta	Haptophyceae (Prymnesiophyceae)	Isochrysidales	
		Coccosphaerales	Coccolithophorida (in part)
		Prymnesiales	
		Pavlovales	
Chrysophyta	Chrysophyceae	Ochromonadales ⎱ Chromulinales ⎰	Chrysomonadida
		Dictyochales	Silicoflagellida
	Chloromonadophyceae		Chloromonadida
Euglenophyta	Euglenophyceae	Eutreptiales[c]	
		Euglenales	
		Rhabdomonadales	Euglenida
		Sphenomonadales	
		Heteronematales	
Dinophyta (Pyrrophyta)	Dinophyceae	Prorocentrales	
		Dinophysiales	
		Gymnodiniales	Dinoflagellida
		Noctilucales	
		Peridiniales	
Cryptophyta	Cryptophyceae	Cryptomonadales	Cryptomonadida

[a] Mainly taken from Parke and Dixon (1976).

[b] Class, Phytomastigophorea; superclass, Mastigophora; phylum, Protozoa. After Honigberg *et al.* (1964).

[c] After Leedale (1967).

with a zoological background, Table I lists the scheme used with the nearest protozoological equivalents (after Honigberg *et al.,* 1964). Table II gives the botanical names of the groups which are basically nonflagellate and which therefore do not have protozoological names.

All phytoflagellates are eukaryotic free-living organisms, and most are unicellular, although a range of colonial types are found in the Chlorophyceae and Chrysophyceae, and palmelloid phases may be found in the Chloromonadophyceae and Cryptophyceae. Every cell is provided with at least one of each of the basic organelles necessary for life in what must inevitably be an aquatic environment. Thus, each cell has at least one flagellum, and there may be as many as six. The basic number is two, but in some species only one emerges from the cell.

There is normally one nucleus, although two are found in some dinoflagellates, and one or more chloroplasts which may be occasionally reduced to leucoplasts or may even be entirely lacking. Each cell has one to several Golgi bodies, a mitochondrial reticulum or several mitochondria, microbodies, endoplasmic reticulum with ribosomes, and microtubules. Additional components which may or may not be present are ejectile organelles, eyespots, and contractile vacuoles. The cell is covered

Table II Classification of the Algal Groups with Flagellate Reproductive Stages

Phylum or division	Class	Main orders	Main life form
Chlorophyta (in part)	Chlorophyceae (in part)	Prasiolales	Parenchymatous
		Chlorococcales	Coccoid
		Ulotrichales	Filamentous
		Ulvales	Parenchymatous
		Chaetophorales	Filamentous
		Cladophorales	Filamentous
		Dasycladales	Siphonous
		Codiales	Siphonous
		Caulerpales	Siphonous
		Oedogoniales	Filamentous
	Charophyceae	Charales	Filamentous
Eustigmatophyta	Eustigmatophyceae		Coccoid or filamentous
Chrysophyta (in part)	Xanthophyceae	Heterochloridales	Ameboid
		Mischococcales	Coccoid
		Tribonematales	Filamentous
		Vaucheriales	Siphonous filaments
	Phaeophyceae	Ectocarpales	Filamentous
		Laminariales	Parenchymatous
		Sphacelariales	Filamentous
		Fucales	Parenchymatous
	Bacillariophyceae	Centrales	Unicellular, filamentous
		Pennales	Unicellular, colonial

by a basic plasma membrane, in addition to which there may be scales or spines, a cell wall, or a complex pellicle or theca. Phagotrophy is found in some phytoflagellates, and in some the cell moves by ameboid or euglenoid movement.

In the account that follows the basic fine structure of each group will first be described, and this is followed by comparative accounts of the flagella, chloroplasts, and nuclei, which appear to be the most valuable structures for use in the subsequent attempt to derive a phylogenetic system for these organisms.

II. FINE STRUCTURE

A. Chlorophyceae

The flagellate members of this class consist of unicells or colonies with numbers varying from about eight to several hundred. The basic cell normally bears two equal-length anteriorly inserted flagella, which are substantially smooth (Figure 8), and contains a large cup-shaped chloroplast, one nucleus, a pyrenoid, an eyespot, and a Golgi region. There are no unusual organelles or structures.

Many members of the class have been studied by electron microscopy, but various species of *Chlamydomonas* more so than any other (e.g., Lembi and Lang, 1965; Ringo, 1967b; Ettl and Green, 1973). The marine flagellate *Dunaliella,* with an almost identical structure apart from the absence of a cell wall, has also been studied (references in Eyden, 1975), as have the apochlorotic chlamydomonadlike cells of *Polytoma* (Sui *et al.,* 1976) and *Polytomella* (Brown *et al.,* 1976). Rather little work has been carried out on the colonial members for, although several received early attention, *Volvox* has been the only one studied in any detail (Pickett-Heaps, 1970; Kochert and Olson, 1970; Olson and Kochert, 1970; Soyer, 1973).

Flagella in the Chlorophyceae and particularly in *Chlamydomonas* were the first algal flagella to be studied in detail, and Ringo (1967a,b) gave the first thorough account of the internal arrangement of the tubules of the axoneme and basal body, and of the stellate structure of the transition zone. The smaller axonemal components, such as subsidiary fibres, attached to the doublet tubules have been described for *Chlamydomonas* (Hopkins, 1970), as have basal body development and flagellar regression and growth during the cell cycle (Cavalier-Smith, 1974). Recently the structure of the flagella and their microtubular root systems have been described for *Polytomella* (Brown *et al.,* 1976).

Cells of many members of this class are surrounded by a cell wall that was traditionally thought to consist of cellulose. Recent work has shown conclusively that the wall in *Chlamydomonas* and some related flagellates consists of a glycoprotein material arranged in the form of a crystalline lattice (Hills *et al.*, 1973; Roberts, 1974).

The chloroplast of the chlorophyll-containing members of this group tends to occupy most of the posterior end of the cell and to contain a large central pyrenoid. The structure of the chloroplast and its development following growth in the dark have been studied in detail in *Chlamydomonas* (Ohad *et al.*, 1967a,b). As in higher plants, the chloroplast envelope consists of two membranes, and the thylakoids are rather irregularly arranged, in places being stacked together to form small grana with many connections from one granum or layer to those adjacent. In apochlorotic members of the group reduced chloroplasts or leucoplasts have been detected, and these contain starch grains, DNA aggregates, and ribosomes, but no thylakoids (Sui *et al.*, 1976).

Most flagellate members of the Chlorophyceae contain eyespots, and these are generally situated within the superficial layers of the chloroplast at one side of the cell [for references, see Dodge (1969a, 1973)]. Normally the eyespot consists of two layers of carotenoid-containing globules, but up to nine layers have been reported in colonial flagellates. In these algae there is no direct connection between the eyespot and the flagella, although it is possible that a microtubular root or the membrane lying over the eyespot passes impulses to the flagella.

Nuclear division and cell division have, so far, been studied mainly in *Chlamydomonas* (see Section II,C). Following mitosis the spindle tubules disappear, but microtubules appear in a band indicating the future cleavage plane. The cell then divides, with microtubules extending from both basal bodies along the sides of the cleavage furrow (Johnson and Porter, 1968). Meiosis takes place in the zygote, and here the first stage (Triemer and Brown, 1977) is the appearance in the nucleus of axial cores and the condensation of the chromosomes, which soon become associated with the nuclear envelope. Later, at pachytene, typical synaptonemal complexes are formed and when these separate at diplotene the chromosomes are in their most condensed state. The subsequent stages appear similar to mitosis, with polar fenestrations of the nuclear envelope allowing development of the spindle. Following anaphase a cleavage furrow forms, and the zygote divides into two before the second division takes place.

In both *Chlamydomonas* (Schötz, 1972) and *Polytomella* (Burton and Moore, 1974) serial sectioning and reconstruction have been carried out to investigate the three-dimensional structure of the mitochondria. This reveals that there is usually only one, though sometimes 4 to 10 mitochondria, rather than the large number deduced from the profiles seen in thin sections. Microbodies have been found in several members of the Chlorophyceae (Silverberg, 1975a,b).

B. Prasinophyceae

The organisms now placed in the class Prasinophyceae were formerly members of the Chlorophyceae and mainly of the order Volvocales. The group consists mainly of unicells which have scaly flagella and distinctive internal features. It seems likely that most members of this group have benthic stages of their life history. A considerable number of genera and species have been studied in the last few years (see list in Dodge, 1973; Belcher *et al.*, 1974, on *Asteromonas*; Moestrup and Thomsen, 1974; Norris and Pearson, 1975; Pennick *et al.*, 1976, on *Pyramimonas*; Manton, 1975, on *Scourfeldia*).

The flagella, which number two to six, are normally inserted into a flagellar pit at the anterior end of the cell. They are uniform in size and structure, are covered with two layers of close-fitting scales, and may have hairlike appendages (Figure 7). In *Pyramimonas* (Norris and Pearson, 1975) the flagellar scales of the inner layer are five- or six-sided with knobs at the periphery and centre, and are arranged in regular spiral rows. The outer scales are very much larger, irregular polygons with no rim but a long central spine. These overlap one another to a considerable extent and are arranged in straight rows. The hairs can be regarded as a third type of scale, as they appear to be easily detached. In *Heteromastix* (Manton *et al.*, 1965) and *Platymonas* (Manton and Parke, 1965) the flagellar scales of the two layers are more equal in size but differ in shape, those of the inner layer being square and those of the outer layer stellate. Internally, the flagella have a normal 9 + 2 arrangement of axonemal tubules, but at the transition zone some variations may exist. A complex stellate structure has been reported for *Heteromastix* (Manton, 1964a; Manton *et al.*, 1965), which is similar to that found in *Chlamydomonas* (Ringo, 1967b), but Moestrup and Thomsen (1974) have suggested that *Pyramimonas* also has a coiled fibre at the base of the free part of the flagellum like that found in some members of the Chrysophyceae (see Section II,A).

A feature of taxonomic importance is provided by the flagellar root system. As usual, there are microtubules which run from the flagellar bases to the cell surface. There is also a banded structure, termed a "synistosome" by Norris and Pearson (1975), which connects the bases. A broad, banded root or rhizoplast runs directly from the bases and splits to continue on either side of the nucleus (Figure 1). The nucleus appears

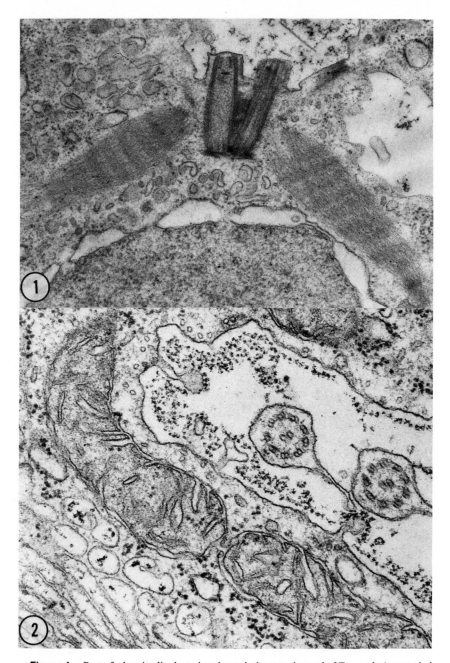

Figure 1. Part of a longitudinal section through the anterior end of *Tetraselmis tetrathele* (Prasinophyceae) at the commencement of division. Note the two flagellar bases from which broad striated roots run toward the nucleus. Magnification: ×32,000.

    Figure 2. The flagellar depression in a recently divided cell of *Tetraselmis*. The small wall particles are collecting inside the depression after they have been formed in Golgi vesicles (lower left). Magnification: ×49,000. (Figures 1 and 2 courtesy of B. R. Oakley.)

to be drawn up into a projection between the branches of the rhizoplast, but in *Pyramimonas parkeae* there is a further modification in the form of a large microbody which is enclosed by the rhizoplast (Norris and Pearson, 1975).

The cell body, like the flagella, of many members of this group is covered with scales, usually in three layers. The inner layer consists of very small, regularly arranged, squarish scales, and over these lie delicate "lacelike" scales which may form two layers (Swale, 1973). The scales are formed in the Golgi cisternae and when mature accumulate in the anteriorly placed scale reservoir. It has been shown that in *Pyramimonas* all six types of scales can be found in one reservoir, but before discharge onto the cell surface groups of similar scales appear to be collected together (Moestrup and Thomsen, 1974). In some members of this class, such as *Platymonas* (=*Tetraselmis*), the cell is covered by a loose-fitting theca. After division, Golgi cisternae produce large numbers of minute stellate bodies which are secreted from the anterior end of the cell, and these fuse together to form the theca (Manton and Parke, 1965) (Figure 2).

The internal structure of the cell is often very similar to that of *Chlamydomonas*. There is a cup-shaped chloroplast with an eyespot on one side normally consisting of one or two layers of pigmented globules, although six layers have been reported from *Pyramimonas montana* (Maiwald, 1971). The chloroplast encloses a large pyrenoid which is usually penetrated by either protrusions of the cytoplasm, in *Platymonas,* or of the nucleus, in *Prasinocladus* (Manton and Parke, 1965).

Interphase nuclei in the Prasinophyceae are situated in the centre or towards the anterior end of the cell. They contain a large nucleolus and rather evenly granular chromatin. Nuclear division has been studied in two species of *Platymonas* (Stewart *et al.,* 1974; Oakley and Dodge, 1974a) and in *Pyramimonas parkeae* (Pearson and Norris, 1975). In the latter the first indication of mitosis is the extension of the single microbody across the cell and the associated division of the chloroplast. The flagella and Golgi also replicate, and the basal body complexes separate to lie at positions anterior to the poles of the nucleus. In *Platymonas* microtubules appear in the cytoplasm at prophase, possibly being initiated by the rhizoplasts; then, whilst the chromatin condenses into a rather amorphous metaphase plate, most of the nuclear envelope disappears. The spindle area, delimited by inflated endoplasmic reticulum, is rather cylindrical with broad ends, and no structures have been found associated with the microtubules, although in *Pyramimonas* they seemed to end at the rhizoplast (Pearson and Norris, 1975). Anaphase remains something of an enigma, but telophase gives rise to daughter groups of chromosomes surrounded by a reformed nuclear envelope. Cytokinesis proceeds immediately, starting at the anterior end of the cell. Ducts and vesicles containing scales appear to coalesce and, no doubt guided by the microtubules which extend from the basal bodies to the posterior of the cell, the cleavage furrow advances until the cell is bisected. In *Pyramimonas* the cells remain motile throughout division, but in *Platymonas* free parts of the flagella are lost at the onset of mitosis and motility is not regained until after cytokinesis. Cell division in *Heteromastix* has recently been described (Mattox and Stewart, 1977) and is almost identical to that of *Platymonas,* apart from the presence of a persistent interzonal spindle during cytokinesis.

Two members of this class, *Pyramimonas parkeae* (Norris and Pearson, 1975) and *P. grossi* (Manton, 1969), possess trichocysts. These consist of a coil of membrane material rather like that found in cryptomonads. On discharge a tubular thread is extruded from the cell.

C. Euglenophyceae

In the Euglenophyceae we have an assemblage of distinctive organisms which combine in various degrees the characteristics of plants and animals. Thus, some appear to be normally autotrophic despite moving by euglenoid motion, whereas others entirely lack chloroplasts. A number of features such as flagella structure and insertion, cell covering or pellicle, and nuclear type unite this group.

Many studies of the general ultrastructure of euglenoids have been published. These include a survey of various mainly nonpigmented genera (Mignot, 1966) and of mainly pigmented ones (Leedale, 1967): *Euglena spirogyra* (Leedale *et al.,* 1965); *Eutreptiella* (Throndsen, 1973); *Menoidium* and *Rhabdomonas* (Leedale and Hibberd, 1974); *Khawkinea* (Schuster and Hershenov, 1974); *Cyclidiopsis* (Mignot, 1975); *Trachelomonas* (Dodge, 1975a).

There is a basic complement of two flagella, but only one of these may be emergent from the reservoir, as in the genus *Euglena.* The free parts of the flagella have an elaborate internal structure composed of a standard axoneme plus a complex paraflagellar rod (Mignot, 1966). Externally the flagella bear a single row of very delicate hairs 2–3 μm long (Leedale, 1967) (Figure 6), and there may be a tuft of hairs at the tip (Dodge, 1975a). Dynesius and Walne (1975) suggest that the hairs are bipartite, with a thicker fibrous base from which the thin hair arises. Mignot *et al.* (1972) used various cytochemical tests to study the composition of flagella hairs from various flagellates and showed that those of euglenoids consisted of glycoproteins. Near the base of the emergent flagellum in *Euglena,* and some other genera, is situated a swelling which contains paracrystalline material with a regular repeating pattern (cf. Walne and Arnott, 1967; Piccinni and Omodeo, 1975; Kivic and Vesk, 1972). The second flagellum is often reduced to a short stump, which does not emerge from the flagellar canal and appears to lack the central axonemal tubules (Leedale and Hibberd, 1974; Dodge, 1975a).

The flagella are inserted at the base of a deep invagination of the anterior end of the cell, termed the reservoir. The distal portion or flagellar canal is narrow and surrounded by both transverse and longitudinal bands of microtubules, some of which are in continuity with those that lie beneath the pellicle (Willey *et al.,* 1973; Dodge, 1975a). A detailed reconstruction of this region has been presented by Leedale and Hibberd (1974). Beneath the canal the reservoir widens out and here it is generally surrounded only by longitudinal microtubules, although in some genera such as *Menoidium* a complex "scroll" of tubular and striated material is also present (Leedale and Hibberd, 1974). Situated at one side of the reservoir is the contractile vacuole, an ephemeral single-membrane lined chamber into which small vesicles discharge. The opposite side of the reservoir is usually occupied by the eyespot, which consists of a number of lipid globules containing red carotenoid pigment. The globules are not normally enclosed by a membrane, as is the case in most other flagellates, and the eyespot does not therefore appear to be homologous with a reduced plastid. In some euglenoids which have bleached chloroplasts the eyespot persists (Palisano and Walne, 1976). Chloramphenicol has been found to inhibit the synthesis of eyespot pigment, although it does not destroy that already present (Kivic and Vesk, 1972).

One of the most characteristic features of the Euglenophyceae is the cell covering or pellicle. In most genera this takes the form of a system of ridges and furrows which spiral or run longitudinally along the cell. The outer surface is formed by the plasmalemma, and immediately beneath this lies a series of complex interlocking proteinaceous strips under which groups of microtubules lie parallel to the strips. Mucilage glands are also present (Leedale, 1964; Mignot, 1965). The form and thickness of the pellicular strips vary with the genera and their mode of existence. For example, in *Trachelomonas,* where the cell lives within a complex ferruginous lorica, the pellicle is delicate and the proteinaceous strips are much reduced (Dodge, 1975a; Leedale, 1975). In *Rhabdomonas* and *Menoidium,* which have rigid cells instead of the normal flexible euglenoid cell, the pellicle is smooth and consists of a continuous layer of amorphous material lying beneath the plasmalemma (Leedale and Hibberd, 1974).

The arrangement of pellicular strips and the close correlation of microtubules with them have been studied in detail for *Euglena gracilis* (Guttman and Ziegler, 1974), using both scanning and transmission electron microscopy. The development of the extracellular lorica in a number of species of *Trachelomonas* has been described by Leedale (1975). These complex structures appear to be formed in part by mucilage secretion from the pellicle and in part by mineral deposition from the surrounding environment.

Many members of the Euglenophyceae possess chloroplasts, and these have been the subject of much study and experimentation. The chloroplast is bounded by a triple-membrane envelope (Leedale, 1967; Ploaie, 1971), a feature shared only with the Dinophyceae, and contains lamellae which generally consist of thylakoids arranged in threes. The development of *Euglena* chloroplasts from the proplastids found in dark-grown cells and bleached cells (Kivic and Vesk, 1974; Palisano and Walne, 1976) has been studied in detail (Klein *et al.*, 1972), and changes during the cell cycle have been reported (Orcival-Lafont and Calvayrac, 1974).

In some species pyrenoids are found as part of the chloroplast. These may be either of the single-stalked type, as in *Trachelomonas* (Dodge, 1975a), or formed within the plastid and suspended by at least two branches of it, as in *E. gracilis* (Klein *et al.*, 1972).

The euglenoid nucleus has long been known to be unusual, and electron microscopy confirms this. In the interphase state (Figure 14) there is a typical double nuclear envelope and a large granular nucleolus. The chromatin is arranged in two types of pattern with a scattered granular matrix and distinct dense, granular bodies which represent the chromosomes (Leedale, 1968). In nuclear division as described by Leedale (1968) for *Euglena,* the nucleus first migrates towards the anterior end of the cell, and then microtubules appear within the nucleus, which becomes somewhat elongated. At metaphase the chromosomes are arranged in a loose equatorial band around the extended nucleolus (often termed the endosome). In anaphase the nucleus elongates further, and bundles of microtubules run from pole to pole, but these have not been seen to make any connection with chromosomes or with any polar structure. The chromosomes segregate slowly to opposite poles, and the nucleolus splits in two. All through division the nuclear envelope remains intact. In *Euglena* there is no obvious involvement of the flagellar bases in the division, but in *Astasia* Sommer and Blum (1965) thought that they probably acted as "division centers."

Euglenoid flagellates contain normal Golgi bodies which are strikingly large and situated near the anterior end of the cell. The mitochondrial profiles are scattered around the cell and have rather distinctive disk-shaped cristae. A few members of the group possess trichocysts, such as the complex tubular structures of *Entosiphon* that appear to consist mainly of polysaccharides (Mignot and Hovasse, 1973).

D. Chloromonadophyceae

This small group of flagellates has a number of distinctive features which have been more definitely established and confirmed by electron microscopy. Studies by Heywood (1972, 1973) and Mignot (1976) have shown that the flagella conform to the heterokont organization as found in the Chrysophyceae and Xanthophyceae. The anterior flagellum bears two rows of tubular hairs or mastigonemes with characteristic tapered bases and cylindrical 17-nm-diameter shafts. Heywood (1972), in a detailed study of these flagellar hairs, showed that they originated in endoplasmic reticulum vesicles which then migrated towards the base of the flagellum, where they appeared to discharge their contents into the base of the flagellar groove. The posterior flagellum is smooth and bears no hairs.

The flagella are anchored in the cell by a rather elaborate root system which has been studied in *Vacuolaria* (Heywood, 1972), *Gonyostomum* (Mignot, 1967), and *Chattonella* (Mignot, 1976). As in many flagellates, there are branches of the root system, composed of microtubules, which run from the base of the flagellum under the cell membrane. The distinctive feature is a rhizoplast composed of a banded, fibrillar root and microtubules, which runs directly to the nucleus and is attached to the anterior end of the nuclear envelope in such a way that the nucleus becomes drawn out and pointed.

Cells of members of this group are entirely naked, being bounded only by a single membrane. Within the cell the cytoplasm is arranged in two distinct zones. The broad outer layer consists of vacuoles among which lie numerous chloroplasts, often with their long axes radially aligned. The inner cytoplasmic zone is much more dense and consists of the nucleus and associated Golgi, mitochondria, and ribosomes.

To describe the structure of the chloroplasts presents a problem because already we know of two different types of construction. First to be described was *Vacuolaria,* which Mignot (1967) showed had a typical chrysophyte organization with parallel lamellar bands consisting of three thylakoids, interconnections between lamellae, a girdle lamella enclosing the ends of the lamellae, and an envelope of four membranes bounding the plastid; Heywood (1977) has recently confirmed this description. In the brackish-water species *Chattonella* (Mignot, 1976) the plastids have some of the features described above but appear to lack girdle lamellae and have pyrenoidal areas taking up the portion of the chloroplast towards the centre of the cell. No eyespots have been reported in this group.

The nucleus is large and more or less central in position. In the interphase condition it contains a nucleolus and scattered granular chromatin (Heywood, 1976). During division the nuclear envelope remains intact until anaphase, when it partly breaks down (Heywood, personal communication). The large chromosomes have distinctive kinetochore attachments to the spindle microtubules (Heywood and Godward, 1972).

A noteworthy feature of the fine structure of members of this group is provided by the ring of closely packed Golgi bodies which ensheath the anterior end of the nucleus (Mignot, 1967; Koch and Schnepf, 1967). The forming faces of the Golgi are adjacent to the nuclear membrane, but towards the cytoplasm numerous vesicles, both rounded and flattened, are apparent. Associated with the Golgi zone and situated between it and the flagellar bases is a contractile vacuole (Schnepf and Koch, 1966). Outer vacuoles from the Golgi were found to fuse to form the contractile vacuoles which, when fully distended, discharged to the outside of the cell. In the peripheral cytoplasm of *Chattonella* there are a number of ejectile organelles termed mucocysts. These are lanceolate bodies containing a fibroreticular material (Mignot, 1976). *Gonyostomum* contains more elaborate trichocysts which discharge to give a mucous filament (Mignot, 1967; Mignot and Hovasse, 1975).

E. Chrysophyceae

The Chrysophyceae, as originally envisaged, contained all organisms with golden-yellow chloroplasts. More recently, organisms with smooth anterior flagella have been removed to the Haptophyceae, which leaves the Chrysophyceae a much more homogeneous class. A considerable number of genera have been studied by electron microscopy and, in addition to the 13 already listed (Dodge, 1973), there have been studies on *Anthophysa* (Belcher and Swale, 1972b), *Phaeaster* (Belcher and Swale, 1971), *Paraphysomonas* (Manton and Leedale, 1961; Pennick and Clarke, 1973; Thomsen, 1975), *Sphaleromantis* (Pienaar, 1976b), *Spumella* (Mignot, 1977), *Syncrypta* (Clarke and Pennick, 1975), and *Uroglena* and *Uroglenopsis* (Wujek, 1976), and a recent survey of the whole class by Hibberd (1976).

There are two flagella inserted in the anterior end of the cell, which is usually depressed or obliquely terminated. The flagella are arranged at an angle of 90°, or greater, to each other such that one is normally directed forwards and the other to the side or backwards. The anterior flagellum is usually long, at least as long as the cell body, whereas the other is normally shorter and may be reduced to a nonemergent stump. The long flagellum is distinctive in bearing two rows of long hairs or mastigonemes which have three main parts: a tapered base, a long tubular shaft with lateral filaments, and fine terminal hairs (Bouck, 1971; Hill and Outka, 1974). In both *Synura* (Hibberd, 1973) and *Mallomonas* (Zimmermann, 1977) it has been shown that there are small oval scales on the long flagellum. These unmineralized scales are loosely attached and difficult to observe in the electron microscope, so it is not as yet clear whether they are found throughout the class. The second flagellum has no hairs but, just after it leaves the cell, bears a swelling which lies in a depression of the cell surface situated just over the eyespot (Figures 3 and 4). The flagellar swelling contains electron-dense material, but this does not normally appear paracrystalline, as in the Euglenophyceae.

Internally both flagella have the normal axonemal structure, but the transition region, immediately above the basal plate, contains a helix of dense material situated between the peripheral doublets and the central tubules (Bouck, 1971; Casper, 1972; Fuchs and Jarosch, 1974; Hibberd, 1976). The flagella are anchored in the cell by two types of roots (Schnepf *et al.*, 1977). One, similar to that found in several other flagellates, consists of a group of microtubules running under the cell surface, and the other root or rhizoplast is a striking banded structure which runs directly from the basal bodies to the nucleus, where it spreads over the nuclear envelope (Bouck and Brown, 1973; Hibberd, 1976). A recent study of *Poterioochromonas* has shown that it has no striated root, but the flagella are anchored by six root fibers (Schnepf *et al.*, 1977). Here, both the chloroplast and a kinetosomal mitochondrion are attached to the root system. In *Dinobryon* a banded root runs from the flagellar bases to the eyespot (Kristiansen and Walne, 1976, 1977).

Chrysophycean cells exhibit a variety of cell coverings. Some, such as the silicoflagellates, are entirely naked, for here the silica spicule is situated within the cell (van Valkenburg, 1971). *Ochromonas* species are also naked, and the function of microtubules in cell support has been investigated in detail (Bouck and Brown, 1973). Many members of the group have silica scales which clothe the cell membrane. These are not formed in the Golgi body, as are scales of the Haptophyceae and Prasinophyceae, but in derivatives of the endoplasmic reticulum (Schnepf and Deichgräber, 1969). The morphology of the scales has been studied for numerous species (for references, see Thomsen, 1975, *Paraphysomonas*; Pienaar, 1976b, *Sphaleromantis*; Zimmermann, 1977; Peterfi and Momen, 1976, *Mallomonas*). In *Syncrypta glomifera* a completely different type of scale has been reported (Clarke and Pennick, 1975) with a hollow ellipsoidal form consisting of a framework of fine fibres.

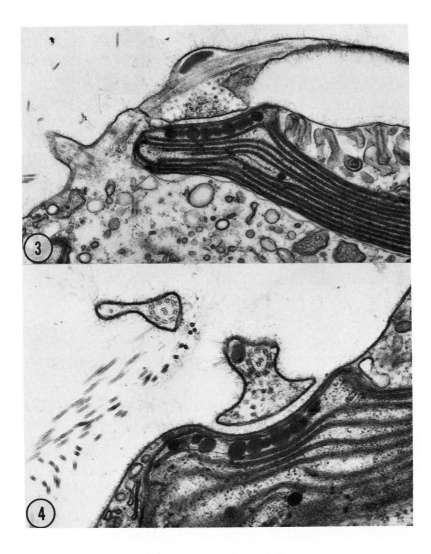

Figure 3. Longitudinal section through the anterior end of *Ochromonas tuberculatus* (Chrysophyceae) showing the flagellar swelling situated above the eyespot which is part of a chloroplast. Magnification: ×20,000.

Figure 4. Section at right angles to Figure 3, showing the flagella and eyespot cut transversely. A number of mastigonemes are seen at the left. Magnification: ×35,000. (Figures 3 and 4 courtesy of D. J. Hibberd; from Hibberd, 1970, reproduced by permission.)

Many members of the group have a cup or lorica loosely surrounding the cell. These are of many types (cf. Kristiansen, 1969, 1972) and appear to normally consist of cellulose fibers (Kramer, 1970; Franke and Herth, 1973).

There are usually two chloroplasts per cell, and these contain three-thylakoid lamellae with a girdle lamella enclosing the ends and also delimiting the DNA zones (Gibbs *et al.*, 1974). The chloroplasts are surrounded by endoplasmic reticulum, and sometimes the inner membrane is expanded to form a periplastidial reticulum that lies between the chloroplast envelope and the endoplasmic reticulum. Pyrenoids are occasionally present and are usually completely immersed in the chloroplast. In some cases the pyrenoid is traversed by thylakoids which are unusually sinuous in *Chrysococcus cordiformis* (Belcher and Swale, 1972a), but in *Phaeaster* (Belcher and Swale, 1971) and *Chrysamoeba* (Hibberd, 1971) the pyrenoid is penetrated by invaginations of the periplastidial space. Most members of the class have an eyespot which consists of a single layer of pigmented droplets situated at the anterior edge of a chloroplast (Figures 3 and 4). Frequently there is a depression over the eyespot associated with the flagellar swelling.

The single interphase nucleus in chrysomonads is of typical appearance, with a regularly poroid envelope (Cole and Wynne, 1973), a large nucleolus, and finely scattered chromatin. Division of the nucleus has, as yet, been studied only in *Ochromonas danica* (Slankis and Gibbs, 1972; Bouck and Brown, 1973) (see Section II,C).

Other organelles and structures found in chrysophycean cells include a single large Golgi body, which is found adjacent to the nucleus at the anterior end of the cell (Cole and Wynne, 1973), leucosin vesicles at the posterior end of the cell, and a contractile vacuole which has rather unusually rectangular collecting vesicles (Aaronson and Behrens, 1974). Uniquely structured ejectile organelles called discobolocysts have been found in one species of *Ochromonas* (Hibberd, 1970).

Some members of this group, in particular the much-studied *O. danica,* perform endocytosis, which probably amounts to phagotrophy. The types of particles which can be ingested and the appearance of the digestive vesicles have been extensively reported (Aaronson, 1973, 1974; Cole and Wynne, 1974), and in *Ochromonas malhamensis* they were found to be entirely determined by what was available (Dubowsky, 1974; Daley *et al.*, 1973).

We should not leave the Chrysophyceae without mention of two unusual flagellates currently placed in this class: *Pedinella hexacostata* (Swale, 1969) and *Apedinella spinifera* (Throndsen, 1971). In both of these the cell structure is reasonably typical of the Chrysophyceae, but the single flagellum is winged and quite distinctive.

F. Haptophyceae (Prymnesiophyceae)

This class was formed by Christensen (1962) to contain the former members of the Chrysophyceae which possess a haptonema in addition to the two flagella. Subsequent ultrastructural studies have revealed a number of other features which are characteristic of the group. Recently, a new typified name, Prymnesiophyceae, has been proposed for the class (Hibberd, 1976) but, owing to the potential confusion this may cause, the original name is used here.

The cell structure has been described for many genera and species. Those reports published before 1973 have already been listed (Dodge, 1973), but since then there have been several studies of *Pavlova* (Billard, 1976; Green, 1973, 1975, 1976b; Van der Veer, 1976; Van der Veer and Lewis, 1977), the description of a new genus, *Imantonia* (Reynolds, 1974; Green and Pienaar, 1977), *Platychrysis* (Chretiennot, 1973), *Diacronema* (Green and Hibberd, 1977), *Hymenomonas* (Pienaar, 1976a), and a new colonial genus *Corymbellus* (Green, 1976a). The much-studied genus *Chrysochromulina* has been reexamined by freeze-etching (Neushul and Northcote, 1976).

Members of the Haptophyceae normally have two flagella which are anteriorly inserted and equal or subequal in length. In most organisms the flagella are smooth, but in the genus *Pavlova* very fine hairs and small, club-shaped appendages are found over the surface of the flagella (Green, 1975; Green and Manton, 1970). The flagella are anchored in the cell by a system of microtubules which run under the cell membrane or adjacent to the chloroplasts (Hibberd, 1976).

The haptonema, formerly known as a third flagellum, varies from about 80 μm long in *Chrysochromulina parva* (Parke *et al.,* 1962) to a residual basal structure in *Imantonia* (Green and Pienaar, 1977), with many grades in between. The basic structure (Manton, 1964b, 1967a) consists of three concentric membranes, with the outer membrane continuous with the cell plasma membrane and the inner two connected to endoplasmic reticulum cisternae within the cell. The membranes enclose six or seven longitudinally arranged microtubules. At its base the haptonema is quite distinct from flagella bases, for the microtubules simply enter the cell as a cluster and are usually increased in number by one or two extras. The base has about the same length as flagellar bases, but it has no roots.

The second major distinguishing feature of the Haptophyceae is the scales that normally cover the cell surface. These have a great variety of forms and ornamentation, although they all seem to be basically derived from an oval flattened plate of organic material, with a thickened margin (Franke and Brown, 1971). Often the scales have a pattern of radiating lines on the inner face and concentric lines on the outer. In the coccolithophorids, scales become mineralized by the deposition of calcium carbonate crystals which are arranged in various patterns. Often there is more than one type of scale on a particular organism, as in the recently described *Chrysochromulina pyramidosa* (Thomsen, 1977), where there are underlying flat circular plates, with wide meshes derived from circular and radiating fibres, and outer scales with a four-strutted pyramidal structure on similar baseplates. All types of scales form within Golgi cisternae, and the process has been described in several organisms (e.g., *Chrysochromulina*, Manton, 1967b; *Hymenomonas carterae*, Outka and Williams, 1971; *Emiliana huxleyi*, Klaveness, 1972, 1976).

Each cell contains a single large Golgi body which is characteristically situated between the nucleus and the flagella bases. It tends to be somewhat fan-shaped as a result of the cisternae being bunched together near the flagella bases and inflated towards the nucleus (Hibberd, 1976). The Golgi body is associated with endoplasmic reticulum cisternae on one side, and scale production takes place progressively towards the other side. It shows considerable changes in appearance, with peculiar dilations related to scale formation cycles, and these have been most carefully analysed in *Hymenomonas pringsheimii* (=*Pleurochrysis scherffelii*), which has a nonmotile benthic phase (Brown, 1969).

The nucleus is usually situated in the posterior half of the cell, and the nuclear envelope is attached to the endoplasmic reticulum envelope surrounding the chloroplasts. In interphase the nucleus contains a nucleolus and scattered clumps of chromatin. Nuclear division has been studied in *Prymnesium parvum* (Manton, 1964c), and here mitosis is preceded by duplication of the flagella, chloroplasts, and Golgi body. The nuclear envelope degenerates, and the chromatin aggregates into a compact metaphase plate. The spindle tubules originate outside the nuclear envelope at prophase, probably from the vicinity of the flagellar bases, although there is no definite evidence that they form poles. Many of the spindle tubules pass through the chromosome clump, and fragments of nuclear envelope (or endoplasmic reticulum) remain scattered throughout the spindle area. After what appears to be a rapid anaphase, in which the clumps of chromosomes move a considerable distance apart, the nuclear envelope reforms, commencing on the "polar" side of each daughter nucleus.

All known members of the Haptophyceae possess chloroplasts, and usually there are two per cell. The envelope consists of four membranes, the outer two being part of the endoplasmic reticulum system of the cell. No elaboration of periplastidial network (cf. Chrysophyceae) or contents (cf. Cryptophyceae) have been reported in the space between the endoplasmic reticulum and the true chloroplast envelope. Within the chloroplast the uniform lamellae are made up of three thylakoids, and there is no girdle lamella. In some species pyrenoids are associated with the chloroplasts, and these are occasionally of the stalked form (e.g., *Pavlova granifera*, Green, 1973), or more usually are completely immersed within the chloroplast (e.g., *H. carterae*, Outka and Williams, 1971). An eyespot is found in *Pavlova*, where it is of the chrysophycean type, being part of a chloroplast and situated in a pit near the flagellar bases (Green and Manton, 1970).

G. Cryptophyceae

The Cryptophyceae is a distinctive group of unicellular organisms, most of which are flagellates. It has no close affinity with any other group, and the ultrastructure of the chloroplasts, flagella, and ejectosomes are all unique. Members of the genera *Cryptomonas, Chroomonas, Chilomonas,* and *Hemiselmis* have all been studied in some detail (see Dodge, 1973; Antia *et al.*, 1973; Faust and Gantt, 1973; Oakley and Dodge, 1976; Santore, 1977; Santore and Greenwood, 1977).

Each cell bears two anterior or lateral flagella which are of more or less the same length. The arrangement of flagellar hairs appears to be rather variable, but in *Cryptomonas* one flagellum bears two rows of hairs and has a swelling near its base bearing a tuft of hairs. The other flagellum has only one row of hairs (Hibberd *et al.*, 1971). The hairs are of the tubular type, as in the Chrysophyceae, and they form within the perinuclear space (Heath *et al.*, 1970). The only detailed study of flagellar bases and roots has been that of Mignot *et al.* (1968), who showed that the basal bodies have the usual arrangement of tubules in triplets at the base, although the distal portion consists of an arrangement of doublets fixed to a thickened ring. The roots, although varying in arrangement with the genus, normally consist of tubules which run beneath the periplast, and a fibrous root which runs near the nucleus.

The cryptophycean cell is covered by a delicate periplast. This consists basically of the plasma membrane, which appears to be coated by a "fuzz" of granular material, but beneath this and closely associated with it lie thin, platelike structures about 10 nm thick, which have been found to have different shapes in the various genera. The first to be described was *Cryptomonas* sp., where the plates have a hexagonal form (Hibberd *et al.*, 1971), and *Chroomonas* sp., where the plate areas are more or less rectangular and arranged in longitudinal rows on the cell (Gantt, 1971). A more detailed study using scanning electron microscopy (Santore, 1977) has demonstrated longitudinal rows of rectangular plates in *Hemiselmis rufescens, Chroomonas salina,* and *Chroomonas* sp. In *Chroomonas* two ejectosomes (see later) are associated with the anterior corners of each plate, and in *Cryptomonas* the ejectosomes are found at the corners of the hexagonal plates. Faust (1974) found that plates which were dissociated from the periplast had a latticelike structure with a periodicity of 20 nm and are most likely composed of protein, as they can be digested with protease.

Cryptomonad chloroplasts possess a number of features which distinguish them from the chloroplasts of all other groups of algae. Perhaps the most striking of these is the normal constitution of the lamellae, which are each made up of two thylakoids (Gibbs, 1962; Dodge, 1969b; Wehrmeyer, 1970b) (Figure 12). Occasional exceptions to this rule have been found, such as single thylakoids or groups of more than two. These seem mostly either to be the result of branching lamellae (Wehrmeyer, 1970b) or may be caused by unusual conditions such as heterotrophic growth (Antia *et al.*, 1973). Like the red and blue-green algae, members of the Cryptophyceae contain phycobilin pigments, but here they are located in a unique position within the thylakoidal cisternae (Gantt *et al.*, 1971). The chloroplast envelope provides the last distinctive feature (Figure 5). As in members of the Chrysophyta, the envelope consists of four membranes, two belonging to the chloroplast and two being derived from the cell endoplasmic reticulum system. Uniquely, there is a substantial space or compartment between the two sets of membranes. Here, starch grains of the food reserve are located on the inner side of the plastids and, more interestingly, adjacent to each chloroplast there is a granular nucleuslike structure which has been termed the cryptonucleus (Greenwood, 1974) or nucleomorph (Figure 5). One important member of the Cryptophyceae, the nonpigmented *Chilomonas paramecium,* has been shown to possess rudimentary chloroplasts or leucoplasts (Sepsenwol, 1973). These consist of vesicles bounded by two pairs of membranes, in which an outer compartment contains starch grains and probably ribosomes, and an inner compartment which contains bodies resembling pigment-containing lipid globules. Many cryptomonads have stalked pyrenoids associated with the plastids, and a few have internal eyespots consisting of a plate of carotenoidal globules extending beyond the pyrenoid (Dodge, 1969b).

The interphase nucleus situated at the posterior end of the cell is of typical construction, with a double nuclear envelope, scattered chromatin, and a localized nucleolus. The process of nuclear division has been described in detail (Oakely and Dodge, 1973, 1976; Oakley and Bisalputra, 1977). It starts with the nucleus moving towards the flagellar bases, which have already replicated. Numerous microtubules are seen to run either side of the nucleus from the flagellar bases and, whilst moving, the nucleus changes into a flattened shape. Microtubules are now arranged transversely on either side of the nucleus and between it and the plastids.

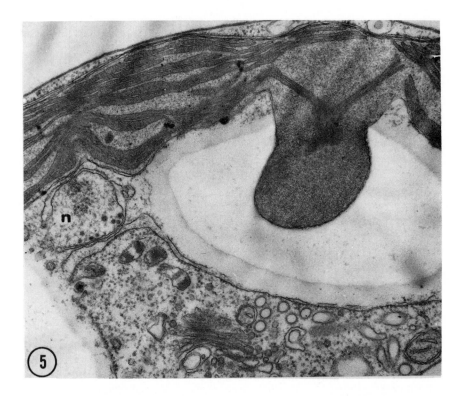

Figure 5. *Hemiselmis salina* (Cryptophyceae). Part of a section through a cell, showing the chloroplast with dense thylakoids mainly arranged in pairs, a pyrenoid with an associated large starch grain situated in the periplastidial compartment, and a nucleomorph (n). Magnification: ×33,000.

The chromatin becomes distinctly condensed and aggregates into a central plate, whilst larger gaps appear in the nuclear envelope and allow microtubules to penetrate into the nucleus. The metaphase condition is distinctive, with a dense central plate of chromatin in which individual chromosomes cannot be distinguished. Some microtubules appear to be attached to the chromatin, whereas others run from pole to pole through cylindrical gaps. The spindle has a squarish shape and at this stage shows no connection with the flagellar bases which are located at its anterior side. Anaphase appears to be rapid, and the chromatin plate divides into two parts which separate towards the poles. The nuclear membrane reforms, using part of the endoplasmic reticulum which remains around the chloroplasts during division. Cytokinesis may commence before mitosis is complete and consists of longitudinal fission of the cell, which commences at its posterior end.

The cells of the Cryptophyceae contain normal Golgi bodies, but the mitochondria, which have been investigated in detail by Santore and Greenwood (1977), have some unusual features. In all cells investigated there appeared to be a single mitochondrion, which in *Hemiselmis* was vermiform whereas in *Cryptomonas* and *Chroomonas* it formed a branched mitochondrial complex. The cristae were found to have a flattened, finger-like form which can be clearly distinguished from the cristae of other organisms.

All members of the group investigated to date have been found to possess ejectile organelles, variously termed ejectosomes, trichocysts, or taeniobolocysts. When discharged, these take the form of a tubular, tapered thread, but within the cell they consist of an oddly shaped reel of membranous material. The structure and discharge of these unique organelles have been investigated by several workers (Anderson, 1962; Mignot *et al.*, 1968, 1970; Wehrmeyer, 1970a).

The majority of cryptomonads exist as free-living cells, but a number have been found living in symbiotic associations. One of the most interesting of these is that involving the ciliate *Mesodinium rubrum*. Earlier studies (Taylor *et al.*, 1969, 1971) suggested that the numerous symbionts within the ciliate cell consisted of cryptomonadlike plastids, each with a pyrenoid and some mitochondria. However, a recent study of this system (Hibberd, 1977) has revealed that there is but one cryptomonad within the *Mesodinium* cell, and this consists of a large nucleus, several mitochondria, endoplasmic reticulum, and two groups of chloroplasts, all linked and bounded by a single membrane. The symbiont contains the typical starch compartment with a nucleomorph and only lacks the flagella, periplast, and ejectosomes of a normal cryptomonad.

H. Dinophyceae

The Dinophyceae is a most distinctive group of organisms consisting mainly of biflagellate unicells. Their unique features include the structure of the flagella, the type of cell covering, the mesokaryotic nucleus, the trichocysts, and the pusule. Many possess chloroplasts, but some are entirely heterotrophic. The majority are free-living, but a few live as parasites and some, that have chloroplasts, as symbionts. Many genera and species have been studied by electron microscopy, and the earlier work has already been the subject of a review (Dodge, 1971b). Recent studies have included descriptions of a new species, *Cachonina illdefina* (Herman and Sweeney, 1976) which has a structure reminiscent of

Heterocapsa triquetra, of a dinoflagellate, *Gyrodinium lebourae* which exists either as a saprophyte or a phagocyte and has an unusual microtubule-containing projection also found in *Crypthecodinium cohnii* (Lee, 1977b), of a small *Gymnodinium* with a reduced number of organelles (Dodge, 1974b), and a detailed study of cell division and thecal formation in *Ceratium tripos* (Wetherbee, 1975a,b,c). A parasitic dinoflagellate, *Haplozoon,* has been studied (Siebert and West, 1974), and the unusual dinoflagellate *Peridinium balticum,* which contains chrysophyte symbionts like those of *Glenodinium foliaceum* (Dodge, 1971a; Jeffrey and Vesk, 1976), has been the subject of much study (Tomas *et al.,* 1973; Tomas and Cox, 1973; Tippit and Pickett-Heaps, 1976).

Of the two flagella, one is normally directed posteriorly, and it may bear fine hairs near its tip and probably contains extraaxonemal material within the sheath. The second flagellum runs around the cell in the transverse groove or girdle. It has a distinctive undulating appearance which derives from the mode of construction involving a taut, striated strand, a stretched sheath and an axoneme thrown into either a helix or undulations. This flagellum bears a single row of long fine hairs. From early electron microscope studies it was interpreted as having the form of a helix (Leadbeater and Dodge, 1967); however, recent studies using scanning microscopy have both supported this idea (Herman and Sweeney, 1977; Wetherbee, personal communication) and suggested that the flagellum might be ribbonlike with the axoneme thrown into undulations distal to the taut striated strand (Taylor, 1975; Leblond and Taylor, 1976; Berdach, 1977). Unfortunately all methods of preparation for electron microscopy involve drastic processing, and it is possible that critical changes have been taking place before examination. In the dinoflagellate *Oxyrrhis,* tiny scales have been found on both flagella (Clarke and Pennick, 1972). The flagellar bases are relatively simple, having no unusual structures, and the root systems invariably consist of both tubular and striated roots but in various arrangements from genus to genus (see, e.g., Dodge and Crawford, 1968, *Amphidinium*; 1970b, *Ceratium*; 1972, *Oxyrrhis*).

The dinoflagellate cell is covered by a complex envelope called the theca or amphiesma. It has a basic uniformity of construction throughout the group, although there are various features which can be identified with particular orders or even genera (Dodge and Crawford, 1970a). There is a continuous outer membrane, which some workers regard as the functional plasma membrane, beneath which lies a close-packed layer of vesicles. In genera such as *Amphidinium* or *Gymnodinium* these vesicles are substantially empty, but in the armoured genera such as *Peridinium* and *Ceratium* each vesicle contains a thick polysaccharide plate with a characteristic shape and ornamentation. In *Ceratium,* where plate formation has been studied (Wetherbee, 1975a,b,c), flattened cytoplasmic vesicles containing precursor material were found to fuse with the base of the thecal vesicles. At an early stage the cell was covered with membranes like those in *Gymnodinium* but with an almost continuous layer of microtubules beneath.

In *Peridinium trochoideum* (Kalley and Bisalputra, 1975) densely stained inclusions which were found throughout the cytoplasm were said to be deposited at the surface to form a continuous wall which later divided up into the characteristic plates. The thecal membranes have also been studied by freeze–fracture techniques (Sweeney, 1976) with a view to discovering any change in particle distribution and size associated with the circadian cycle. It was found that, although the protoplasmic face of the membrane beneath the plates was constant with time, the outer face (EF) varied, with twice as many particles per unit area at 18 hr than at 6 hr. Numerous studies of the structure and arrangement of thecal plates have been carried out with scanning electron microscopy, and these have given a much better understanding of how the plates articulate together and their various types of ornamentation (e.g., Dürr and Netzel, 1974, *Gonyaulax polyedra*; Loeblich and Loeblich, 1975, *Gonyaulax* spp.; Taylor, 1971, 1973, *Ornithocercus*). In the naked dinoflagellate *Oxyrrhis marina* very delicate circular scales have been found on the outer surface of the thecal membranes (Clarke and Pennick, 1976), and in the thecate *Heterocapsa triquetra* elaborate scales cover the outer cell membrane (Pennick and Clarke, 1977).

Studies of dinoflagellate chloroplasts have shown that there are two main types of construction with a number of variations (Dodge, 1975b). The type most characteristic of the group consists of thylakoids arranged parallel to each other and mainly in threes (Figure 10). There is very little interconnection between adjacent lamellae, and only in some organisms do one or two thylakoids form a peripheral lamella. The chloroplast is bounded by a three-membrane envelope (Dodge, 1968). This type of chloroplast may contain a simple lenticular internal pyrenoid or one of a variety of types of stalked pyrenoid (Dodge and Crawford, 1971). A quite distinctive type of chloroplast is found in *Glenodinium foliaceum* (Dodge, 1975b; Jeffrey and Vesk, 1976) and *P. balticum* (Tomas and Cox, 1973). Here there is a distinct girdle lamella, there are connections between adjacent lamellae, and an internal pyrenoid is present. These two species contain the carotenoid pigment fucoxanthin and have a supernumerary nucleus in addition to the typical mesokaryotic type. As the chloroplasts and extra nucleus appear to be completely enclosed by a single-membrane envelope, it has been suggested that they represent a symbiont from the Chrysophyceae (Tomas and Cox, 1973). There are also some dinoflagellates of the genus *Gymnodinium*, which have chloroplasts with internal pyrenoids and fucoxanthin pigment but no girdle lamella and no second nucleus in the cell (Dodge, 1975b).

One study has been made of circadian changes in the structure of the chloroplasts of *Gonyaulax polyedra* (Herman and Sweeney, 1975). Here, it was found that the spacing between the lamellae in the parts of the chloroplasts near the centre of the cell was wider during the day than at night. Ribosomes were absent from these spaces when they were expanded, but plentiful when contracted.

Much work has been carried out on the unique nuclei of dinoflagellates. Early studies (reviewed in Dodge, 1971b) had shown that the permanently condensed chromosomes were essentially lacking in histone, and that nuclear division took place within an entire nuclear envelope which was invaginated to form cytoplasmic channels through the nucleus. Studies on the chromosome structure, in particular those involving the use of spread chromosomes, have suggested that the chromosome consists of a bundle of circular chromatids twisted together to give the characteristic banded appearance in section (Haapala and Soyer, 1973, 1974; Soyer and Haapala, 1974; Oakley and Dodge, 1979).

More recent studies of mitosis have shown that there are connections between the microtubules which lie in the cytoplasmic channels and dividing chromosomes which are attached to depressions in the walls of these channels (Oakely and Dodge, 1974b, 1977; Cachon and Cachon, 1974) (Figure 15). Mitosis in the parasitic dinoflagellate *Haplozoon* appears to follow the same lines as in free-living organisms (Siebert and West, 1974), as also does that in the desmokont section of the class (Prorocentrales), for microtubule-containing channels have been found in two members, *Prorocentrum triestinum* (Dodge and Bibby, 1973) and *P. mariae-lebourae* (Loeblich and Hedberg, 1976). The supernumerary nucleus of *Peridinium balticum* has been found to divide by what appears to be amitotic division involving simple cleavage (Tippit and Pickett-Heaps, 1976). Cytokinesis has been studied in *Amphidinium* (Oakley and Dodge, 1977) and *Ceratium* (Wetherbee, 1975b). In this process, which seems to take place by simple constriction, a row of microtubules lies beneath the cell membrane in the plane of division and is involved in forming the cleavage furrow.

A structure unique to the dinoflagellates is the pusule, which probably functions in osmoregulation. It consists of an invagination of the plasma membrane from the base of the flagellar canal, and within the cell this becomes associated with internal vacuolar membranes to form distinctive pusule vesicles and tubules. The pusules take many forms (Dodge, 1972), varying from simple collections of pusule vesicles situated around the flagellar canals in *Gymnodinium* species to elaborate systems of tubules, vesicles, and vacuolar reticulum found in the freshwater dinoflagellate *Wolosznyskia coronata* (Crawford and Dodge, 1974) and the parasite *Oodinium* (Cachon *et al.*, 1970).

Most dinoflagellates contain numerous trichocysts. In the undischarged state they consist of a neck attached to the cell theca and a body containing paracrystalline proteinaceous material. When discharged, the trichocyst shoots out a long, cross-banded thread which is square or rhomboidal in section (Bouck and Sweeney, 1966; Dodge, 1973). Apart from the fact that they lack a pointed tip, these trichocysts bear a strong resemblance to those found in ciliates. A few dinoflagellates, notably genera such as *Nematodinium* and *Polykrikos* (Mornin and Francis, 1967; Greuet, 1972), contain very elaborate nematocysts reminiscent of those found in the Coelenterata. A few of these larger dinoflagellates such as *Warnowia* possess elaborate ocelli with both lens- and retinalike receptor (Greuet, 1977).

I. Miscellaneous Flagellates and Related Organisms

Although not strictly phytoflagellates, there are a number of algal groups in which flagellate stages such as zoospores and gametes are produced. These are clearly outside the scope of this chapter but their presence should be considered in any general phylogenetic scheme. For example, there are many green algae where the main vegetative phase is filamentous (e.g., *Ulothrix* and *Cladophora*) or parenchymatous (e.g., *Ulva*), and which produce both zoospores and gametes. These could fit the description of flagellates in the Volvocales (Volvocida) (Bråten, 1971). There are other green algae belonging to the order Oedogoniales, which have more elaborate multiflagellate zoospores (Hoffman, 1970), which suggests that despite their green chloroplasts they are somewhat removed from the Volvocales, and some authors place them in a separate algal class, the Oedogoniophyceae. In the order Chaetophorales some genera have zoospores with scaly flagella (e.g., *Chaetosphaeridium,* Moestrup, 1974) suggesting that they are more closely related to the Prasinophyceae than to the Chlorophyceae.

Among the algal classes in which no dominant flagellated vegetative stages are found are the Phaeophyceae, Xanthophyceae, and Eustigmatophyceae; however, all three have zoospores, and the first-named also has motile gametes. In the first two classes motile stages are very similar to the equivalent stages in the Chrysophyceae, for the anterior flagellum bears stiff hairs whilst the posterior flagellum is smooth. Within the cell in both these classes the chloroplast is surrounded by endoplasmic reticulum, and there are girdle lamellae enclosing the remaining three-thylakoid lamellae. The Eustigmatophyceae, which was formed to embrace certain organisms formerly placed in the Xanthophyceae (Hibberd and Leedale, 1972), has zoospores with an anterior hairy flagellum with a flagellar swelling, and a posterior flagellum which is relatively short. The chloroplast is surrounded by endoplasmic reticulum, but there are no girdle lamellae. The eyespot, which in the Xanthophyceae and Phaeophyceae is formed as part of a chloroplast, consists simply of an aggregation of pigmented globules rather like that seen in the Euglenophyceae. In the Bacillariophyceae the only known flagellated stages are male gametes in some centric diatoms, which bear one hairy anterior flagellum lacking the central two tubules of the axoneme (Manton and von Stosch, 1966; Heath and Darley, 1972). However, a distinctive silica cell wall and nuclear division involving a central spindle have been observed (Tippit et al., 1975), which are quite unique among algae.

There are also a number of flagellates—mainly apochlorotic forms—which have long been designated, "of uncertain systematic position." Pride of place here must be taken by *Cyanophora paradoxa*. This small flagellate has two hairy flagella, one directed forwards and the other backwards (Thompson, 1973). It contains two or three cyanelles which probably represent symbiotic blue-green algae (Hall and Claus, 1963; Pickett-Heaps, 1972b), and its cell surface is somewhat reminiscent of that seen in the Dinophyceae (Mignot et al., 1969). It has variously been placed in the Cryptophyceae and the Dinophyceae. Mignot and Brugerolle (1974) have studied the ultrastructure of *Colponema* which, although it lacks cyanelles, has a marked similarity to the structure of *Cyanophora*. On the basis of the new evidence these authors suggest that these two problematical genera should be placed in a distinctive group between the Chloromonadophyceae and Dinophyceae.

III. PHYLOGENETIC INDICATORS

Having briefly described the ultrastructure of the various phytoflagellate groups, we will now examine their phylogenetic relationships. Here we face a number of problems. Firstly it is necessary to decide how much weight to give to the currently resurrected theories regarding the symbiotic origin of organelles such as chloroplasts, mitochondria, and flagella (Margulis, 1970; Taylor, 1974), or should we instead follow Allsopp (1969) and Cavalier-Smith (1975) (see also Taylor, 1976b) in relying on a more traditional, essentially autogenous, evolutionary approach? Secondly, we face the problem of deciding whether to use only specific organelles in our phylogenetic studies or to try to include all structures. For the sake of simplicity, in the account which follows, we will concentrate on what appear to be the three major structural indicators of phylogeny: flagella, chloroplasts, and nuclei. The range of structure found in these three organelles is first described before we turn to a general discussion and conclusions. It should here be mentioned that Casper (1974) has produced an extensive phylogenetic survey of all microorganisms (algae, fungi, and protozoa) based essentially on chloroplast pigments and flagellar structure, and that there is a recent general review by Taylor (1976a) and brief reviews based mainly on ultrastructure (Dodge, 1973, 1974a).

A. Flagella

The basic number of flagella is two in all phytoflagellates, for even when there is only one emergent flagellum, as in *Euglena* and *Cromulina,* there is still a vestigial second flagellum. *Monomastix* (Prasinophyceae) takes this one step further by having only one emergent flagellum and the other reduced to a basal body. Where more than two flagella are present, usually in the Prasinophyceae and Chlorophyceae, they are always in multiples of two. There is no group that could clearly be said to be representative of an earlier uniflagellate condition, although the nonflagellate Rhodophyceae could well represent a direct line from an ancestral condition.

In most groups the flagella are inserted at or near the anterior end of the cell, although at times, as in the Cryptophyceae and more particularly in zoospores of some brown algae, the insertion becomes more and more lateral. In the majority of dinoflagellates the insertion is midventral (Peridiniales and Gymnodiniales), although it may also be lateral (Dinophysiales) or apical (Prorocentrales). To some extent the type of insertion varies with the type of flagella. Thus the pairs of "smooth" flagella (Figure 8) in the Chlorophyceae and Haptophyceae are anteriorly inserted and directed. The heterokont flagella of the Chrysophyceae and Chloromonadophyceae (plus the Phaeophyceae and Xanthophyceae) are anterior but with a tendency towards lateral for the smooth flagellum which beats posteriorly. The Cryptophyceae vary from genus to genus, but in *Cryptomonas* both flagella, which bear mastigonemes like the anterior flagellum of the Chrysophyceae, are directed and beat in a forwards direction. In *Hemiselmis* they are more or less laterally inserted and beat forwards to laterally. In the Euglenophyceae we also have variety, for when only one flagellum is emergent it beats forwards, but when two are seen the second normally beats backwards. The Dinophyceae add more variety to the types of behaviour for, in species with a ventral insertion, one beats backwards and the other around the cell in the girdle, but in species with lateral or anterior insertion both beat freely, one in a sine wave motion and the other with a helical or irregular undulation.

Flagellar appendages have also provided taxonomic and phylogenetic markers, particularly since the advent of electron microscopy. Unfortunately, the picture is now less clear than it was when Manton (1965) first surveyed the scene, because the larger number of organisms which have now been examined have revealed much variety and many anomalies. The types of appendages fall into the following categories: (1) stiff tubular hairs (mastigonemes) usually having fine appendages at the tips and may have easily lost fibrous lateral appendages (Bouck, 1971), usually found in two lateral rows (Chrysophyceae and Chloromonadophyceae) but in only one row on one flagellum of some cryptophytes; (2) fine hairs which may be long and form a single row (Dinophyceae and Euglenophyceae) (Figure 6), or may be in two rows as on the dinophycean longitudinal flagellum, or may be very sparse as in *Chlamydomonas* and on the posterior flagellum of *Ochromonas*; (3) brittle hairs which are easily detached, as in the Prasinophyceae (Figure 7) and possibly in the supposed chrysophyte *Pedinella*; (4) scales, which were initially associated almost exclusively with the very scaly flagella in the Prasinophyceae but more recently have been demonstrated on the flagella of *Synura* (Chrysophyceae) (Hibberd, 1973), *Oxyrrhis* (Dinophyceae) (Clarke and Pennick, 1972), and spermatozoids of a number of green algae such as *Coleochaetae* and *Chara* (Moestrup, 1970). Clublike scales or appendages are present on several species of *Pavlova* (Haptophyceae) (Green, 1976b). Thus, scales are now a very problematical factor.

In many organisms the transition zone of the flagellum is relatively simple, the central two strands of the axoneme terminate at a disk, and the peripheral doublets enter the cell. Some years ago it was shown by Lang (1963) that in some green algae there was also a complex stellate structure within the axoneme at the transition zone. Later, a similar structure was reported in *Prymnesium* (Haptophyceae) (Manton and Leedale, 1963) and *Heteromastix* (Prasinophyceae) (Manton, 1964a) as well as in many other members of the Chlorophyceae. In the Chrysophyceae (Casper, 1972) and Eustigmatophyceae (Hibberd, 1976) (and probably also in the Xanthophyceae) instead of the stellate body there is a spiral structure or fibril which is quite distinctive.

Other internal features which have been discovered include the paraflagellar rod, which runs parallel to the axoneme in euglenids and the striated strand which is associated with an expanded sheath in the dinoflagellate transverse flagellum. Swellings, often containing regularly packed material, are found near the base of the anterior flagellum in the Eustigmatophyceae, on the emergent or anterior flagellum in the Euglenophyceae, but on the posterior flagellum in the Chrysophyceae, Xanthophyceae, and Phaeophyceae. None is known from the Chlorophyceae, Prasinophyceae, Haptophyceae, Dinophyceae, or Cryptophyceae.

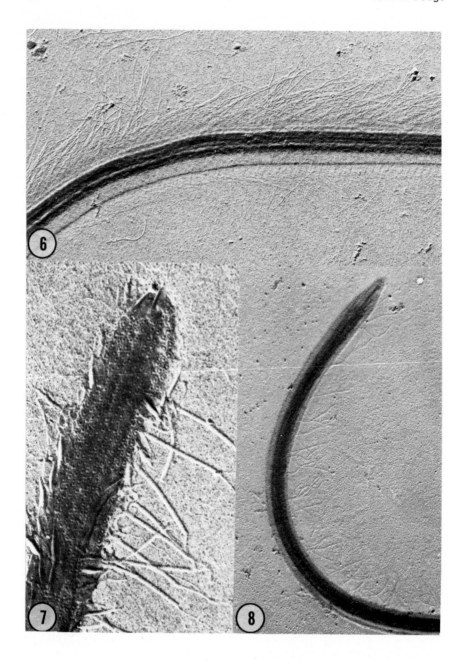

Figures 6–8. Some types of flagella.

The basal bodies and root structures at present provide only scanty evidence. For example, in the Volvocales the pairs of flagella are joined together by a striated band, but the roots are microtubular and run beneath the cell wall. By way of contrast, in the Prasinophyceae a striated band (synistosome) may be present, but the flagella appear to be mainly anchored by a broad striated root which passes close to the nucleus. Flagellar roots in dinoflagellates form a variety of patterns, although they always consist of both microtubular and striated portions.

One cannot leave the subject of flagella without mentioning the unique appendage called the haptonema. This bears a resemblance to the flagellum in its location and size, but its construction is quite different and relatively primitive. Could it be an example of an earlier type of flagellum that has survived?

To draw phylogenetic conclusions from this assemblage of flagellar structural details is not easy. The present situation seems to represent mainly divergent lines with few indications of an earlier common ancestry (Figure 9). For example, the Chlorophyceae show, by their general lack of hairs and the anterior insertion, similarity to the Haptophyceae, where there is also evidence of a stellate transition zone. The distinctive haptonema is the problem in any close relationship here. The Prasinophyceae, with scaly flagella, could be derived from a stock common to both the Chlorophyceae and Haptophyceae, as the flagella here are normally anteriorly inserted. The Dinophyceae and Euglenophyceae, with their long fine hairs and extraaxonemal rods or strands, have a certain similarity and could share a common ancestor. The remaining groups (together with the other algal groups not generally included in this chapter) probably all have a common origin. They all have mastigonemes, most have some kind of flagellar swelling, and almost all are heterokont. The Eustigmatophyceae and Cryptophyceae appear to have diverged before the others, although it could be argued that, as both the Chrysophyceae and Eustigmatophyceae share a similar spiral transition zone, they could be fairly closely related.

Figures 6–8. (*Continued*).

Figure 6. Part of the emergent flagellum of *Trachelomonas* (Euglenophyceae), showing the rather wide flagellar sheath and the unilateral array of long, fine hairs. Magnification: ×11,000.

Figure 7. The tip of one of the flagella of *Pyramimonas* (Prasinophyceae), which is clothed with small scales and bears a number of stiff hairs. Magnification: ×38,000.

Figure 8. One of the two flagella of *Chlamydomonas* (Chlorophyceae), which bears only a few delicate hairs. Magnification: ×17,500. (Figures 6–8 after Dodge, 1973.)

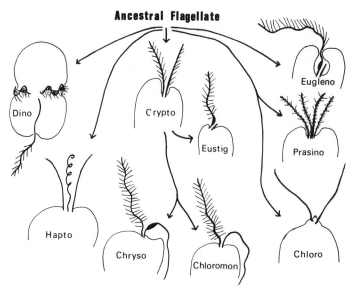

Figure 9. Diagram to illustrate the possible phylogenetic relationships of the various flagellar types.

B. Chloroplasts and Associated Structures

The ultrastructure of the chloroplasts of algae has already been the subject of three reviews (Gibbs, 1970; Dodge, 1973; Bisalputra, 1974) so, apart from summarizing the earlier findings, this chapter is mainly concerned with introducing new information which may have a bearing on the phylogenetic relations of the flagellate groups.

In the phytoflagellates we find five distinct types of chloroplast organization, but before describing them it is necessary to mention blue-green algae and the chloroplasts of the Rhodophyceae (red algae), where there are no flagellate species. In blue-green algae there are no organelles as such, and the thylakoids lie singly, often parallel to the cell wall. Phycobilin pigments are located in granular phycobilisomes attached to the outer faces of the thylakoids. In the Rhodophyceae the chloroplasts are bounded by a two-membrane envelope and contain single thylakoids arranged more or less in parallel or sometimes concentrically (type I of Bisalputra, 1974). A peripheral thylakoid may lie inside the chloroplast envelope, and an internal pyrenoid may be present. Starch grains occur in the cytoplasm of the cell and never within the chloroplast. The chloroplasts contain chlorophyll a, and in addition the phycobilin pigments phycocyanin and phycoerythrin are located in phycobilisomes, as in blue-green algae.

In the flagellates one type of chloroplast organization (type II of Bisalputra) is found only in the Cryptophyceae. Here the chloroplast is enclosed by four membranes. The outer two are part of the endoplasmic reticulum system and often have ribosomes attached to the outer face. The inner two membranes constitute the chloroplast envelope proper. Between the two pairs of membranes is a periplastidial space (Figure 5) of variable width containing starch grains plus ribosomes and a small nucleuslike structure termed the nucleomorph. An internal, multiple-stalked pyrenoid may be present, and one or two species contain an eyespot which is situated on an extension of the chloroplast (Dodge, 1969b). The chloroplasts contain chlorophylls a and c plus phycobilin pigments located within the thylakoids (Gantt et al., 1971), which are basically arranged in loose pairs (Figure 12).

The next type of chloroplast organization (type IIIb of Bisalputra) is found in chrysophyte algae (Chrysophyceae and Chloromonadophyceae, plus the nonflagellate groups Bacillariophyceae, Phaeophyceae, and Xanthophyceae). Here, as in the Cryptophyceae, the chloroplasts are surrounded by both endoplasmic reticulum and a chloroplast envelope; however, there is very little space between the two structures, and it only contains occasional tubular or membranous elements which may form a periplastidial reticulum (Falk and Kleinig, 1968). In these organisms there is normally no solid food reserve, and the cells store oil or the liquid polysaccharide chrysolaminarin in the cytoplasm. Within the chloroplasts, the thylakoids are almost always arranged in threes, and there may be some interconnection between adjacent lamellae. There is always a girdle lamella running around the edges of the chloroplast, and this invariably encloses the genophore (chloroplast DNA) which is in the form of a ring (Bisalputra and Bisalputra, 1969). The main chloroplast pigments here are chlorophyll a and c, together with a range of carotenoids. Eyespots are situated within the chloroplast, usually as a single row of pigmented granules (Figures 3 and 4). Pyrenoids are not always present but, when they are, tend to be of the internal intralamellar type, except in the Phaeophyceae, where they are usually stalked. Chloroplasts conforming exactly to this chrysophyte type are known for two dinoflagellates, *Glenodinium foliaceum* and *Peridinium balticum* (see Section II, H).

The next type of chloroplast organization is found only in the Haptophyceae. Here, the envelope and endoplasmic reticulum system surrounding the chloroplast is exactly as described above, but internally there is no girdle lamella. Pyrenoids are either stalked or internal, and food reserves and pigments are fairly similar to those of the Chrysophyceae. A few members of this group have eyespots that are rather elaborate internal extensions of the chloroplast (e.g., *Pavlova*, Green and Manton, 1970).

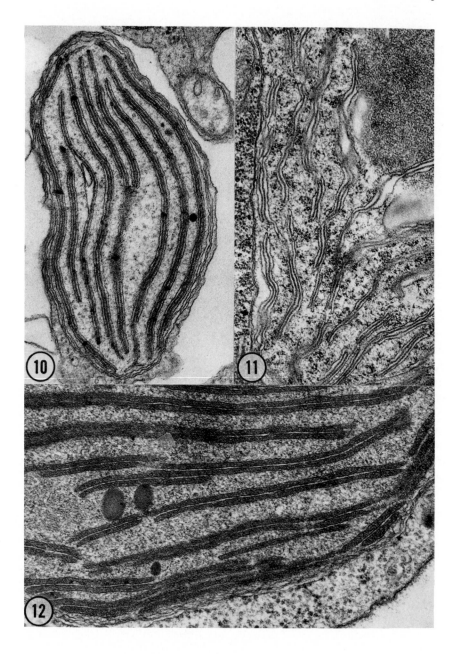

Figures 10–12. Some types of chloroplasts.

A slight variation on the above type of chloroplast structure unites most of the pigmented members of the Dinophyceae and Euglenophyceae. In these groups the chloroplast envelope consists of three membranes, the outer one of which (contrary to what is reported in Bisalputra, 1974, p. 137) probably does not consist of endoplasmic reticulum. Internally these plastids consist of lamellae made up of three thylakoids which are usually more closely appressed than in the chrysophyte groups (Figure 10). Many types of pyrenoids are present, five in the Dinophyceae alone (Dodge and Crawford, 1971). On the type of pigments, food reserves, and eyespots this group divides neatly in two. The Euglenophyceae have chlorophylls a and b, store paramylon grains in the cytoplasm, and have eyespots situated near the anterior end of the cell, which are quite detached from the chloroplasts. The Dinophyceae have chlorophylls a and c plus a number of distinctive carotenoids, store starch grains in the cytoplasm, and have a variety of eyespot types, some of which are associated with chloroplasts.

The final type of chloroplast construction (type V of Bisalputra) is found in the green algae Chlorophyceae and Prasinophyceae. In these groups the chloroplast envelope consists of two membranes (as in all higher plants), and there is no association with the endoplasmic reticulum. Within the plastid, thylakoids are rather randomly arranged, so that they may remain single or form into bands and grana in which adjacent membranes are fused together (Figure 11). There is considerable interconnection between grana. These chloroplasts contain a large, central pyrenoid, and starch is stored both around the pyrenoid and between the thylakoids but never in the cytoplasm. When eyespots are present, they consist of one to eight layers of carotenoid-containing granules, usually situated on one side of the chloroplast. The pigments here are chlorophylls a and b, together with carotenoids similar to those found in higher plants.

The possible phylogenetic sequence of the types of chloroplast structures is illustrated in Figure 13. General phylogenetic schemes will be discussed in Section IV. Suffice it here to mention that there is strong evidence that chloroplasts as organelles were derived originally from

---

Figures 10–12. (*Continued*).

Figure 10. A chloroplast from a peridinin-containing dinoflagellate (*Amphidinium britanicum*), showing thylakoids in threes and a triple-membrane envelope. Magnification: ×48,000.

Figure 11. Part of the chloroplast of an unidentified green flagellate (Chlorophyceae) in which the thylakoids are variously arranged and the envelope (left) consists of two membranes. Magnification: ×39,000.

Figure 12. Chloroplast of *Chroomonas* (Cryptophyceae) where the phycobilin-containing thylakoids are arranged in pairs. Magnification: ×49,000.

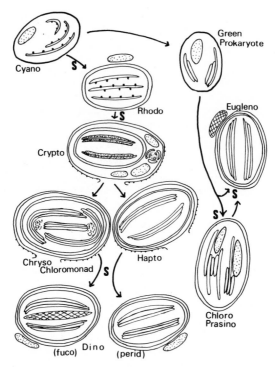

Figure 13. Diagram illustrating the possible phylogenetic relationships of the various types of chloroplasts. Symbiosis is presumed to have taken place where S appears.

blue-green algae taken up symbiotically, rather as the colourless flagellate *Cyanophora paradoxa* currently employs blue-green cyanelles (for references, see Taylor, 1974).

C. Nuclei and Nuclear Division

Various aspects of nuclear structure and division in the algae and protozoa have been reviewed in recent years (Pickett-Heaps, 1969, 1972a; Leedale, 1970; Hollande, 1972, 1974; Kubai, 1975; Dodge, 1973; Taylor, 1976a). Ultrastructural information in this area has been much more difficult to obtain than for the flagella or chloroplasts, and in some groups only one organism has been studied in detail. This discussion must therefore be regarded as being much more tentative than that of the other main phylogenetic markers.

Two of the phytoflagellate groups, the Dinophyceae and Euglenophyceae, stand out as having nuclei which are quite distinctive from those of all the others, although they show a number of features which might link them to other protozoal or fungal groups. The most primitive of these two is generally taken to be the Dinophyceae (Ris, 1975). Here the combination of virtually histoneless, permanently condensed chromosomes which are probably circular in basic construction (Haapala and Soyer, 1973; Oakley and Dodge, 1979), together with an intranuclear mitosis in which the chromosomes are attached to the nuclear membrane (Figure 15), is very reminiscent of certain aspects of genophore structure and division in prokaryotic organisms. There are, however, several advanced features such as the multiplicity of chromosomes and the clear mitotic division involving a cytoplasmic spindle and kinetochorelike structures (Oakley and Dodge, 1974b). These features have some similarities to those reported in the dubious dinoflagellate *Syndinium* (Ris and Kubai, 1974), although here centrioles are involved in the formation and movement of the spindle. There are also parallels with some hypermastigine flagellates (Hollande, 1974; Kubai, 1973), where there is an extranuclear spindle and the chromosomes are attached to pluglike kinetochores fixed to the nuclear envelope.

The second unique type of nucleus is found in the Euglenophyceae. Here, the chromosomes are also permanently condensed (Figure 14), but unfortunately we have no information on their chemistry or ultrastructure. Their mitosis is unique amongst the algae in being entirely intranuclear, with the spindle microtubules forming inside a permanently closed nuclear envelope (Leedale, 1968; Chaly *et al.*, 1977). The nearest approach to this type of division appears to be in the trypanosomes (Vickerman and Preston, 1970) and in amoeboflagellates such as *Naegleria* (Schuster, 1975). A number of aquatic fungi also have intranuclear mitosis (for references, see Leedale, 1970), but in these organisms distinct centrioles are present at the poles, whereas in euglenids no such spindle-organizing structure has been definitely reported.

All the remaining phytoflagellate groups (and indeed all other algae) have typical eukaryotic nuclei with only relatively minor differences in nuclear division mainly concerned with the type of spindle organizer and whether the mitosis is open or semiopen. In the Cryptophyceae mitosis has been described in detail for *Chroomonas* (Oakley and Dodge, 1976) (Section II,G). Here, the initial microtubule-organizing centre (MTOC) is provided by the flagellar bases, but once the nucleus has migrated to the anterior end of the cell and prophase commences the bases do not appear to be involved in spindle orientation or formation. Mitosis is open and there are both interzonal microtubules passing right through the chromatin mass and also others which end in it. The situation in the Haptophyceae (Manton, 1964c) is fairly similar (see Section II,F), although

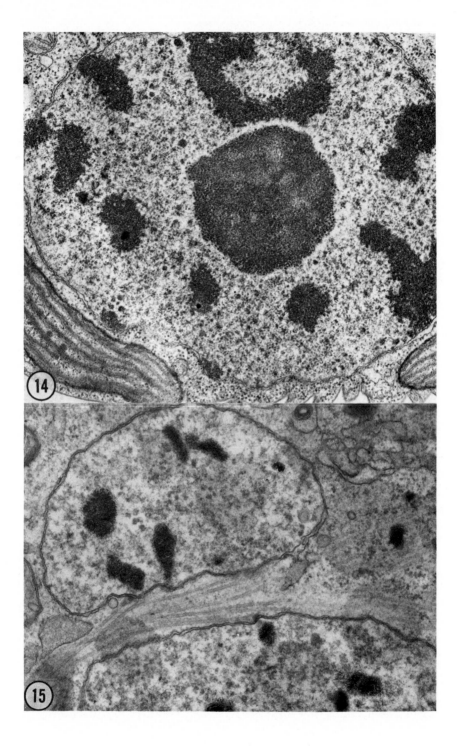

14

15

there appears to be a considerably greater extension of the spindle during anaphase.

Rather surprisingly, a comparable type of division is found in the Prasinophyceae, where the extranuclear spindle microtubules appear to be organized either by the banded flagellar root (rhizoplast) (Figure 1) or a derivative of it. The spindle is almost completely open (Stewart *et al.*, 1974; Pearson and Norris, 1975), and relatively simple kinetochores are present (Oakley and Dodge, 1974b).

In the Chrysophyceae (Slankis and Gibbs, 1972; Bouck and Brown, 1973) a striated flagellar root, which runs to the nucleus, replicates when the flagella duplicate at the start of division. One rhizoplast then becomes associated with each pole of the nucleus, and numerous microtubules appear in this area. Small apertures form in the nuclear envelope (polar fenestrae), allowing microtubules to enter the nucleus, where some run from pole to pole whilst others contact the chromosomes. Most of the nuclear envelope now disappears, and the spindle extends, the rhizoplast structures remaining at the poles throughout. We have, as yet, few details of mitosis in the Chloromonadophyceae, but it appears that the nuclear envelope remains intact throughout most of the division up to anaphase, apart from polar pores (Heywood and Godward, 1972). Basal bodies acting as centrioles appear to be situated at the poles, and the chromosomes, which in the light microscope are very long and centric, are attached to the spindle by distinct kinetochores. It should be mentioned that this appears rather similar to the situation in the Phaeophyceae where the few reports available suggest that true centrioles are present at the poles and that the nuclear envelope remains substantially intact (Bouck, cited in Leedale 1970; Neushul and Dahl, 1972). This is also essentially the type of division found in the Xanthophyceae (Ott and Brown, 1972).

In the Chlorophyceae, division of the flagellate *Chlamydomonas* has been studied in some detail (Johnson and Porter, 1968; Coss, 1974; Triemar and Brown, 1974). Here the initial duplication of the flagella and the movement of the nucleus towards the anterior end of the cell appear to happen essentially as in the Cryptophyceae, Haptophyceae, Chrysophyceae, and Prasinophyceae. The nucleus extends into a spindle shape, and microtubules pass into it through polar gaps. There are no obvious MTOCs at the poles, and the main part of the nuclear envelope remains

---

Figure 14. The interphase nucleus of *Trachelomonas* (Euglenophyceae) showing the large central nucleolus and condensed chromosomes. Magnification: ×28,000.

Figure 15. Part of an anaphase nucleus of *Amphidinium carterae* (Dinophyceae), showing cytoplasmic microtubules passing through a tunnel to the walls of which the small chromosomes are attached. Magnification: ×32,000. (Courtesy of B. R. Oakley.)

intact, giving what Taylor (1967a) has called semiopen division. Further extension of the nucleus is associated with anaphase separation of the chromatids, but no kinetochores have been observed.

For the sake of subsequent discussion the nuclear division of the two remaining algal groups should be briefly mentioned (no details are available for the Eustigmatophyceae as yet). In the Rhodophyceae the dividing nucleus in *Membranoptera* (McDonald, 1972) becomes spindle-shaped, and at its ends polar rings develop. These are short cylinders which appear to act as a focus for the spindle microtubules. The nuclear envelope remains intact, apart from polar gaps which allow the entry of microtubules. At anaphase in *Polysiphonia* (Scott *et al.*, 1977) an interzonal spindle develops. In centric diatoms (Bacillariophyceae) the nuclear division has been studied in detail in *Lithodesmium* (Manton *et al.*, 1969) and *Melosira* (Tippit *et al.*, 1975). In these organisms there is a permanent spindle precursor (or "persistent polar complex" of Tippett *et al.*) situated at one side of the nucleus. It extends and splits into two plates at the start of division, and numerous microtubules form in association with it to give a central spindle consisting of two overlapping half-spindles which then sink into the nucleus as the nuclear envelope breaks down. The chromatin appears to move polewards along the microtubules, but there is very little extension of the spindle after metaphase. Clearly this form of nuclear division is unique so far as algae and phytoflagellates are concerned, although certain aspects of it are reminiscent of what takes place in some fungi and protozoa (for references, see Tippit *et al.*, 1975).

IV. DISCUSSION AND PHYLOGENETIC CONCLUSIONS

Over the past few years several schemes have been proposed to explain the evolution of the algae and related organisms. In the first of these, Raven (1970) postulated that several independent symbiotic events may have taken place in the development of the present spectrum of chloroplast types. He suggested that blue-green algae, having evolved a system of photosynthesis based on chlorophyll a, might then have developed along three lines. The first of these has continued as present-day blue-green algae, the second became "yellow prokaryotes," and the third evolved chlorophyll b and became "green prokaryotes." All three would have been involved in symbiotic relationships with primitive eukaryotic cells, the first giving rise to plastids in the Rhodophyceae and Cryptophyceae, the second giving rise to the plastids of the Chrysophyceae, Phaeophyceae, and Dinophyceae, whereas the third was used by the Chlorophyceae and Euglenophyceae. At the time this theory seemed rather speculative, but in part it has now been justified by the finding of a green prokaryote, *Prochloron,* which contains chlorophyll b, and stacked thylakoids in what appears to be a perfectly good blue-green algal cell

(Lewin, 1976, 1977; Whatley, 1977). It is perhaps only a matter of time before a yellow prokaryote is discovered.

An alternative theory was presented by Lee (1972). He was concerned that there *were* no known yellow or green prokaryotes, and he thought that pyrenoid phylogeny had not been adequately considered. He also pointed out that chlorophyll c was involved in two of Raven's lines. Lee suggested that *Cyanophora paradoxa,* a cryptophytelike flagellate with blue-green cyanelles, probably provided the key to the ancestry of most algal groups. He postulated that the chloroplast first evolved in a cryptophycean alga and from this the other groups developed along three lines. One gave rise to the green groups by the loss of phycobilins and the acquisition of chlorophyll b. The second phylogenetic line also required loss of the phycobilins, but with the development of chlorophyll c, to give the Chrysophyceae and its allies and the Dinophyceae. The third line contains the present-day Cryptophyceae and by the eventual loss of flagella gave rise to the Rhodophyceae. This scheme involves only one basic symbiotic act but considerable evolutionary change. Recently Lee (1977a) proposed a completely new scheme in which algal groups with chloroplast endoplasmic reticulum are derived from a primary symbiotic association between a blue-green alga and a flagellate, as in his earlier scheme. Then a secondary symbiosis took place between this organism and a ciliate with both a macro- and a micronucleus, which resulted in four membranes around the symbiont. Later, some descendents of the double symbiotic organism lost the outer membrane of the micronucleus, and this membrane was replaced by the outer membrane of the symbiont. Complete loss of the micronucleus (and most of the cilia) then gave rise to the Dinophyceae and Euglenophyceae, with their distinctive nuclei, and loss of the macronucleus gave rise to the line in which we now find the Chrysophyceae and Chloromonadophyceae. This seems a rather roundabout way of deriving the extant groups of flagellates and algae.

One further insight into the relationships between the groups was provided by Greenwood's (1974) brief discussion of the Cryptophyceae. He published, for the first time, details of the "cryptonucleus" (later called "nucleomorph") found in the periplastidial compartment of these organisms. He also pointed out that, although starch was present in the cytoplasm of the rhodophytes and dinoflagellates and in the periplastidial compartment of the Cryptophyceae, it was not found at all in the chrysophyte groups (Chromophyta). Consequently it was suggested that cryptophytes could have retained a primitive condition which had been lost, together with the nucleomorph, during evolution along the chrysophyte line. He concluded that the characteristics of the Cryptophyceae placed them at the base of the chrysophyte line (or Chromophyta) and between this line and the Rhodophyceae.

By putting together aspects of most of the above schemes it is possible

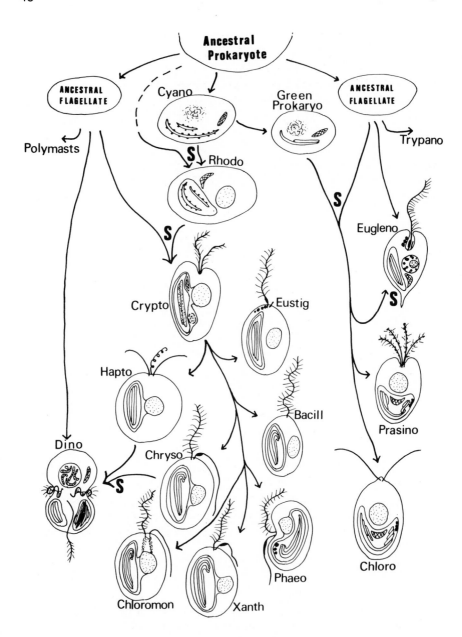

Figure 16. A diagrammatic summary of the phylogenetic relationship of phytoflagellates and other groups as described in the text. S indicates that a symbiotic event is necessary.

to propose a new hypothesis which utilizes the recent evidence (Figure 16). The starting point in this scheme is an ancestral prokaryote which gave rise to at least two types of organisms. One of these, the blue-green alga, is well known; its origin by the invagination of membranes and acquisition of the ability to photosynthesize is well documented (Cavalier-Smith, 1975). Another derivative remained heterotrophic but developed or acquired flagella by an association with spirochetelike bacteria (Margulis, 1970). It thus became an ancestral flagellate, and the nuclear material was gradually separated from the cytoplasm and enclosed by an envelope, mitochondria were obtained (Margulis, 1970), and basic types of cell organization like those found in various nonpigmented flagellates evolved. Here, for the moment, we leave this line and return to the blue-green alga. Coccoid algal cells were captured phagocytically by derivatives of the ancestral prokaryote and instead of being digested were retained as symbionts. As a result of this, the blue-green alga gradually became reduced to little more than a system of thylakoids as the host took over the protective and food storage roles—a situation like that found in *Cyanophora* today. This association eventually gave rise to red algae (Rhodophyceae), in which considerable morphological development took place.

At some stage a single-celled red alga was taken in by an ancestral flagellate, and this form gave rise to an early version of the Cryptophyceae. The combination of cell membranes of the red alga and the phagocytic vesicle of the flagellate produced the two membranes of the endoplasmic reticulum sheath, and the remnants of the cytoplasm and nucleus of the symbiont remained within the periplastidial compartment. Changes which are more difficult to explain are the movement of phycobilins into the thylakoids instead of remaining on their surface in phycobilisomes, and the acquisition of chlorophyll c. Directly from this cryptomonadlike ancestor have evolved the various groups of the chrysophyte line. In all these groups the periplastidial compartment became much reduced, and the chloroplasts lost their phycobilin pigments and acquired the ability to stack and form girdle lamellae. The chloroplast DNA became concentrated into genophores. The basic storage metabolism changed to being based on fats and chrysolaminarin. One flagellum retained the tubular mastigonemes of the Cryptophyceae in the Chrysophyceae and Chloromonadophyceae (as well as in the Xanthophyceae, Phaeophyceae, and Bacillariophyceae), and a second flagellum, when present, became differently orientated and virtually smooth. The Eustigmatophyceae probably developed along a separate line from fairly early times for, although it shares a hairy flagellum and chloroplast endoplasmic reticulum, it has a number of distinctive features. The Haptophyceae present an even greater problem, for although their chloroplast is virtually identical with that of the Chrysophyceae (save the girdle lamella), their flagella system is very different. One can only speculate that, again, this was an early offshoot which lost the ability to make flagellar hairs but developed a microtubular appendage which became the haptonema. At present there is a great range of haptonema lengths (see Section III,A), which could be either a developmental sequence or could represent progressive reduction in this organelle. Throughout this section—Cryptophyceae—Chrysophyceae or Haptophyceae—the various types of nuclear division are sufficiently similar for them to have a common ancestry.

Returning to the blue-green algae, we now have the green prokaryote *Prochloron,* which by symbiotic association with an ancestral flagellate could have given rise to the Chlorophyceae and Prasinophyceae. This explains the two-membrane envelope of plastids in these groups (one belonging to the prokaryote and one to the phagocytic vesicle). Here, the phycobilin pigments were lost at the prokaryotic stage, but the ability to store starch in the symbiont (plastid) was retained. The split between the Chlorophyceae (i.e., Volvocales) and Prasinophyceae could either be a result of two symbiotic events involving different ancestral flagellates or, more likely, part of the vast diversification which has taken place in the green algae to give filaments, nonmotile cells, parenchymatous organisms, siphonaceous algae and, indeed, higher plants (for comprehensive reviews, see Pickett-Heaps, 1975; Stewart and Mattox, 1975). On the subject of the relationships with other extant groups there is some similarity between the nuclear division of the Prasinophyceae and Cryptophyceae, which may be accounted for by both having a similar ancestral flagellate. Less easy to explain, except by suggesting that it is a result of convergent evolution, is the similar flagellar system (two smooth anterior flagella) in both the Haptophyceae and Chlorophyceae.

Because of their similar pigment composition, euglenids have often been linked with the green algae, despite their vast differences in all other characteristics. It is a fact that many members of the Euglenophyceae are completely lacking in chloroplasts, and on some grounds the group seems rather closely related to the trypanosomes. This prompts the suggestion that they are derived directly from the ancestral flagellates and may have only relatively recently taken up chloroplasts which they obtained from green algae after partial digestion. This could explain the pigments and the triple-membrane chloroplast envelope; it does, however, leave as a major problem the big change from starch storage inside the plastid to paramylon storage in the cytoplasm. It is possible that the ancestral flagellate already had a paramylon system in action before plastids were acquired. The unique flagellar and nuclear structures in this group indicate that, apart from the plastids, there is no close link with any of the other pigmented groups.

Finally we are left with the dinoflagellates, a group with a primitive nucleus and distinctive structure. Once again we must go back to the ancestral flagellates for, in this group, more than half the species are nonphotosynthetic and have no vestigial plastids. It seems likely that chloroplast acquisition was, therefore, a relatively recent event and, judging from the various types of plastid (Dodge, 1975b), they have probably been acquired from several sources (Loeblich, 1976). First, in the main peridinin-containing dinoflagellates, where there is no girdle lamella, the plastids were possibly obtained from a member of the Haptophyceae with loss of the endoplasmic reticulum sheath but acquisition of an extra membrane from the phagocytic vesicle. Those of the second type, which contain fucoxanthin plus internal lenticular pyrenoids, were probably obtained from the Chrysophyceae, as also were the plastids of the third group, so far represented only by two species (*Peridinium balticum* and *Glenodinium foliaceum*), which have resulted from such a recent symbiotic event that the nuclei and some cytoplasm of the symbiont have also been retained and the plastids are effectively surrounded by a total of five membranes. It is particularly fascinating to note that this must be the second time these two organisms have obtained chloroplasts, for they also contain eyespots bounded by three membranes (Dodge and Crawford, 1969), which almost certainly represent reduced chloroplasts from an earlier symbiosis. Once again, the completely unique flagellar system and nucleus of dinoflagellates indicate no close links with any other extant groups of phytoflagellates, although there may be a distant link with the polymastigotes.

As stated above, this phylogenetic scheme employs many ideas from earlier schemes but it also brings in more recent discoveries. However, a scheme is only as good as the facts on which it is based, and it is highly likely that, as research unfolds more of the details of the ultrastructure of phytoflagellates, the scheme will need revision from time to time. Clearly the phytoflagellates provide a fascinating study in variation, and they have had a most interesting developmental history.

ACKNOWLEDGMENTS

I wish to thank Drs. D. J. Hibberd and B. R. Oakley for allowing me to use their micrographs, and Mrs. D. Hughes for her diligent work in typing the manuscript.

REFERENCES

Aaronson, S. (1973). *Microbios* **7**, 231–233.
Aaronson, S. (1974). *J. Gen. Microbiol.* **83**, 21–29.
Aaronson, S., and Behrens, U. (1974). *J. Cell Sci.* **14**, 1–9.
Allsopp, A. (1969). *New Phytol.* **68**, 591–612.
Anderson, E. (1962). *J. Protozool.* **9**, 380–395.
Antia, N. J., Kalley, J. P., McDonald, J., and Bisalputra, T. (1973). *J. Protozool.* **20**, 377–384.

Belcher, J. H., and Swale, E. M. F. (1971). *Br. Phycol. J.* **6**, 157–169.
Belcher, J. H., and Swale, E. M. F. (1972a). *Br. Phycol. J.* **7**, 53–59.
Belcher, J. H., and Swale, E. M. F. (1972b). *Br. Phycol. J.* **7**, 335–346.
Belcher, J. H., Pennick, N. C., and Clarke, K. J. (1974). *Br. Phycol. J.* **9**, 101–106.
Berdach, J. T. (1977). *J. Phycol.* **13**, 243–251.
Billard, C. (1976). *Bull. Soc. Phycol. Fr.* **21**, 18–27.
Bisalputra, T. (1974). *In* "The Physiology and Biochemistry of Algae" (W. D. P. Stewart, ed.), pp. 124–160. Blackwell, Oxford.
Bisalputra, T., and Bisalputra, A. A. (1969). *J. Ultrastruct. Res.* **29**, 151–170.
Bouck, G. B. (1971). *J. Cell Biol.* **50**, 362–384.
Bouck, G. B., and Brown, D. L. (1973). *J. Cell Biol.* **56**, 340–359.
Bouck, G. B., and Sweeney, B. M. (1966). *Protoplasma* **61**, 205–223.
Bråten, T. (1971). *J. Cell Sci.* **9**, 621–635.
Brown, D. L., Massalski, A., and Patenaude, R. (1976). *J. Cell Biol.* **69**, 106–125.
Brown, R. M. (1969). *J. Cell Biol.* **41**, 109–123.
Burton, M. D., and Moore, J. (1974). *J. Ultrastruct. Res.* **48**, 414–419.
Cachon, J., and Cachon, M. (1974). *C. R. Acad. Sci., Ser. D* **278**, 1735–1737.
Cachon, J., Cachon, M., and Greuet, C. (1970). *Protistologica* **6**, 467–476.
Casper, S. J. (1972). *Arch. Protistenk.* **114**, 65–82.
Casper, S. J. (1974). "Grundzüge eines naturlichen Systems der Mikroorganismen." Fischer, Jena.
Cavalier-Smith, T. (1974). *J. Cell Sci.* **16**, 529–556.
Cavalier-Smith, T. (1975). *Nature (London)* **256**, 463–468.
Chaly, N., Lord, A., and Lafontaine, J. G. (1977). *J. Cell Sci.* **27**, 23–45.
Chretiennot, M.-J. (1973). *J. Mar. Biol. Assoc. U.K.* **53**, 905–914.
Christensen, T. (1962). *In* "Botanik" (T. W. Bocher *et al.*, eds.), Vol. 2, Nr. 2. Munksgaard, Copenhagen.
Clarke, K. J., and Pennick, N. C. (1972). *Br. Phycol. J.* **7**, 357–360.
Clarke, K. J., and Pennick, N. C. (1975). *Br. Phycol. J.* **10**, 363–370.
Clarke, K. J., and Pennick, N. C. (1976). *Br. Phycol. J.* **11**, 345–348.
Cole, G. T., and Wynne, M. J. (1973). *Cytobios* **8**, 161–173.
Cole, G. T., and Wynne, M. J. (1974). *J. Phycol.* **10**, 397–410.
Coss, R. A. (1974). *J. Cell Biol.* **63**, 325–329.
Crawford, R. M., and Dodge, J. D. (1974). *Nova Hedwigia Z. Kryptogamenkd.* **22**, 699–719.
Daley, R. J., Morris, G. P., and Brown, S. R. (1973). *J. Protozool.* **20**, 58–61.
Dodge, J. D. (1968). *J. Cell Sci.* **3**, 41–48.
Dodge, J. D. (1969a). *Br. Phycol. J.* **4**, 199–210.
Dodge, J. D. (1969b). *Arch. Microbiol.* **69**, 266–280.
Dodge, J. D. (1971a). *Protoplasma* **73**, 145–157.
Dodge, J. D. (1971b). *Bot. Rev.* **37**, 481–508.
Dodge, J. D. (1972). *Protoplasma* **75**, 285–302.
Dodge, J. D. (1973). "The Fine Structure of Algal Cells." Academic Press, New York.
Dodge, J. D. (1974a). *Sci. Prog. Oxf.* **61**, 257–274.
Dodge, J. D. (1974b). *J. Mar. Biol. Assoc. U.K.* **54**, 171–177.
Dodge, J. D. (1975a). *Arch. Protistenk.* **117**, 65–77.
Dodge, J. D. (1975b). *Phycologia* **14**, 253–263.
Dodge, J. D., and Bibby, B. T. (1973). *Bot. J. Linn. Soc.* **67**, 175–187.
Dodge, J. D., and Crawford, R. M. (1968). *Protistologica* **4**, 231–242.
Dodge, J. D., and Crawford, R. M. (1969). *J. Cell Sci.* **5**, 479–493.
Dodge, J. D., and Crawford, R. M. (1970a). *Bot. J. Linn. Soc.* **63**, 53–67.
Dodge, J. D., and Crawford, R. M. (1970b). *J. Phycol.* **6**, 137–149.
Dodge, J. D., and Crawford, R. M. (1971). *Bot. J. Linn. Soc.* **64**, 105–115.

Dodge, J. D., and Crawford, R. M. (1972). *Protistologica* **7**, 399–409.
Dubowsky, N. (1974). *J. Protozool.* **21**, 295–298.
Dürr, G., and Netzel, H. (1974). *Cell Tissue Res.* **150**, 21–41.
Dynesius, R. A., and Walne, P. L. (1975). *J. Phycol.* **11**, 125–130.
Ettl, H., and Green, J. C. (1973). *J. Mar. Biol. Assoc. U.K.* **53**, 975–985.
Eyden, B. P. (1975). *J. Protozool.* **22**, 336–344.
Falk, H., and Kleinig, H. (1968). *Arch. Mikrobiol.* **61**, 347–362.
Faust, M. A. (1974). *J. Phycol.* **10**, 121–124.
Faust, M. A., and Gantt, E. (1973). *J. Phycol.* **9**, 489–495.
Franke, W. W., and Brown, R. M. (1971). *Arch. Mikrobiol.* **77**, 12–19.
Franke, W. W., and Herth, W. (1973). *Arch. Mikrobiol.* **91**, 323–344.
Fuchs, B., and Jarosch, R. (1974). *Protoplasma* **79**, 215–223.
Gantt, E. (1971). *J. Phycol.* **7**, 177–184.
Gantt, E., Edwards, M. R., and Provasoli, L. (1971). *J. Cell Biol.* **48**, 280–290.
Gibbs, S. P. (1962). *J. Ultrastruct. Res.* **7**, 418–435.
Gibbs, S. P. (1970). *Ann. N.Y. Acad. Sci.* **175**, 454–473.
Gibbs, S. P., Cheng, D., and Slankis, T. (1974). *J. Cell Sci.* **16**, 557–577.
Green, J. C. (1973). *Br. Phycol. J.* **8**, 1–12.
Green, J. C. (1975). *J. Mar. Biol. Assoc. U.K.* **55**, 785–793.
Green, J. C. (1976a). *J. Mar. Biol. Assoc. U.K.* **56**, 31–38.
Green, J. C. (1976b). *J. Mar. Biol. Assoc. U.K.* **56**, 595–602.
Green, J. C., and Hibberd, D. J. (1977). *J. Mar. Biol. Assoc. U.K.* **57**, 1125–1136.
Green, J. C., and Manton, I. (1970). *J. Mar. Biol. Assoc. U.K.* **50**, 1113–1130.
Green, J. C., and Pienaar, R. N. (1977). *J. Mar. Biol. Assoc. U.K.* **57**, 7–17.
Greenwood, A. D. (1974). *Proc. Int. Congr. Electron Microsc., 8th, Canberra* **2**, 566–567.
Greuet, C. (1972). *C. R. Acad. Sci., Ser. D* **275**, 1239–1242.
Greuet, C. (1977). *Protistologica* **13**, 127–143.
Guttman, H. N., and Ziegler, H. (1974). *Cytobiologie* **9**, 10–22.
Haapala, O. K., and Soyer, M. O. (1973). *Nature (London), New Biol.* **244**, 195–197.
Haapala, O. K., and Soyer, M. O. (1974). *Hereditas* **76**, 83–90.
Hall, W. T., and Claus, G. (1963). *J. Cell Biol.* **19**, 551–563.
Heath, I. B., and Darley, W. M. (1972). *J. Phycol.* **8**, 51–59.
Heath, I. B., Greenwood, A. D., and Griffiths, H. B. (1970). *J. Cell Sci.* **7**, 445–461.
Herman, E. M., and Sweeney, B. M. (1975). *J. Ultrastruct. Res.* **50**, 347–354.
Herman, E. M., and Sweeney, B. M. (1976). *J. Phycol.* **12**, 198–205.
Herman, E. M., and Sweeney, B. M. (1977). *Phycologia* **16**, 115–118.
Heywood, P. (1972). *J. Ultrastruct. Res.* **39**, 608–623.
Heywood, P. (1973). *Br. Phycol. J.* **8**, 43–46.
Heywood, P. (1976). *Cytobios* **17**, 79–86.
Heywood, P. (1977). *J. Phycol.* **13**, 68–72.
Heywood, P., and Godward, M. B. E. (1972). *Chromosoma* **39**, 333–339.
Hibberd, D. J. (1970). *Br. Phycol. J.* **5**, 119–143.
Hibberd, D. J. (1971). *Br. Phycol. J.* **6**, 207–223.
Hibberd, D. J. (1973). *Arch. Mikrobiol.* **89**, 291–304.
Hibberd, D. J. (1976). *Bot. J. Linn. Soc.* **72**, 55–80.
Hibberd, D. J. (1977). *J. Mar. Biol. Assoc. U.K.* **57**, 45–61.
Hibberd, D. J., and Leedale, G. F. (1972). *Ann. Bot.* **36**, 49–71.
Hibberd, D. J., Greenwood, A. D., and Griffiths, H. B. (1971). *Br. Phycol. J.* **6**, 61–72.
Hill, F. G., and Outka, D. E. (1974). *J. Protozool.* **21**, 299–312.
Hills, G. J., Gurney-Smith, M., and Roberts, K. (1973). *J. Ultrastruct. Res.* **43**, 179–192.
Hoffman, L. R. (1970). *Can. J. Bot.* **48**, 189–196.

Hollande, A. (1972). *Ann. Biol.* **11** (fasc 9–10), 427–466.
Hollande, A. (1974). *Protistologica* **10**, 413–451.
Honigberg, B. M., Balamuth, W., Bovee, E. C., Corliss, J. O., Gojdics, M., Hall, R. P., Kudo, R. R., Levine, N. D., Loeblich, A. R., Jr., Weiser, J., and Wenrich, D. J. (1964). *J. Protozool.* **11**, 7–20.
Hopkins, J. M. (1970). *J. Cell Sci.* **7**, 823–839.
Jeffrey, S. W., and Vesk, M. (1976). *J. Phycol.* **12**, 450–455.
Johnson, U. G., and Porter, K. R. (1968). *J. Cell Biol.* **38**, 403–425.
Kalley, J. P., and Bisalputra, T. (1975). *Can. J. Bot.* **53**, 483–494.
Kivic, P. A., and Vesk, M. (1972). *Planta* **105**, 1–14.
Kivic, P. A., and Vesk, M. (1974). *Can. J. Bot.* **52**, 695–699.
Klaveness, D. (1972). *Protistologica* **8**, 335–346.
Klaveness, D. (1976). *Protistologica* **12**, 217–224.
Klein, S., Schiff, J. A., and Holowinsky, A. W. (1972). *Dev. Biol.* **28**, 253–273.
Koch, W., and Schnepf, E. (1967). *Arch. Mikrobiol.* **57**, 196–198.
Kochert, G., and Olson, L. W. (1970). *Arch. Mikrobiol.* **74**, 19–30.
Kramer, D. (1970). *Z. Naturforsch., Teil B* **256**, 1017–1020.
Kristiansen, J. (1969). *Bot. Tidsskr.* **64**, 162–168.
Kristiansen, J. (1972). *Sven. Bot. Tidskr.* **66**, 184–190.
Kristiansen, J., and Walne, P. L. (1976). *Protoplasma* **89**, 371–374.
Kristiansen, J., and Walne, P. L. (1977). *Br. Phycol. J.* **12**, 329–341.
Kubai, D. F. (1973). *J. Cell Sci.* **13**, 511–552.
Kubai, D. F. (1975). *Int. Rev. Cytol.* **43**, 167–227.
Lang, N. J. (1963). *J. Cell Biol.* **19**, 631–634.
Leadbeater, B., and Dodge, J. D. (1967). *J. Gen. Microbiol.* **46**, 305–314.
Leblond, P. H., and Taylor, F. J. R. (1976). *Biosystems* **8**, 33–39.
Lee, R. E. (1972). *Nature (London)* **237**, 44–46.
Lee, R. E. (1977a). *S. Afr. J. Sci.* **73**, 179–182.
Lee, R. E. (1977b). *J. Mar. Biol. Assoc. U.K.* **57**, 303–315.
Leedale, G. F. (1964). *Br. Phycol. Bull.* **2**, 291–306.
Leedale, G. F. (1967). "Euglenoid Flagellates." Prentice-Hall, Englewood Cliffs, New Jersey.
Leedale, G. F. (1968). *In* "The Biology of Euglena" (D. E. Buetow, ed.), Vol. 1, pp. 185–242. Academic Press, New York.
Leedale, G. F. (1970). *Ann. N.Y. Acad. Sci.* **175**, 429–453.
Leedale, G. F. (1975). *Br. Phycol. J.* **10**, 17–41.
Leedale, G. F., and Hibberd, D. J. (1974). *Arch. Protistenk.* **116**, 319–345.
Leedale, G. F., Meeuse, B. J. D., and Pringsheim, E. G. (1965). *Arch. Mikrobiol.* **50**, 68–102.
Lembi, C. A., and Lang, N. J. (1965). *Am. J. Bot.* **52**, 464–477.
Lewin, R. A. (1976). *Nature (London)* **261**, 697–698.
Lewin, R. A. (1977). *Phycologia* **16**, 217.
Loeblich, A. R. (1976). *J. Protozool.* **23**, 13–28.
Loeblich, A. R., and Hedberg, M. F. (1976). *Bot. Mar.* **19**, 255–257.
Loeblich, L. A., and Loeblich, A. R. (1975). *Proc. Int. Conf. Toxic Dinoflagellate Blooms, 1st, Boston, Mass.* pp. 207–24.
McDonald, K. (1972). *J. Phycol.* **8**, 156–166.
Maiwald, M. (1971). *Arch. Protistenk.* **113**, 334–344.
Manton, I. (1964a). *J. R. Microsc. Soc.* **82**, 279–285.
Manton, I. (1964b). *Arch. Mikrobiol.* **49**, 315–330.
Manton, I. (1964c). *J. R. Microsc. Soc.* **83**, 317–325.

Manton, I. (1965). *Adv. Bot. Res.* **2**, 1–34.
Manton, I. (1967a). *J. Cell Sci.* **2**, 265–272.
Manton, I. (1967b). *J. Cell Sci.* **2**, 411–418.
Manton, I. (1969). *Oesterr. Bot. Z.* **116**, 378–392.
Manton, I. (1975). *Arch. Protistenk.* **117**, 358–368.
Manton, I., and Leedale, G. F. (1961). *Phycologia* **1**, 37–57.
Manton, I., and Leedale, G. F. (1963). *Arch. Mikrobiol.* **45**, 285–303.
Manton, I., and Parke, M. (1965). *J. Mar. Biol. Assoc. U.K.* **45**, 743–754.
Manton, I., and von Stosch, H. A. (1966). *J. R. Microsc. Soc.* **85**, 119–134.
Manton, I., Rayns, D. G., Ettl, H., and Parke, M. (1965). *J. Mar. Biol. Assoc. U.K.* **45**, 241–255.
Manton, I., Kowallik, K., and von Stosch, H. A. (1969). *J. Microsc. (Oxford)* **89**, 295–320.
Margulis, L. (1970). "Origin of Eukaryotic Cells." Yale Univ. Press, New Haven, Connecticut.
Mattox, K. R., and Stewart, K. D. (1977). *Am. J. Bot.* **64**, 931–945.
Mignot, J. P. (1965). *Protistologica* **1**, 5–15.
Mignot, J. P. (1966). *Protistologica* **2**, 51–117.
Mignot, J. P. (1967). *Protistologica* **3**, 5–23.
Mignot, J. P. (1975). *Protistologica* **11**, 177–185.
Mignot, J. P. (1976). *Protistologica* **12**, 279–293.
Mignot, J. P. (1977). *Protistologica* **13**, 219–231.
Mignot, J. P., and Brugerolle, G. (1974). *Protistologica* **11**, 429–444.
Mignot, J. P., and Hovasse, R. (1973). *Protistologica* **9**, 373–391.
Mignot, J. P., and Hovasse, R. (1975). *Ann. Stn. Biol. Besse Chandesse* 1974–1975, pp. 201–211.
Mignot, J. P., Joyon, L., and Pringsheim, E. G. (1968). *Protistologica* **4**, 493–506.
Mignot, J. P., Joyon, L., and Pringsheim, E. G. (1969). *J. Protozool.* **16**, 138–145.
Mignot, J. P., Hovasse, R., and Joyon, L. (1970). *J. Microsc. (Paris)* **9**, 127–132.
Mignot, J. P., Brugerolle, G., and Metenier, G. (1972). *J. Microsc. (Paris)* **14**, 327–342.
Moestrup, Ø. (1970). *Planta* **93**, 295–308.
Moestrup, Ø. (1974). *Biol. J. Linn. Soc.* **6**, 111–125.
Moestrup, Ø., and Thomsen, H. A. (1974). *Protoplasma* **81**, 247–269.
Mornin, L., and Francis, D. (1967). *J. Microsc. (Paris)* **6**, 759–772.
Neushul, M., and Dahl, A. L. (1972). *Am. J. Bot.* **59**, 401–410.
Neushul, M., and Northcote, D. H. (1976). *Am. J. Bot.* **63**, 1225–1236.
Norris, R. E., and Pearson, B. R. (1975). *Arch. Protistenk.* **117**, 192–213.
Oakley, B. R., and Bisalputra, T. (1977). *Can. J. Bot.* **22**, 2789–2800.
Oakley, B. R., and Dodge, J. D. (1973). *Nature (London)* **244**, 521–522.
Oakley, B. R., and Dodge, J. D. (1974a). *Br. Phycol. J.* **9**, 222.
Oakley, B. R., and Dodge, J. D. (1974b). *J. Cell Biol.* **63**, 322–325.
Oakley, B. R., and Dodge, J. D. (1976). *Protoplasma* **88**, 241–254.
Oakley, B. R., and Dodge, J. D. (1977). *Cytobios* **17**, 35–46.
Oakley, B. R., and Dodge, J. D. (1979). *Chromosoma (Berl.)* **70**, 277–291.
Ohad, I., Siekevitz, P., and Palade, G. E. (1967a). *J. Cell Biol.* **35**, 521–552.
Ohad, I., Siekevitz, P., and Palade, G. E. (1967b). *J. Cell Biol.* **35**, 553–584.
Olson, L. W., and Kochert, G. (1970). *Arch. Mikrobiol.* **74**, 31–40.
Orcival-Lafont, A.-M., and Calvayrac, R. (1974). *J. Phycol.* **10**, 300–307.
Ott, D. W., and Brown, R. M. (1972). *Br. Phycol. J.* **7**, 361–374.
Outka, D. E., and Williams, D. C. (1971). *J. Protozool.* **18**, 285–297.
Palisano, J. R., and Walne, P. L. (1976). *Nova Hedwigia Z. Kryptogamenkd.* **27**, 455–481.

Parke, M., and Dixon, P. S. (1976). *J. Mar. Biol. Assoc. U.K.* **56**, 527–594.
Parke, M., Lund, J. W. G., and Manton, I. (1962). *Arch. Mikrobiol.* **42**, 333–352.
Pearson, B. A., and Norris, R. E. (1975). *J. Phycol.* **11**, 113–124.
Pennick, N. C., and Clarke, K. J. (1973). *Br. Phycol. J.* **8**, 147–151.
Pennick, N. C., and Clarke, K. J. (1977). *Br. Phycol. J.* **12**, 63–66.
Pennick, N. C., Clarke, K. J., and Cann, J. P. (1976). *Arch. Protistenk.* **118**, 221–226.
Peterfi, L. S., and Momen, L. (1976). *Nova Hedwigia Z. Kryptogamenkd.* **27**, 353–392.
Piccinni, E., and Omodeo, P. (1975). *Boll. Zool.* **42**, 57–59.
Pickett-Heaps, J. D. (1969). *Cytobios* **3**, 257–280.
Pickett-Heaps, J. D. (1970). *Planta* **90**, 174–190.
Pickett-Heaps, J. D. (1972a). *Cytobios* **5**, 59–77.
Pickett-Heaps, J. D. (1972b). *New Phytol.* **71**, 561–567.
Pickett-Heaps, J. D. (1975). "Green Algae. Structure, Reproduction and Evolution in Selected Genera." Sinauer Assoc., Sunderland, Massachusetts.
Pienaar, R. N. (1976a). *J. Mar. Biol. Assoc. U.K.* **56**, 1–11.
Pienaar, R. N. (1976b). *Br. Phycol. J.* **11**, 83–92.
Ploaie, P. G. (1971). *Rev. Roum. Biol. Ser. Bot.* **16**, 179–183.
Raven, P. H. (1970). *Science* **169**, 641–646.
Reynolds, N. (1974). *Br. Phycol. J.* **9**, 429–434.
Ringo, D. L. (1967a). *J. Ultrastruct. Res.* **17**, 266–277.
Ringo, D. L. (1967b). *J. Cell Biol.* **33**, 543–571.
Ris, H. (1975). *Biosystems* **7**, 298–304.
Ris, H., and Kubai, D. F. (1974). *J. Cell Biol.* **60**, 702–720.
Roberts, K. (1974). *Philos. Trans. R. Soc. London, Ser. B* **268**, 129–146.
Santore, U. J. (1977). *Br. Phycol. J.* **12**, 255–270.
Santore, U. J., and Greenwood, A. D. (1977). *Arch. Mikrobiol.* **112**, 207–218.
Schnepf, E., and Deichgräber, G. (1969). *Protoplasma* **68**, 85–106.
Schnepf, E., and Koch, W. (1966). *Arch. Mikrobiol.* **54**, 229–236.
Schnepf, E., Deichgräber, G., Röderer, G., and Herth, W. (1977). *Protoplasma* **92**, 87–107.
Schötz, F. (1972). *Planta* **102**, 152–159.
Schuster, F. L. (1975). *Tissue Cell* **7**, 1–12.
Schuster, F. L., and Hershenov, B. (1974). *J. Protozool.* **21**, 33–39.
Scott, J., Schornstein, K., and Thomas, J. (1977). *J. Phycol.* **13**, Suppl., p. 61.
Sepsenwol, S. (1973). *Exp. Cell Res.* **76**, 395–409.
Siebert, A. E., and West, J. A. (1974). *Protoplasma* **87**, 17–35.
Silverberg, B. A. (1975a). *Protoplasma* **83**, 269–295.
Silverberg, B. A. (1975b). *Protoplasma* **85**, 373–376.
Slankis, T., and Gibbs, S. P. (1972). *J. Phycol.* **8**, 243–256.
Sommer, J. R., and Blum, J. J. (1965). *Exp. Cell Res.* **39**, 504–527.
Soyer, M. O. (1973). *Ann. Sci. Nat. Zool. Paris* **15**, 231–258.
Soyer, M. O., and Haapala, O. K. (1974). *J. Micros. (Paris)* **19**, 137–146.
Stewart, K. D., and Mattox, K. R. (1975). *Bot. Rev.* **41**, 104–135.
Stewart, K. D., Mattox, K. R., and Chandler, C. D. (1974). *J. Phycol.* **10**, 65–79.
Sui, C.-H., Swift, H., and Chiang, K. S. (1976). *J. Cell Biol.* **69**, 352–370.
Swale, E. M. F. (1969). *Br. Phycol. J.* **4**, 65–86.
Swale, E. M. F. (1973). *Br. Phycol. J.* **8**, 95–99.
Sweeney, B. M. (1976). *J. Cell Biol.* **68**, 451–461.
Taylor, F. J. R. (1971). *J. Phycol.* **7**, 249–258.
Taylor, F. J. R. (1973). *J. Phycol.* **9**, 1–10.
Taylor, F. J. R. (1974). *Taxon* **23**, 229–258.

Taylor, F. J. R. (1975). *Phycologia* **14**, 45–47.
Taylor, F. J. R. (1976a). *J. Protozool.* **23**, 28–40.
Taylor, F. J. R. (1976b). *Taxon* **25**, 377–390.
Taylor, F. J. R., Blackbourn, D. J., and Blackbourn, J. (1969). *Nature (London)* **224**, 819–821.
Taylor, F. J. R., Blackbourn, D. J., and Blackbourn, J. (1971). *J. Fish. Res. Board Can.* **28**, 391–407.
Thompson, A. (1973). *J. S. Afr. Bot.* **39**, 35–39.
Thomsen, H. A. (1975). *Br. Phycol. J.* **10**, 113–127.
Thomsen, H. A. (1977). *Bot. Not.* **130**, 147–153.
Throndsen, J. (1971). *Norw. J. Bot.* **18**, 47–64.
Throndsen, J. (1973). *Norw. J. Bot.* **20**, 271–280.
Tippit, D. H., and Pickett-Heaps, J. D. (1976). *J. Cell Sci.* **21**, 273–289.
Tippit, D. H., McDonald, K. L., and Pickett-Heaps, J. D. (1975). *Cytobiologie* **12**, 52–73.
Tomas, R. N., and Cox, E. R. (1973). *J. Phycol.* **9**, 304–323.
Tomas, R. N., Cox, E. R., and Steidinger, K. A. (1973). *J. Phycol.* **9**, 91–98.
Triemer, R. E., and Brown, R. M. (1974). *J. Phycol.* **10**, 419–433.
Triemer, R. E., and Brown, R. M. (1977). *Br. Phycol. J.* **12**, 23–44.
Van der Veer, J. (1976). *J. Mar. Biol. Assoc. U.K.* **56**, 21–30.
Van der Veer, J., and Lewis, R. J. (1977). *Acta Bot. Neerl.* **26**, 159–176.
van Valkenburg, S. D. (1971). *J. Phycol.* **7**, 113–118.
Vickerman, K., and Preston, T. M. (1970). *J. Cell Sci.* **6**, 365–383.
Walne, P. L., and Arnott, H. J. (1967). *Planta* **77**, 325–353.
Wehrmeyer, W. (1970a). *Protoplasma* **70**, 295–315.
Wehrmeyer, W. (1970b). *Arch. Mikrobiol.* **71**, 367–383.
Wetherbee, R. (1975a). *J. Ultrastruct. Res.* **50**, 58–64.
Wetherbee, R. (1975b). *J. Ultrastruct. Res.* **50**, 65–76.
Wetherbee, R. (1975c). *J. Ultrastruct. Res.* **50**, 77–87.
Whatley, J. M. (1977). *New Phytol.* **79**, 309–313.
Willey, R. L., Durban, E. M., and Bowen, W. R. (1973). *J. Phycol.* **9**, 211–215.
Wujek, D. E. (1976). *Cytologia* **41**, 665–670.
Zimmermann, B. (1977). *Br. Phycol. J.* **12**, 287–290.

POSTSCRIPT

In a recent paper [*Can. J. Bot.* **56**, 2883 (1978)] Gibbs has discussed the origin of the *Euglena* chloroplast (see Section IV of this chapter). She comes to the conclusion that the chloroplasts have been derived from endosymbiotic green algae. It is also suggested that dinoflagellate chloroplasts may have evolved from symbiotic eucaryotic algae. Whatley, John, and Whatley [*Proc. Roy. Soc. Lond. Ser. B*, **204**, 165 (1979)] suggest that chloroplasts in both the Euglenophyceae and Dinophyceae may be derived from isolated chloroplasts, rather as is suggested in the present article.

Structural Features of Protozoan Cytochromes

3

GRAHAM W. PETTIGREW

I. INTRODUCTION

Cytochromes are components of electron transport processes in both prokaryotic and eukaryotic organisms. The c-type cytochromes, that is, those containing covalently bound heme, are for reasons of size, solubility, and ease of purification the best characterized class. These cytochromes are involved in diverse types of electron transport and show corresponding diversity in properties (Kamen and Horio, 1970), but recent amino acid sequence determination, principally by Ambler and co-workers (1976), has indicated that certain similarities in structure are apparent in functionally very dissimilar cytochromes. These findings have been extended by recent X-ray crystallographic studies (Salemme et al., 1973; Dickerson et al., 1976; Korszun and Salemme, 1978; Almassy and Dickerson, 1978) which show that the tertiary structures of cytochromes c

BIOCHEMISTRY AND PHYSIOLOGY OF PROTOZOA
SECOND EDITION, VOL. 1

retain common features even though primary structure similarities are extremely tenuous.

This comparative structural work has two principal contributions to make. Firstly, details of changes that occur in a protein during evolution provide circumstantial evidence indicating which areas of a molecule are functionally important. Thus the comparative biochemistry of cytochromes affects proposals concerning mechanism. Secondly, the differences observed between two proteins reflect in some way the time of separation of their respective genes and thus can give information on the relationships between the organisms of which these genes are a part.

Protozoa are uncharted territory as far as evolutionary relationships are concerned (Corliss, 1974; Hutner and Corliss, 1976), and yet they promise to answer some of the most interesting phylogenetic questions. These involve problems of the origin of Metazoa, the divergence of animals and plants, the relationships of algae to Protozoa, and the origin of eukaryotic organelles and eukaryotes themselves from presumed prokaryotic ancestors. If the potential prizes are great, the effort to date as far as amino acid sequences are concerned has been unimpressive, and only a very few proteins from selected protozoan species have been studied. This chapter concerns itself with the small number of protozoan c-type cytochromes that have been investigated and discusses whether these and future protein sequences can contribute to the elucidation of eukaryotic phylogeny.

II. THE ISOLATION OF MITOCHONDRIAL CYTOCHROMES c FROM PROTOZOA

Mitochondrial cytochromes c (that is, cytochromes which donate electrons to cytochrome oxidase in eukaryotic cells) have been isolated from relatively few Protozoa. This is in contrast to the large number of such cytochromes purified from other eukaryotes (Margoliash and Schejter, 1966). Mitochondrial cytochrome c has been isolated from *Euglena gracilis* (Gross and Wolken, 1960; Perini *et al.*, 1964; Meyer and Cusanovich, 1972; Lin *et al.*, 1973; Pettigrew *et al.*, 1975b), *Crithidia oncopelti* (Pettigrew *et al.*, 1975b), *Crithidia fasciculata* (Kusel *et al.*, 1969; Hill *et al.*, 1971a), and some trypanosome species (Hill *et al.*, 1971b). Cytochrome c_{553} from *Tetrahymena pyriformis* which, according to the above definition is also a member of the class of mitochondrial cytochromes c (Lloyd and Chance, 1972), but which has atypical properties, has been isolated by Yamanaka *et al.* (1968) and Tarr and Fitch (1976).

All the mitochondrial cytochromes c (with the exception of *T. pyriformis* cytochrome c_{553}) are basic proteins and can be collected on a cation ex-

Table I Purification of Mitochondrial Cytochromes c from Protozoa[a]

Species and source	α peak	Chromatographic conditions	Isoelectric point	Yield (μmoles/kg)	Reference
Crithidia oncopelti (B. Newton, Molteno Institute, Cambridge, England)	557	CM-cellulose, 40 mM phosphate, pH 7	—	20	Pettigrew et al. (1975b)
Crithidia fasciculata (ATCC 11745)	555	Amberlite CG-50, 0.19 M Na$^+$, pH 8	9.9	2.7	Kusel et al. (1969)
		Amberlite IRC-50, 0.16 M Na$^+$, pH 8	8.8	1.5	Hill et al. (1971a)
		CM-cellulose, 60 mM phosphate, pH 7	—	5.3	Pettigrew (unpublished data)
Euglena gracilis strain Z	558	CM-cellulose, 80 mM NaCl, 20 mM Tris–HCl, pH 7.3	—	3	Meyer and Cusanovich (1972)
		Amberlite CG-50, 0.28 M Na$^+$, pH 7	9.55	—	Lin et al. (1973)
		CM-cellulose, 40 mM phosphate, pH 7	—	2	Pettigrew et al. (1975b)
Acanthamoeba castellanii (K. M. G. Adam, University of Edinburgh, Edinburgh, Scotland)	550	CM-cellulose, 40 mM phosphate, pH 7	—	1.5	Pettigrew (unpublished data)
Tetrahymena pyriformis (K. Kuroda, University of Osaka, Osaka, Japan)	553	DEAE-cellulose, ionic strength not given	—	4[b]	Yamanaka et al. (1968)
		DE-Sephadex A-25, 0.03 M Na$^+$, pH 8.4	6.5	1.2	Tarr and Fitch (1976)
Horse heart	550	Amberlite IRC-50, 0.22 M Na$^+$, pH 8	10.0	20	Margoliash and Walasek (1967)
		CM-cellulose, 80 mM phosphate, pH 7	—	—	Pettigrew (unpublished data)

[a] The sources of the organisms are from the first reference listed in each case. Chromatographic conditions are for elution of cytochromes from ion-exchange columns and are recorded to give an approximate idea of the ionic strength necessary to achieve chromatography. In gradient elution the actual concentration of salt in the main cytochrome fractions varies with the column size and capacity and the volume and concentrations of the gradient components.

[b] Yield per kilogram of acetone powder. All other figures are for packed wet cells.

changer from desalted or diluted cell extracts (Table I). Gradient elution can then be employed to separate the cytochrome from contaminating proteins. Pettigrew *et al.* (1975b) made use of the redox properties of cytochromes during purification. Chromatography was carried out first with the cytochrome in the reduced state and then in the oxidized state (Figure 1). The oxidized form was eluted more slowly from the column and could therefore be separated from contaminants which comigrated with the cytochrome in the initial chromatography of the reduced form. The chromatography conditions for the basic protozoan cytochromes *c* studied are very similar and indicate that they are rather less basic than,

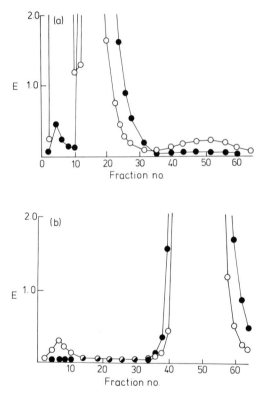

Figure 1. Chromatography of ferrocytochrome c_{557} (a) and ferricytochrome c_{557} (b) on CM-cellulose. (a) *Crithidia oncopelti* cytochrome c_{557} in 10 mM sodium phosphate, pH 7, containing 0.5 mM sodium ascorbate was adsorbed onto a CM-cellulose bed (Whatman CM52, 10 × 1 cm). The cytochrome was eluted with 40 mM sodium phosphate, pH 7. (b) After dilution (three times) and oxidation with a few crystals of potassium ferricyanide, the cytochrome was readsorbed on an identical CM-cellulose bed and again eluted with 40 mM sodium phosphate, pH 7. Open circles, E_{280}; solid circles, E_{412}. (From Pettigrew *et al.*, 1975b, reproduced by permission.)

for example, horse cytochrome c. This is borne out by studies of their isoelectric point (Table I) and also by their overall charge calculated from amino acid sequence data.

The purity index $E_{\alpha \text{ peak reduced}}/E_{280_{ox}}$ is used to monitor purification (Margoliash and Walasek, 1967). Only *Tetrahymena* cytochrome c_{553} gave values of 1.2–1.3, corresponding to those obtained from mitochondrial cytochromes c from higher organisms (Margoliash and Walasek, 1967). The lower values of the final purity index achieved for *C. oncopelti* cyto-chrome c_{557}, *C. fasciculata* cytochrome c_{555}, and *Euglena* cytochrome c_{558} can be rationalized in terms of a lower α-peak extinction and also, in the case of *Euglena* cytochrome c_{558}, the presence of a second tryptophan residue (and therefore a higher molar extinction at 280 nm).

Although useful for monitoring the removal of contaminant proteins during purification, the purity index does not allow assessment of the presence of modified forms of cytochromes. In the case of mammalian cytochromes c modified species include polymeric and deamidated forms and can be detected by their ability to bind carbon monoxide or by electrophoresis (Margoliash and Walasek, 1967). The ability to bind carbon monoxide is of limited usefulness in the purification of these protozoan cytochromes, since some appear to bind carbon monoxide in a slow reaction (Pettigrew *et al.*, 1975a).

During the purification of cytochrome c_{557} from *C. oncopelti*, a minor cytochrome band was observed, the amino acid analysis and NH_2-terminus of which were consistent with the loss of residues -10 to -5 from the NH_2-terminus (Pettigrew *et al.*, 1975b). It is not known whether this arginyl–glutamyl bond is particularly susceptible to hydrolysis, or whether other bonds were broken and the resultant species were not observed. A similar minor species was noted during the purification of *C. fasciculata* cytochrome c_{555} (Hill *et al.*, 1971a).

Our preliminary unpublished results for *Acanthamoeba castellanii* indicate that a cytochrome c is present which has chromatographic properties similar to those of other protozoan cytochromes c but which has an α peak$_{(max)}$ at 550 nm, confirming the spectroscopic studies of whole mitochondria by Edwards *et al.* (1977).

Tetrahymena cytochrome c_{553} is atypical among the mitochondrial cytochromes c in that it is acidic and can be collected on anion-exchange columns such as DEAE-cellulose (Yamanaka *et al.*, 1968; Tarr and Fitch, 1976). Tarr and Fitch carefully looked for a basic cytochrome in *Tetrahymena* extracts but could find none, and conversely no acidic cytochrome comparable to cytochrome c_{553} has been found in extensive studies of other eukaryotes.

The yields obtained from different species are shown in Table I.

Crithidia oncopelti cells contain much more cytochrome *c* than the others, with yields equal to those of cytochrome *c* in mammalian heart muscles (15–20 μmoles/kg; Margoliash and Walasek, 1967).

III. PROPERTIES OF MITOCHONDRIAL CYTOCHROMES c FROM PROTOZOA

A. Reactivity with Cytochrome Oxidase and NADH–Cytochrome c Reductase from Mammalian Sources

Yamanaka and Okunuki (1968) detected differences in the reactivity of cytochromes *c* with bovine cytochrome oxidase. They proposed that such differences may reflect how closely cytochromes, and the organisms from which they derive, are related. The subsequent work of Smith *et al.* (1973a) indicated that such differences were due to the presence of inhibiting ions in different preparations and that cytochromes dialysed near their isoelectric point were very similar in their reactivity with bovine cytochrome oxidase. Recently, Ferguson-Miller *et al.* (1976) and Errede *et al.* (1976) studied reactivity over a wide range of cytochrome *c* concentrations and demonstrated differences between different mitochondrial cytochromes *c*. They emphasize that determinations of reactivity over such a concentration range are necessary for meaningful comparisons between cytochromes, because a cytochrome which shows a higher rate constant than another at one concentration may have a lower rate constant relative to the other at a different concentration. The mitochondrial cytochromes *c* studied included *Euglena* cytochrome c_{558}, and in Figure 2 the Eadie–Hofstee plot is compared with those for yeast and horse cytochromes *c*. Ferguson-Miller *et al.* interpret these results in terms of high- and low-affinity binding sites on the oxidase for horse cytochrome *c* and only a low-affinity binding in the case of *Euglena* cytochrome c_{558}. According to these authors, the apparent K_m values derived from such plots are interpretable as dissociation constants for productive complex formation between enzyme and substrate. However, Errede *et al.* emphasize that alternative kinetic schemes are equally consistent with the results, in particular, a mechanism involving no active complexes but rather the formation of unproductive "dead-end" complexes.

The results in Table II are the results of Errede (1976) for the protozoan cytochromes considered in this chapter. These are for single cytochrome concentrations, but the *C. fasciculata* cytochrome c_{555} was more extensively compared because of the results of Hill *et al.* (1971a), who reported a lower activity of this cytochrome with cytochrome oxidase. According

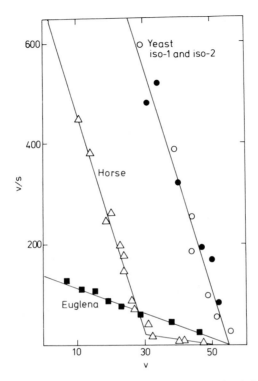

Figure 2. Comparison of the activity of yeast iso-1, yeast iso-2, horse, and *Euglena* cytochromes with Keilin–Hartree particle cytochrome c oxidase. Squares, *Euglena;* triangles, horse; open circles, yeast iso-1; solid circles, yeast iso-2. Conditions were 25 mM Tris–cacodylate, pH 7.8, 250 mM sucrose, 7 mM sodium ascorbate, and 0.7 mM TMPD at 25°C. Final concentrations were Keilin–Hartree particle protein, 0.07 mg/ml; cytochrome c, 0.05–3.2 μM. (From Ferguson-Miller *et al.*, 1976, reproduced by permission.)

to the interpretation of Ferguson-Miller *et al.*, *C. fasciculata* c_{555} resembles *Euglena* c_{558} in having a lower affinity for site 1 on the oxidase (that is, the high-affinity site). However, the two *Crithidia* cytochromes show reactivity very similar to that of horse cytochrome c at concentrations near those employed by Hill *et al.* Their claim, therefore, that *C. fasciculata* cytochrome c_{555} reacts at only 20% of the rate of horse heart cytochrome c is not supported by these studies.

The reverse experiment, in which protozoan cytochrome oxidases are tested for reactivity with mammalian cytochrome c, has been attempted for *Euglena*. (R. H. Brown, personal communication). When horse cytochrome c was added to depleted *Euglena* mitochondria, no activity was detected, whereas *Euglena* cytochrome c_{558} restored normal respiration. This remarkable observation suggests that the differences observed in the

Table II Reactivity of Protozoan Cytochromes with Bovine Cytochrome Oxidase[a]

Cytochrome	Rate relative to that of horse cytochrome c
Horse cytochrome c	100
Crithidia oncopelti c_{557}	103[b]
Crithidia fasciculata c_{555}	156
Euglena gracilis c_{558}	303[b]
Tetrahymena pyriformis c_{553}	0.7

[a] Conditions were 100 mM 2-N-(morphilino)ethane-sulfonic acid, pH 6.0, 10 μM EDTA, 5 μM cytochrome, 25°C. These results are from Table 7 of Errede (1976) and personal communication of unpublished results (Errede, personal communication).

[b] *Crithidia oncopelti* cytochrome c_{557} and *E. gracilis* cytochrome c_{558} were assayed in a different set of experiments using a polarographic method in 20 mM potassium phosphate, pH 7, 10 μM cytochrome, 1 mM ascorbate, 20 μM N,N,N',N'-tetramethyl-p-phenyl-enediamine (TMPD), 30°C, and expressed relative to the reactivity of horse cytochrome c under the same conditions.

bovine cytochrome oxidase assays between horse and *Euglena* cytochromes may reflect a fundamental difference in the molecules.

It was found by Ryley (1952) and confirmed by later workers (Lloyd and Chance, 1972) that the same type of experiment with *Tetrahymena* mitochondria again showed that added horse cytochrome c could not restore respiration. In this case, however, the purified *Tetrahymena* cytochrome c_{553} was unable to react with bovine cytochrome oxidase (Yamanaka and Okunuki, 1968). In view of the report that the regions of interaction of cytochrome c with its oxidase and its reductase are separate (Smith *et al.*, 1973b), it was of interest to know whether *Tetrahymena* cytochrome c_{553} could be reduced by bovine NADH–cytochrome c reductase. The results of Errede (1976) (Table III) show that it cannot, the very low amount of reduction that took place being only partially inhibited by antimycin A.

Thus *Tetrahymena* cytochrome c_{553} is clearly atypical in its lack of reactivity with either mammalian oxidase or reductase, although its functional role as the immediate electron donor to cytochrome oxidase in *Tetrahymena* was established by Lloyd and Chance (1972). Possible evolutionary implications of this are discussed in Section IV. As for the other protozoan cytochromes considered, it is difficult to assess the significance of the differences observed in these assays. A reasonable conclusion may

Table III Reactivity of Protozoan Cytochromes with Bovine
NADH–Cytochrome c Reductase[a]

Cytochrome	Reaction rate relative to that of horse cytochrome c		Antimycin A inhibition (%)
	Zero-order region	First-order region	
Horse cytochrome c	100	100	97
Crithidia oncopelti c_{557}	97	61	98
Crithidia fasciculata c_{555}	101	84	98
Euglena gracilis c_{558}	102	102	97
Tetrahymena pyriformis c_{553}	0.08		20

[a] Conditions were 50 mM HEPES buffer, pH 7.5, 20 μM EDTA, 250 μM KCN, 100 μM NADH, 2.9 μg complex I–III, 25°C. The zero-order and first-order regions of the time course of the reaction refer to the analysis of Errede (1976). The percentage inhibition by antimycin A is a measure of the specificity of reduction.

be that the similarity in reactivity of *C. oncopelti* c_{557}, *C. fasciculata* c_{555}, *Euglena* c_{558}, and horse cytochrome c with mammalian oxidase and reductase preparations suggests that the structural features that set these protozoan cytochromes apart (discussed in Section IV) do not affect the functional properties of the proteins.

B. Spectroscopic Studies and the Heme-Binding Site

1. Optical Absorption Spectra

The optical absorption spectra of cytochromes c derive from electronic transitions in the heme. These transitions are influenced by chemical substituents on the tetrapyrrole ring and by the environment created by the enclosing protein which also provides the fifth and sixth coordination ligands of the iron.

Mitochondrial cytochromes c form a very homogeneous class with respect to their optical absorption spectra, but the protozoan cytochromes considered here are markedly different. The spectra of *C. oncopelti* cytochrome c_{557} are shown in Figure 3. All maxima are shifted towards the red end of the spectrum relative to other mitochondrial cytochromes c. This shift in the case of the α peak is compared in Figure 4 with those of other protozoan cytochromes. Hill *et al.* (1971b) reported that the α-peak absorption maxima of several other trypanosomatid cytochromes fall between 555 and 558 nm.

All these spectra are atypical when compared to those of other mitochondrial cytochromes, although complex α peaks shifted to various

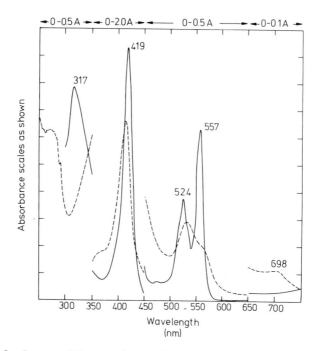

Figure 3. Spectra of *C. oncopelti* cytochrome c_{557} in the wavelength region 250–750 nm. Fully oxidized cytochrome (dashed line) was obtained by the addition of potassium ferricyanide and removal of the excess reagent by passage through a small column of Sephadex G-25. Reduction of the cytochrome (solid line) was with 0.5 mM sodium ascorbate. Spectra were recorded at different absorbance range settings in 10 mM sodium phosphate, pH 7, at room temperature.

degrees are not unusual among bacterial cytochromes c (Kamen and Horio, 1970) and algal cytochromes f (Yakushiji, 1971; Katoh, 1959). They all, however, show the near-infrared band in the oxidized state, which is thought to derive from methionine coordination of the heme iron (Schechter and Saludjian, 1967). The details of the spectra are summarized in Table IV.

2. Spectrum of the Pyridine Ferrohemochrome

This spectrum is used as a diagnostic aid in determining the nature of the prosthetic group in hemoproteins (Enzyme Nomenclature Recommendations, 1965). Perini *et al.* (1964) found that the α-peak maximum of the pyridine ferrohemochrome of *Euglena* cytochrome c_{558} was at 553 nm, intermediate between the corresponding peaks for c-type cytochromes in general (550 nm) and proteins containing protoheme IX (557 nm). Kusel *et al.* (1969) reported this same anomaly for *C. fasciculata* cytochrome c_{555},

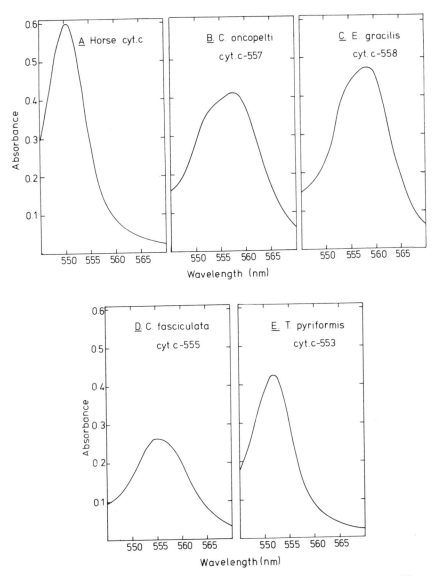

Figure 4. The α peaks of protozoan cytochromes. The spectra of the α peak for different protozoan cytochromes were recorded and compared with that for horse cytochrome c.

and Meyer and Cusanovich (1972) noted the similarity of these pyridine hemochrome spectra to that of 2(4)-vinyl-4(2)-ethyl-substituted heme.

There is now substantial evidence that the heme groups of the two *Crithidia* cytochromes and *Euglena* cytochrome c_{558} are attached through

Table IV Features of the Absorption Spectra of Protozoan Cytochromes between 250 and 750 nm[a]

Source	Ferrocytochrome maxima (nm)	Ferricytochrome maxima (nm)	Pyridine hemochrome α-peak maximum (nm)	α/280	Reference
Crithidia oncopelti[b]	α: 557$_{as}$ (24.7) β: 524 (15.0) 419 (140) 317 (30)	NIR: 698 (0.9) 532 (12.5) 412 (100) 275 (24.1)	553 —	1.02	Pettigrew et al. (1975b)
Crithidia fasciculata[c]	α: 555 (29.7) β: 525 (16.8) 420 (154)	NIR: not determined 533 (12.2) 413 (112)	553 —	1.04	Kusel et al. (1969)
Euglena gracilis[b]	α: 558$_{as}$ (26.3) β: 526 (15.5) 422 (158) 318 (31)	NIR: 702 (1.1) 530 (10.5) 412 (101) 280 (30.0)	553 — — —	0.88	Perini et al. (1964) Pettigrew et al. (1975b)
Acanthamoeba castellanii[b]	α: 550 β: 521	— —	— —	0.93	Pettigrew (unpublished data)
Tetrahymena pyriformis[d]	α: 553 (27.4) β: 523 414	410	550 —	1.17	Yamanaka et al. (1968)
Horse heart[e]	α: 550 (27.7) β: 521 (15.9) 416 (129) —	NIR: 695 (0.81) 528 (11.2) 410 (106) 280 (23.2)	550 — — —	1.2	Margoliash and Frohwirth (1959)

[a] Extinction coefficients are shown in parentheses where available; as, asymmetrical peak; NIR, near-infrared absorption band centered around 695 nm. Spectra were recorded at room temperature.
[b] 10 mM sodium phosphate, pH 7.
[c] 50 mM sodium phosphate, pH 7.
[d] 10 mM Tris–HCl, pH 8.5.
[e] 0.1 M sodium phosphate, pH 6.8.

only one thioether bond to the apoprotein and that one vinyl group remains unsaturated. Thus the α peaks of the pyridine ferrohemochromes of cytochromes c_{557} and c_{558} can be shifted to 551 nm by heating in hydrazine hydrate (Pettigrew et al., 1975b), which reacts with free vinyl groups. The heme of these cytochromes cannot be removed by acid acetone treatment but is released by mercuric chloride in dilute acid. The isolated hemin shows chromatographic and spectroscopic properties intermediate between those of hematohemin and protoheme IX. Also, the amino acid sequences of the two Crithidia cytochromes and Euglena cytochrome c_{558} (see Section IV) have only one cysteine, at residue 17. Residue 14, always a heme-bonding cysteine in other cytochromes c, is alanine.

Recently Miller and Rapoport (1977) isolated the porphyrin of cytochrome c_{558} and established rigorously that it is 2-vinyl-4-ethyldeuteroporphyrin IX. The structure of the heme-binding site of Euglena cytochrome c_{558} is therefore shown in Figure 5. By analogy, other c-type cytochromes presumably have Cys 14 attached to the 2-position.

Thus the shift to longer wavelengths of the spectra of these protozoan cytochromes can be explained by extended delocalization of the heme-conjugated electrons to the free vinyl side chain. This is also reflected in the resonance Raman spectra of these cytochromes, which show characteristics intermediate between those of cytochromes c and b_5 (Adar, 1977).

The shift observed for Tetrahymena cytochrome c_{553} cannot be explained in this way, for this cytochrome shows a normal pyridine ferrohemochrome spectrum (Yamanaka et al., 1968) and a normal heme-binding site. In this shift it resembles mitochondrial cytochrome c_1, algal cytochromes f, and some bacterial cytochromes. Mitochondrial cytochromes c with red-shifted spectra are not a distinguishing feature of Pro-

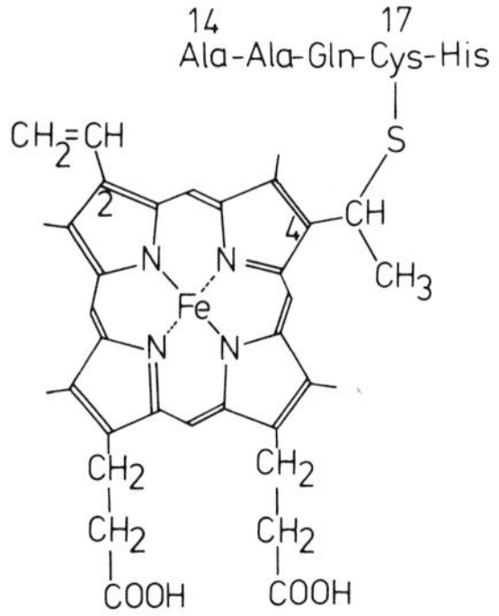

Figure 5. Attachment of the heme to the protein in Euglena cytochrome c_{558}.

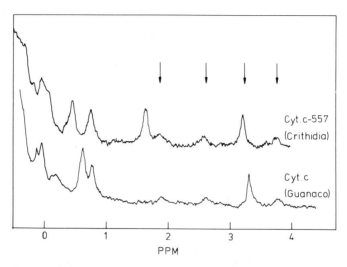

Figure 6. High-field part of the 220-MHz proton NMR spectra at 35°C of ferrocyto-chrome c from guanaco and $C.$ $oncopelti.$ The spectrum of ferrocytochrome c from guanaco is representative of the NMR spectra of all the vertebrate cytochromes c studied. The arrows indicate the positions of the resonances of the sixth heme ligand, Met 80, at 3.3 ppm (methyl) and 1.9, 2.6, and 3.75 ppm (protons) in the vertebrate cytochrome $c,$ and at 3.2 ppm (methyl) and 1.85, 2.55, and 3.75 ppm (protons) in $C.$ $oncopelti$ cytochrome $c_{557}.$ The spectra and resonance positions are from Keller et $al.$ (1973).

tozoa, since $A.$ $castellanii$ contains a basic cytochrome c with an α peak coincident with that of horse cytochrome c at 550 nm (Edwards et $al.,$ 1977; Pettigrew, unpublished observations) (Table IV).

3. Proton NMR Spectra

The ^{1}H NMR spectra of cytochrome c show resonances shifted outside the normal spectral region. These shifts are due to interactions with the unpaired electron on the iron in the oxidized protein and to local heme ring current effects in the reduced protein (Wuthrich, 1969; McDonald et $al.,$ 1969). Shifted resonances include those of the protons of the heme group, the thioether bridges, and the axial ligands, and these resonances reflect the electronic structure of the heme group. Comparison of such shifted resonances in the NMR spectra of guanaco and $C.$ $oncopelti$ fer-rocytochromes c (Figure 6) indicates that the coordination of the iron in the protozoan cytochrome is essentially identical to that in other mitochondrial cytochromes $c.$ The small differences in chemical shifts of these methionine resonances correspond to increases in the iron–sulfur bond length of only a few hundredths of an angstrom in cytochrome $c_{557}.$

Table V　Midpoint Oxidation–Reduction Potentials of Protozoan Cytochromes c^a

Cytochrome	E_m(mV)	Reference
Crithidia oncopelti c_{557}	254	Pettigrew *et al.* (1975a)
Crithidia fasciculata c_{555}	280	Kusel *et al.* (1969)
	253	Pettigrew (unpublished data)
Euglena gracilis c_{558}	307	Perini *et al.* (1964)
	244	Pettigrew *et al.* (1975a)
Tetrahymena pyriformis c_{553}	245	Yamanaka *et al.* (1968)
Horse cytochrome c	261	Margalit and Schejter (1973)

[a] For the results of Pettigrew *et al.* (1975a) and Pettigrew (unpublished data) the conditions of Margalit and Schejter, 1973 were used. Thus these values are for single equilibrium measurements at 25°C in air assuming $n = 1$. Cytochrome (5×10^{-6} M) in 1 mM sodium phosphate, pH 7 (3 ml) was fully oxidized by the addition of 50 nmoles potassium ferricyanide (10 μl), and the spectrum of the α band region was recorded. A redox equilibrium was then established by the addition of 1500 nmoles potassium ferrocyanide (10 μl), and the spectrum was recorded again. The poised potential, E_h, at these concentrations and a total ionic strength of 0.008 mole/liter, is 290 mV (E_m for the ferroferricyanide couple being 378 mV under these conditions). Finally, sodium dithionite was added, and the completely reduced spectrum was obtained.

C. Physicochemical Properties of the Heme Crevice

The heme group of cytochromes can be used as a natural built-in probe of the protein structure. Properties such as the oxidation–reduction potential and the binding of ligands to the iron are determined not only by the coordination of the iron but also by the local environment of the heme provided by folding of the peptide chain.

Mitochondrial cytochromes c are highly conserved in redox potential, only varying between 260 and 265 mV in five species studied (Margalit and Schejter, 1973). The redox potentials of the protozoan cytochromes considered here are shown in Table V. There is disagreement in the results for *C. fasciculata* cytochrome c_{555} and *Euglena* cytochrome c_{558}. Whether these discrepancies are due to differences in the cytochrome preparations or to different methods used for determination of the redox potential is not known. They probably cannot be explained by the different ionic strength conditions or by errors due to the assumption of 430 mV as the midpoint potential for the ferroferricyanide couple, because all workers agree on control values for horse cytochrome c.

When the results of Pettigrew *et al.* (1975a) and unpublished observations are compared, which were obtained under the conditions of Margalit and Schejter (1973), *Euglena* cytochrome c_{558} has a lower potential than the two *Crithidia* cytochromes, which have slightly lower potentials than those of the mitochondrial cytochromes c studied by Margalit and Schejter. However, the deviation is not large (compare, for example, the 100-mV range encompassed by the algal cytochromes f (Yakushiji, 1971)), and the redox potential can still be considered a remarkably conserved property of the mitochondrial cytochromes c in view of the observed variability of 75% of their amino acid sequence.

In other properties also, the heme crevice of the single thioether protozoan cytochromes is little different from that of other cytochromes c. Attempts to perturb the heme crevice by heat, alkaline pH, and ligands (Pettigrew *et al.*, 1975a) gave results slightly different from those for horse cytochrome c, but no more different than the corresponding results for yeast cytochrome c. Thus the loss of a covalent linkage to the heme does not appear to result in increased heme accessibility or greater susceptibility to unfolding.

IV. AMINO ACID SEQUENCE STUDIES ON MITOCHONDRIAL CYTOCHROMES c

A. Evolutionary Implications

The amino acid sequences of C. *oncopelti* cytochrome c_{557} (Pettigrew *et al.*, 1975b), C. *fasciculata* cytochrome c_{555} (Hill and Pettigrew, 1975), $E.$ *gracilis* cytochrome c_{558} (Lin *et al.*, 1973; Pettigrew, 1973), and $T.$ *pyriformis* cytochrome c_{553} (Tarr and Fitch, 1976) are compared in Figure 7 with cytochrome c from horse (Margoliash *et al.*, 1961), wheat (Stevens *et al.*, 1967), and *Neurospora* (Heller and Smith, 1966), and cytochrome c_2 from the photosynthetic prokaryote *Rhodomicrobium vannielii* (Ambler *et al.*, 1976). For the purposes of discussion the most divergent sequence, that of *Tetrahymena* cytochrome c_{553}, is considered separately.

1. Crithidia and Euglena Cytochromes c

These protozoan cytochrome c sequences have certain unusual features which distinguish them from other mitochondrial cytochromes c. Thus the conserved pattern of the heme-binding site

$$Cys\text{-}X\text{-}Y\text{-}Cys\text{-}His$$

which is found, not only in mitochondrial cytochromes c, but in almost all bacterial c-type cytochromes studied, is altered by the replacement of Cys 14 by an alanine. Of course it is not known whether a change to cysteine

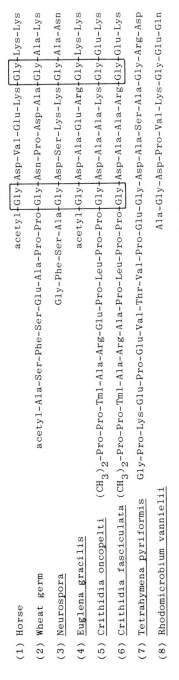

1   5

(1) Horse   acetyl-Gly-Asp-Val-Glu-Lys-Gly-Lys-Lys

(2) Wheat germ   acetyl-Ala-Ser-Phe-Ser-Glu-Ala-Pro-Pro-Gly-Asn-Pro-Asp-Ala-Gly-Ala-Lys

(3) Neurospora   Gly-Phe-Ser-Ala-Gly-Asp-Ser-Lys-Lys-Gly-Ala-Asn

(4) Euglena gracilis   acetyl-Gly-Ala-Glu-Arg-Gly-Lys-Lys

(5) Crithidia oncopelti $(CH_3)_2$-Pro-Pro-Tml-Ala-Arg-Glu-Pro-Leu-Pro-Pro-Gly-Asp-Ala-Ala-Lys-Gly-Glu-Lys

(6) Crithidia fasciculata $(CH_3)_2$-Pro-Pro-Tml-Ala-Arg-Ala-Pro-Leu-Pro-Pro-Gly-Asp-Ala-Ala-Arg-Gly-Glu-Lys

(7) Tetrahymena pyriformis Gly-Pro-Lys-Glu-Pro-Glu-Val-Thr-Val-Pro-Glu-Val-Gly-Asp-Ala-Ser-Ala-Gly-Arg-Asp

(8) Rhodomicrobium vannielii   Ala-Gly-Asp-Pro-Val-Lys-Gly-Glu-Gln

Figure 7. Comparison of the amino acid sequences of selected cytochromes c. The four known amino acid sequences of protozoan cytochromes are aligned with representative sequences from the three eukaryotic kingdoms. Some regions of the sequence of *C. fasciculata* cytochrome c_{555} are not fully characterized and are assigned only on the basis of amino acid composition and homology with the other sequences. These regions are shown in parentheses. The positions common to all mitochondrial cytochromes c with the exception of *T. pyriformis* are boxed in solid lines. Positions which have accepted variation only to the extent of the conserved pairs of amino acids Lys/Arg, Ser/Thr, Asp/Glu, Tyr/Phe, and Ile/Leu are shown boxed in broken lines. The NH_2-terminus of *C. oncopelti* c_{557} is *N*-dimethylproline. Tml, Trimethyllysine. The amino acid sequences are from Margoliash *et al.* (1961) (horse heart); Stevens *et al.* (1967) (wheat germ); Heller and Smith (1966), and corrections by Lederer and Simon (1974) (*Neurospora*); Lin *et al.* (1973), and Pettigrew (1973) (*Euglena*); Pettigrew *et al.* (1975b) (*C. oncopelti*); Hill and Pettigrew (1975) (*C. fasciculata*); Tarr and Fitch (1976) (*Tetrahymena*), and Ambler *et al.* (1976) (*R. vannielii*). (Continued on pp. 76 and 77.)

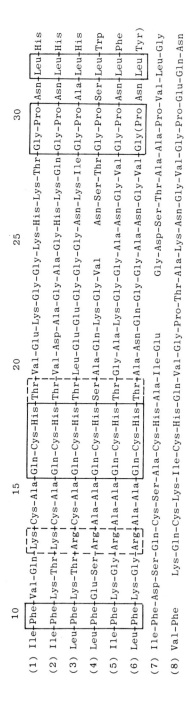

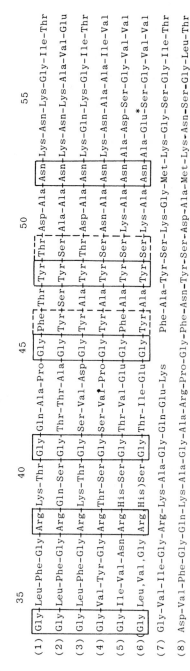

60 65 70 75 80

(1) Trp-Lys-Glu-Glu-Thr-Leu-Met-Glu-Tyr-Leu-Glu-Asn-Pro-Lys-Lys-Tyr-Ile-Pro-Gly-Thr-Lys-Met-Ile-Phe-Ala

(2) Trp-Glu-Glu-Asn-Thr-Leu-Tyr-Asp-Tyr-Leu-Leu-Asn-Pro-Lys-Lys-Tyr-Ile-Pro-Gly-Thr-Lys-Met-Val-Phe-Pro

(3) Trp-Asp-Glu-Asn-Thr-Leu-Phe-Glu-Tyr-Leu-Glu-Asn-Pro-Lys-Lys-Tyr-Ile-Pro-Gly-Thr-Lys-Met-Ala-Phe-Gly

(4) Trp-Glu-Glu-Gly-Thr-Leu-His-Lys-Phe-Leu-Glu-Asn-Pro-Lys-Lys-Tyr-Val-Pro-Gly-Thr-Lys-Met-Ala-Phe-Ala

(5) Trp-Thr-Pro-Glu-Val-Leu-Asp-Val-Tyr-Leu-Glu-Asn-Pro-Tml-Lys-Phe-Met-Pro-Gly-Thr-Lys-Met-Ser-Phe-Ala

(6) Trp-Thr-Pro-Asp-Val-Leu-Asp-Val-Tyr-Leu-Glu-Asn-Pro-Tml-Lys-Met-Pro-Gly-Thr-Lys-Met-Ser-Phe-Ala

(7) Trp-Asn-Glu-Lys-His-Leu-Phe-Val-Leu-Phe-Leu-Asn-Pro-Ser-Lys-Val-His-Val-Pro-Gly-Thr-Lys-Met-Ala-Phe-Ala

(8) Trp-Asp-Glu-Ala-Thr-Leu-Asp-Lys-Tyr-Leu-Asp-Asn-Pro-Lys-Asn-Pro-Lys-Ala-Val-Ala-Val-Pro-Gly-Thr-Lys-Met-Val-Phe-Val

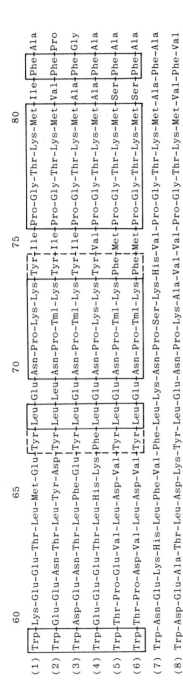
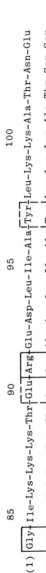

85 90 95 100

(1) Gly-Ile-Lys-Lys-Thr-Glu-Arg-Glu-Asp-Leu-Ile-Ala-Tyr-Leu-Lys-Lys-Ala-Thr-Asn-Glu

(2) Gly-Leu-Tml-Lys-Pro-Gln-Asp-Arg-Ala-Asp-Leu-Ile-Ala-Tyr-Leu-Lys-Lys-Ala-Thr-Ser-Ser

(3) Gly-Leu-Lys-Lys-Asp-Lys-Asp-Arg-Asn-Asp-Leu-Ile-Thr-Phe-Met-Lys-Glu-Ala-Thr-Ala

(4) Gly-Ile-Tml-Ala-Lys-Lys-Asp-Arg-Gln-Asp-Ile-Ile-Ala-Tyr-Met-Lys-Thr-Leu-Lys-Asp

(5) Gly-Ile-Lys-Lys-Pro-Gln-Glu-Arg-Ala-Asp-Leu-Ile-Ala-Tyr-Leu-Lys-Glu-Ala-Thr-Leu-Asn-Lys-Lys

(6) Gly-Met-Lys-Lys-Pro-Gln-Glu-Arg-Ala-Asp-Leu-Ile-Ala-Tyr-Leu-Lys-Glu-Ala-Thr-Leu-Lys-Ser-Val

(7) Gly-Leu-Pro-Ala-Asp-Arg-Ala-Asp-Leu-Ile-Ala-Tyr-Leu-Lys-Ser-Val

(8) Gly-Leu-Lys-Asn-Pro-Gln-Asp-Arg-Ala-Asp-Val-Ile-Ala-Tyr-Leu-Lys-Gln-Leu-Ser-Gly-Lys

Figure 7. (*Continued*).

took place in the line leading to the other eukaryotes or if there was a change to alanine in the line leading to these Protozoa. That is, did the nearest common ancestor of these Protozoa and the Metazoa contain a cysteine or an alanine? The preliminary results for *Acanthamoeba* cytochrome c_{550} indicate that a single thioether bridge is not a general feature of protozoan mitochondrial cytochromes c, and since distantly related bacterial cytochromes (discussed below) have both cysteine residues, the more likely alternative is that there was a loss of cysteine in the line leading to these flagellates. This change has occurred so seldom in the history of c-type cytochromes that one might argue that it has taken place only once in the Protozoa and that the two *Crithidia* cytochromes and *Euglena* cytochrome c_{558} are related.

However, comparison of the rest of the sequence does not really support this conclusion. Thus the changes at conserved residues Tyr 67 and Lys 87, not found in *C. oncopelti* cytochrome c_{557}, are unique to *Euglena* cytochrome c_{558}. Changes at Lys 27 and Ile 75 are found only in these two cytochromes but have given rise to different residues. The methylated proline NH_2-terminus of *C. oncopelti* cytochrome c_{557} is unique, while *Euglena* cytochrome c_{558} resembles vertebrate cytochromes c in having an acetylated glycine at the NH_2-terminus. Nor does the presence of methylated lysine side chains distinguish these protozoan cytochromes from other mitochondrial cytochromes c. *Crithidia oncopelti* cytochrome c_{557} is unique in having trimethyllysine near the NH_2-terminus (residue minus 8), and the methylation of Lys 72 is also observed in plant and fungal cytochromes c. *Euglena* cytochrome c_{558} has neither of these methylated lysines but has trimethyllysine at residue 86 like plant cytochromes c.

Furthermore, in an overall count of the number of differences between sequences (Table VI) there is little indication that *Euglena* and *Crithidia* cytochromes are more related to each other than they are to other cytochromes c. They appear to derive from earlier offshoots than the divergence of the animals, plants, and fungi, which for the three sequences considered show an average of 38% difference. The 45% difference observed between the *Euglena* and *Crithidia* cytochromes is slightly lower than the 49% overall average difference between the three protozoan cytochromes and horse, wheat, and *Neurospora* cytochromes, but horse cytochrome c is no more different from *Euglena* and *C. oncopelti* cytochromes than these proteins are from each other. The high value for the comparison horse versus *C. fasciculata* relative to horse versus *C. oncopelti* is an indication of the random fluctuations present in these figures.

In our opinion these results offer no support for the proposal that trypanosome flagellates may have derived from a euglenoid line during evolution of the arthropods (Bovee, 1971), but it does not disprove the

Table VI Percentage Difference Matrix of Selected Cytochromes c^a

Cytochrome source	(a)	(b)	(c)	(d)	(e)	(f)	(g)
(a) Horse	—	—	—	—	—	—	—
(b) Wheat germ	37	—	—	—	—	—	—
(c) *Neurospora crassa*	34	42	—	—	—	—	—
(d) *Euglena gracilis*	44	50	45	—	—	—	—
(e) *Crithidia oncopelti*	45	49	55	46	—	—	—
(f) *Crithidia fasciculata*	50	49	53	44	14	—	—
(g) *Tetrahymena pyriformis*	53	61	56	53	58	59	—
(h) *Rhodomicrobium vannielii*	51	53	55	56	54	54	55

^{a} Figures were obtained by comparison of all positions common to a pair of sequences. An internal gap was scored as one difference. Differences in length at the NH_2- or COOH-terminus were also scored as one difference. Thus in the comparison horse versus *Neurospora* there are 103 positions in alignment, 34 of which contain different residues. *Neurospora* cytochrome c is also longer at the NH_2-terminus (1 difference) and shorter at the COOH-terminus (1 difference) producing a final score of 36/105 differences. Trimethyllysine and lysine were scored as identical.

hypothesis. Were *E. gracilis* a descendant of that euglenoid line we would have expected a percentage difference between *Crithidia* and *Euglena* cytochromes lower than the 38% difference between eukaryotic kingdoms. However, we would only obtain such a result if we happened to choose an extant descendant of that particular line. Otherwise the percentage difference obtained would reflect an original divergence among the euglenoids.

In contrast to these conclusions, Lin *et al.* (1973) argue that an analysis of the inferred mRNA sequences supports a closer relationship for *Crithidia* and *Euglena* (Table VII). In the dendrogram of Dayhoff (1976),

Table VII Minimum Pairwise Disjointness of Selected Cytochromes c^a

Cytochrome source	(a)	(b)	(c)	(d)	(e)
(a) Pig	—	—	—	—	—
(b) *Drosophila*	17	—	—	—	—
(c) Cotton	39	43	—	—	—
(d) *Neurospora*	35	30	34	—	—
(e) *Euglena*	38	42	41	33	—
(f) *Crithidia*	43	38	33	34	26

^{a} The minimum pairwise disjointness is the number of nucleotides by which the mRNAs for cytochrome c from two species must differ in those nucleotide positions for which there are two or more different nucleotides appearing two or more times each. From Lin *et al.* (1973, Table 3).

however, *Crithidia, Euglena,* and other eukaryotes diverge together, and earlier than the divergence of the eukaryotic kingdoms.

It is probable that at the levels of dissimilarity involved the usefulness of cytochrome *c* as a marker for evolutionary relationships becomes very limited. Because of the uncertainty over multiple, back, and parallel mutations, the level of confidence in assigning remote divergence points must be lower than for more recent events. As Sneath (1974) emphasizes, the methods used to generate dendrograms of relationships give the most parsimonious solution, and often several solutions are almost equally parsimonious. Even if the minimal tree can be selected, its superiority over others may not be statistically significant. If this is an overly pessimistic view of the usefulness of cytochrome *c* sequences, it should be emphasized that the problem could be solved by choosing a more slowly evolving protein than cytochrome *c,* so that evolutionary changes could be assigned with more certainty. Also, cytochrome *c* may contribute at a more detailed level. Thus the two *Crithidia* cytochromes differ at 16 positions, a level of dissimilarity equal to that between mammalian and reptilian cytochromes *c* (Dickerson, 1971). Yet *C. oncopelti* and *C. fasciculata* are very similar in morphology, a principal difference being the presence of a proposed endosymbiotic bacterium in the former (Newton, 1968). Here, cytochrome *c* sequences can help quantitate relatedness where morphological characters have limited usefulness.

However, it is difficult to counter the biologist who may ask, Why give protein sequences special consideration when the analysis of sequence results gives rise to ambiguities and disagreement among their proponents?

2. Tetrahymena Cytochrome c

Although homologous to mitochondrial cytochromes *c, Tetrahymena* cytochrome c_{553} is markedly different from all members of this family in its acidity and its inability to interact with bovine cytochrome oxidase or NADH–cytochrome *c* reductase. This is to some extent reflected in the overall matrix in Table VI, where *Tetrahymena* c_{553} appears to have diverged earliest of all the eukaryotic cytochromes considered. There are anomalies; the cytochromes *c* of horse and *Tetrahymena* are more similar than those of *Crithidia* and *Neurospora.* But if individual sequence positions are considered, *Tetrahymena* cytochrome c_{553} appears to be distinct from the other mitochondrial cytochromes *c.* Thus it is unique in having substitutions at the otherwise conserved residues Lys/Arg 13, Gln 16, Gly 29, Gly 45, Asn 52, Lys 72, Tyr/Phe 74, and Lys 86. Three of these involve basic residues at the front face of the molecule, which have been implicated in interaction with cytochrome oxidase (discussed in Section IV). In contrast, *Euglena* cytochrome c_{558} is unique at only one position (Phe 67), *C. oncopelti* c_{557} at none, and *Neurospora* cytochrome c at one (Phe 97).

This is an interesting result from the point of view of the early evolution of the eukaryotes. Was the mitochondrial electron transport chain in the process of change during the time of divergence of the ciliates? Or did the distinctive terminal respiration of *Tetrahymena* evolve from an already established mitochondrial cytochrome system? In any event, the results contribute to the well-established picture of the ciliates as a remote and distinctive group of organisms (Hutner and Corliss, 1976).

In the endosymbiotic theory of the origin of mitochondria (Margulis, 1970) one possible scenario is that different but related types of aerobic bacteria became endosymbionts in primitive eukaryotes and that the ciliates derive from one such distinctive event while other eukaryotes derive from another.

Are there extant free-living aerobic bacteria which might be related to these putative endosymbionts? John and Whatley (1975) have argued in favour of *Paracoccus denitrificans* fulfilling such a role on the basis of its electron transport system. However, the amino acid sequence of *Paracoccus* cytochrome c_{550} has been shown to be similar to that of the cytochromes c_2 found in photosynthetic nonsulfur bacteria (Timkovich *et al.*, 1976), and in this family of cytochromes there are members which resemble mitochondrial cytochromes c more closely than *Paracoccus* cytochrome c_{550} does. One such cytochrome c is that from *R. vannielii* (Ambler *et al.*, 1976), the sequence of which is shown in Figure 7. The matrix of differences (Table VI) indicates that this sequence is only marginally more different from mitochondrial cytochromes c than is, for instance, *Euglena* cytochrome c_{558}. This led Ambler *et al.* (1976) to suggest that there may be a structural continuum between the cytochromes c_2 and mitochondrial cytochromes c. However, like *Tetrahymena* cytochrome c_{553}, *R. vannielii* cytochrome c_2 can be distinguished from mitochondrial cytochromes c by consideration of individual sequence positions. The bacterial cytochrome differs at the otherwise unvaried positions Lys/Arg 13, Gln 16, Leu 32, Arg 38, Asn 52, Lys 73, and Tyr/Phe 74. Thus it is still possible to define mitochondrial cytochrome c by structural features.

The question of the origin of mitochondrial cytochrome c remains open. *Rhodomicrobium vannielii* cytochrome c_2 shows considerably homology, but it reacts very poorly with bovine cytochrome oxidase (Errede, 1976), has a redox potential at pH 7 of 355 mV (Pettigrew *et al.*, 1975c), and functions in photosynthetic electron transport. It is possible that organisms descended from those giving rise to mitochondria no longer exist, and also that intermediate stages preceding the final development of the mitochondrial electron transport system are not found in present-day eukaryotes. But the alternative, that a type of molecular fossil record still exists, is exciting and should stimulate a more intensive study of bacterial aerobes and protozoan mitochondria.

B. Implications for Cytochrome Function

Although the protozoan cytochrome c sequences now known probably do not allow any definite conclusion regarding phylogenetic relationships, they have been important in extending the knowledge of what constitutes a functional cytochrome c. Residues which had been proposed to be essential for cytochrome c function were found to vary in *C. oncopelti* cytochrome c_{557} and *Euglena* cytochrome c_{558}. Most notable is the presence of alanine rather than cysteine at residue 14. However, as discussed above, this loss of a heme-binding residue does not appear to affect markedly the properties of the proteins.

Euglena cytochrome c_{558} is the only mitochondrial cytochrome c (excluding *Tetrahymena* cytochrome c_{553}) which contains an internal gap. It is a remarkable feature of mitochondrial cytochromes c that changes in length are only tolerated at the NH_2- and COOH-termini. The gap in *Euglena* c_{558} occurs in the loop between residues 20 and 27 and possibly results in a shortened loop with little effect on the topography of the rest of the polypeptide chain.

The presence of phenylalanine at residue 67 in *Euglena* cytochrome c_{558} was originally one of the main pieces of evidence against a free-radical electron transfer mechanism, through the so-called left channel of aromatic side chains (Takano *et al.*, 1973). It is unlikely that a phenylalanine radical could be stable enough to be a reaction intermediate.

The NH_2-terminus of *C. oncopelti* cytochrome c_{557} could not be characterized by conventional sequence methods (Pettigrew *et al.*, 1975b). NMR analysis of isolated NH_2-terminal peptides and their hydrolysis products showed that the NH_2-terminal amino acid was *N*-dimethylproline (Pettigrew and Smith, 1977). It is not known whether the enzyme involved in the methylation is the same as that which methylates lysine side chains (*C. oncopelti* cytochrome c_{557} has methyllysine residues at positions -9 and 72). The dimethylproline is attached to an NH_2-terminal ''extension'' of 10 residues which vertebrate cytochromes c lack. Such an extension is also found in plant cytochromes c (eight residues and acetylated) and yeast and insect cytochromes (which have fewer residues and a free α-amino group), but its folded structure is not known. It is tempting to speculate that the high proline content

$(CH_3)_2$-Pro-Pro-Tml-Ala-Arg-Glu-Pro-Leu-Pro-Pro

may give rise to an extended collagen-type helix of the kind proposed by Almassy and Dickerson (1978) as part of the structure of *Pseudomonas* cytochrome c_{551}. If such an extension projects out from the surface, the methylation of the NH_2-terminal proline and Lys -9 may simply reflect easy accessibility of the nitrogens to the methylase, rather than specificity.

There is now persuasive evidence (Redfield and Gupta, 1971; Salemme *et al.*, 1973; Brautigan and Ferguson-Miller, 1976; Smith *et al.*, 1977; Rieder and Bosshard, 1977; Pettigrew, 1978) that the region of the molecule at which electron transfer takes place is the front face in Dickerson's terminology (Dickerson *et al.*, 1971).

Lysines have been implicated in the interaction of cytochrome *c* with cytochrome oxidase, and Figure 8 is a comparison of the lysine distribu-

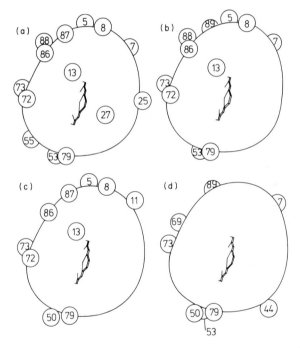

Figure 8. Diagrammatic representation of the front surface of selected cytochrome *c* molecules on the basis of the X-ray crystallographic studies of Dickerson *et al.* (1971) and a Labquip Nicholson molecular model of horse ferricytochrome *c*. (a) Horse cytochrome *c*; (b) *Euglena* cytochrome c_{558}; (c) *C. oncopelti* cytochrome c_{557}; (d) *Tetrahymena* cytochrome c_{553}. These diagrams are based on the assumption that the tertiary structures of the protozoan cytochromes are identical to that of horse cytochrome *c*. The partially exposed pyrrole ring II is shown side-on. Circles indicate the approximate position of ε-amino groups of lysine or guanidino groups of arginine. (From Pettigrew, 1978, reproduced by permission.)

tion on the front face in different c-type cytochromes. Smith *et al.* (1977) showed that modification of Lys 8, 13, 27, 72, or 79 reduced reactivity with cytochrome oxidase. Pettigrew (1978) obtained evidence that members of the groups of residues (5, 7, 8), (86, 87, 88), and (72, 73), and residues 13 and 27 were protected by interaction with yeast cytochrome c peroxidase.

Tetrahymena cytochrome c_{553} has fewer lysines on this front surface and lacks the most prominent residues of this face (8), (13), (27), (72), (86), and (87). In fact, *Tetrahymena* cytochrome c_{553} has seven fewer basic residues than horse cytochrome c, and six of them can be accounted for by these losses from the front of the molecule.

Crithidia oncopelti cytochrome c_{557}, which reacts with bovine cytochrome oxidase, shows a front-face pattern very similar to that of horse cytochrome c but lacks basic residues at positions 25, 27, and 88. The protein contains five fewer positive side chains than horse cytochrome c, and of these three are accounted for by these front-face residues. Ferguson-Miller *et al.* (1976) and Smith *et al.* (1977) note that the lower affinity of *Euglena* cytochrome c_{558} for site 1 on the cytochrome oxidase may be due to the missing lysine side chains at positions 25 and 27, and this comment has also been applied to *C. fasciculata* cytochrome c_{555} (Errede, 1976).

Thus, if *Tetrahymena* c_{553} represents a primitive form of mitochondrial cytochrome c, one may speculate that there has been evolution towards a highly basic cytochrome from a more acidic ancestor and that this involved increasing the positive charge density mainly on one surface of the molecule so that specific, high-affinity electrostatic interactions could be achieved with a cytochrome oxidase.

V. CYTOCHROMES f

The so-called algal cytochromes f are small, c-type cytochromes that function in photosynthetic electron transport, probably as the immediate electron donors to photooxidized photosystem 1. They are dealt with only briefly here, as they are not found in nonphotosynthetic organisms, but their sequences have provided a very convincing bridge across the eukaryote–prokaryote division. From the work of Wood (1977) it now seems likely that the "true" cytochrome f [i.e., the membrane-bound cytochrome studied by Davenport and Hill (1951)] is a distinct species and in algae donates electrons to one or both of the small soluble redox components, plastocyanin and algal cytochrome f [see also Bohme *et al.* (1978)]. Wood (1977) has demonstrated the presence of a membrane-bound cytochrome c_{554} in *Euglena* chloroplasts. Thus algal f-type cytochromes should be renamed, since they are functionally and probably structurally distinct from the membrane-bound protein.

These findings of course make the electron transport chain of chloroplasts even more similar to that of mitochondria:

$$b\text{-type cytochromes} \longrightarrow \begin{array}{c} \text{membrane-bound} \\ c\text{-type cytochrome} \end{array} \xrightarrow{\text{equipotential}} \begin{array}{c} \text{small, soluble} \\ \text{redox protein} \end{array} \longrightarrow \text{oxidase}$$

Unlike mitochondrial cytochromes c, algal f-type cytochromes are acidic proteins and can be collected from cell extracts on anion-exchange columns. A large number of these cytochromes have been purified (Katoh, 1959; Yakushiji, 1971), and they have redox potentials in the range 300–400 mV, α peaks shifted slightly towards the red compared to mitochondrial cytochrome c, and high ratios of Soret to α-peak extinction, and they do not react with eukaryotic cytochrome oxidase (Yamanaka and Okunuki, 1974).

The sequences of several prokaryotic and eukaryotic f-type cytochromes are now known, and there is considerable sequence similarity within the group. A matrix of amino acid differences is shown in Table VIII, and on this basis *Alaria* (Phaeophyta), *Porphyra* (Rhodophyta), and *Bumilleriopsis* (Xanthophyceae) appear related, while the others— *Monochrysis* (Chrysophyta), *Euglena*, and the two blue-green algae *Spirulina* and *Plectonema*—appear about equally distant (although it is interesting that *Euglena* c_{552} is the most divergent sequence of the group, being less like the other eukaryotic cytochromes than are the two blue-green algal cytochromes). A possible dendrogram of sequence relationships is shown in Figure 9. Ambler and Bartsch (1975) note that these results are open to several interpretations: "(1) The whole genomes of the

Table VIII Percentage Difference Matrix for Algal f-Type Cytochromes[a]

Cytochrome source	(a)	(b)	(c)	(d)	(e)	(f)
(a) *Euglena gracilis*	—	—	—	—	—	—
(b) *Monocrysis lutheri*	61	—	—	—	—	—
(c) *Porphyra tenera*	62	50	—	—	—	—
(d) *Alaria esculenta*	59	51	29	—	—	—
(e) *Bumilleriopsis filiformis*	58	43	22	21	—	—
(f) *Plectonema boryanum*	57	48	43	48	38	—
(g) *Spirulina maxima*	55	52	47	50	45	44

[a] Scoring of differences was done as described in Table VI, footnote *a*. The amino acid sequences were taken from Pettigrew (1974) (*Euglena*); Laycock (1972) (*Monochrysis*); Ambler and Bartsch (1975) (*Porphyra* and *Spirulina*); Laycock (1975) (*Alaria*); Aitken (1976) (*Plectonema*); and Ambler, Kunert, and Boger (personal communication) (*Bumilleriopsis*).

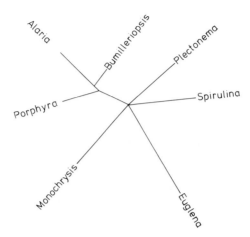

Figure 9. A possible diagram of the relationships between the algal *f*-type cytochrome sequences. The diagram was constructed from the following considerations of the matrix of differences. *Porphyra, Alaria,* and *Bumilleriopsis* clearly form a group relative to the others. However, of these three, the *Alaria* sequence is consistently most distant from the other sequences while the *Bumilleriopsis* sequence is closest to the other sequences. This suggests unequal numbers of changes in these three lines of descent, and this is indicated on the diagram where the lengths of the arms are proportional to the numbers of assigned changes. *Spirulina* and *Plectonema* sequences are not much closer to each other than each is to the *Porphyra* sequence, while the *Euglena* sequence appears the most divergent of the group although it is closer to the sequences from the blue-green algae than to any others.

eukaryotic cells are evolutionarily derived from a prokaryote closely related to the blue-green algae. Or (2) the eukaryotic chloroplasts are derived from a prokaryote related to the blue-green algae, but the remainder of the eukaryotic genomes are derived from different precursors. Or (3) transfer of the cytochrome *f* genes (or of a cluster of genes) has taken place in one direction or other between the blue-green algal line and a eukaryotic ancestor.''

The close similarity of the photosythetic apparatus but the considerable differences in other structures and biochemical functions between blue-green algae and autotrophic eukaryotes argues strongly against possibility 1. Several different lines of molecular evidence support a relationship between the blue-green algae and the chloroplast rather than the whole eukaryotic cell (e.g., RNA sequences, Zablen *et al.,* 1975; plastocyanins, Aitken, 1975). Hypothesis 3 becomes indistinguishable from hypothesis 2 if a large cluster of genes (i.e., the cyanobacterial genome) is transferred. Thus in our view there is very strong molecular evidence now in favour of an endosymbiotic origin for the chloroplast.

VI. CONCLUSIONS

This chapter has discussed the structural features of protozoan cytochromes c from the point of view of the origin and function of the mitochondrial electron transport chain. The absence of some basic residues on the front face of protozoan cytochromes, in particular *Tetrahymena* cytochrome c_{553}, may indicate that eukaryotes inherited an electron transport system to molecular oxygen which was still in the process of change, and that this change may have involved the joint evolution of the molecular surfaces of interaction of cytochrome c and cytochrome aa_3.

The presumed prokaryotic origins of the mitochondrial electron transport chain cannot yet be defined. The bacterial cytochrome closest to mitochondrial cytochromes c in sequence is the cytochrome c_2 from a photosynthetic anaerobe, an organism which bears little resemblance to a putative precursor of the mitochondrion or the eukaryotic cell. If such photosynthetic anaerobes can be considered primitive, this supports the suggestion of George (1964) that the cytochrome chain predates cytochrome aa_3 and also that functionally diverse electron transport chains have evolved from a common ancestor (Dickerson *et al.*, 1976).

The recent X-ray crystallographic studies on *Chlorobium* cytochrome c_{555} (Korszun and Salemme, 1978) give further support to this hypothesis. This cytochrome shows some sequence homology with the algal f-type cytochromes discussed here (Van Beeumen *et al.*, 1976) but only a very slight similarity to mitochondrial cytochromes c or cytochromes c_2. The tertiary structures of these cytochromes have, however, retained principal features in common when almost all similarities in primary structure have been lost.

Thus the two small c-type cytochromes of eukaryotes, functioning in quite different electron transport chains, probably have a remote common ancestor, and it is possible that the electron transport chains themselves are also related [but note the misgivings of Ambler (1977) and Thatcher (1976) regarding the possibility that gene transfer events have been frequent contributors to the course of evolution].

At a more detailed level the protozoan cytochrome c sequences are too few at present to draw any conclusions regarding the relationships of Protozoa or even to judge whether further sequence analysis of this particular protein is justified. In our opinion, cytochrome c may have incorporated changes too rapidly for it to be useful in the definition of the very early events in the divergence of Protozoa. But in view of the structural novelties which characterize the protozoan cytochromes studied, further examination of this group would be valuable and would give cytochrome c the chance to prove itself of phylogenetic usefulness in an area where classical approaches require assistance.

ACKNOWLEDGMENTS

Drs. R. P. Ambler and B. J. Errede kindly allowed use of their unpublished data. I am indebted to Drs. R. P. Ambler, T. E. Meyer, M. D. Kamen, R. G. Bartsch, B. J. Errede, and G. M. Smith for their advice and encouragement during the course of some of the work described here.

REFERENCES

Adar, F. (1977). *Arch. Biochem. Biophys.* **181,** 5–7.
Aitken, A. (1975). *Biochem. J.* **149,** 675–683.
Aitken, A. (1976). *Nature (London)* **263,** 793–796.
Almassy, R. J., and Dickerson, R. E. (1978). *Proc. Natl. Acad. Sci. U.S.A.* **75,** 2674–2678.
Ambler, R. P. (1977). *CNRS Int. Symp. Electron Transp. Microorganisms, Marseilles.*
Ambler, R. P., and Bartsch, R. G. (1975). *Nature (London)* **253,** 285–288.
Ambler, R. P., Meyer, T. E., and Kamen, M. D. (1976). *Proc. Natl. Acad. Sci. U.S.A.* **73,** 472–475.
Bohme, H., Kunert, K. J., and Boger, P. (1978). *Biochim. Biophys. Acta* **501,** 275–285.
Bovee, E. C. (1971). *J. Protozool.* **18,** Suppl., p. 14.
Brautigan, D. L., and Ferguson-Miller, S. (1976). *Fed. Proc., Fed. Am. Soc. Exp. Biol.* **35,** Abstr. No. 1219.
Corliss, J. O. (1974). *Taxon* **23,** 497–522.
Davenport, H. E., and Hill, R. (1951). *Proc. R. Soc. London, Ser. B* **139,** 327–345.
Dayhoff, M. O. (1976). "Atlas of Protein Sequence and Structure," Vol. 5, Suppl. No. 2, p. 24. Natl. Biomed. Res. Found., Washington, D.C.
Dickerson, R. E. (1971). *J. Mol. Evol.* **1,** 26–45.
Dickerson, R. E., Takano, T., Eisenberg, D., Kallai, O. B. Samson, L., Cooper, A., and Margoliash, E. (1971). *J. Biol. Chem.* **246,** 1511–1536.
Dickerson, R. E., Timkovich, R., and Almassy, R. J. (1976). *J. Mol. Biol.* **100,** 473–491.
Edwards, S. W., Chagla, A. H., Griffiths, A. J., and Lloyd, D. (1977). *Biochem. J.* **168,** 113–121.
"Enzyme Nomenclature Recommendations" (1965). pp. 18–24. Elsevier, Amsterdam.
Errede, B. J. (1976). Ph.D. Thesis, Univ. of California, San Diego.
Errede, B. J., Haight, G. P., and Kamen, M. D. (1976). *Proc. Natl. Acad. Sci. U.S.A.* **73,** 113–117.
Ferguson-Miller, S., Brautigan, D., and Margoliash, E. (1976). *J. Biol. Chem.* **251,** 1104–1115.
George, P. (1964). *In* "Oxidases and Related Redox Systems" (T. E. King, H. S. Mason, and M. Morrison, eds.), Vol. 1, pp. 3–36. Univ. Park Press, Baltimore, Maryland.
Gross, J., and Wolken, J. (1960). *Science* **132,** 357–358.
Heller, J., and Smith, E. L. (1966). *J. Biol. Chem.* **241,** 3165–3180.
Hill, G. C., and Pettigrew, G. W. (1975). *Eur. J. Biochem.* **57,** 265–271.

Hill, G. C., Chan, S. K., and Smith, L. (1971a). *Biochim. Biophys. Acta* **253**, 78–87.
Hill, G. C., Gutteridge, W. E., and Mathewson, N. W. (1971b). *Biochim. Biophys. Acta* **243**, 225–229.
Hutner, S. H., and Corliss, J. O. (1976). *J. Protozool.* **23**, 48–56.
John, P., and Whatley, F. R. (1975). *Nature (London)* **254**, 495–498.
Kamen, M. D., and Horio, T. (1970). *Annu. Rev. Biochem.* **39**, 673–700.
Katoh, S. (1959). *J. Biochem. (Tokyo)* **46**, 629–632.
Keller, R. M., Pettigrew, G. W., and Wuthrich, K. (1973). *FEBS Lett.* **36**, 151–156.
Korszun, Z. R., and Salemme, F. R. (1978). *Proc. Natl. Acad. Sci. U.S.A.* **74**, 5244–5247.
Kusel, J. P., Suriano, J. R., and Weber, M. M. (1969). *Arch. Biochem. Biophys.* **133**, 293–304.
Laycock, M. V. (1972). *Can. J. Biochem.* **50**, 1311–1325.
Laycock, M. V. (1975). *Biochem. J.* **149**, 271–279.
Lederer, F., and Simon, A. M. (1974). *Biochem. Biophys. Res. Commun.* **56**, 317–323.
Lin, D. K., Niece, R. L., and Fitch, W. M. (1973). *Nature (London)* **241**, 533–535.
Lloyd, D., and Chance, B. (1972). *Biochem. J.* **128**, 1171–1182.
McDonald, C. C., Phillips, W. D., and Vinogradov, S. N. (1969). *Biochem. Biophys. Res. Commun.* **36**, 442–449.
Margalit, R., and Schejter, A. (1973). *Eur. J. Biochem.* **32**, 492–499.
Margoliash, E., and Frohwirth, N. (1959). *Biochem. J.* **71**, 570–572.
Margoliash, E., and Schejter, A. (1966). *Adv. Protein Chem.* **21**, 113–286.
Margoliash, E., and Walasek, O. F. (1967). *In* "Oxidation and Phosphorylation" (R. W. Estabrook and M. E. Pullman, eds.), Methods in Enzymology, Vol. 10, pp. 339–348. Academic Press, New York.
Margoliash, E., Smith, E. L., Kreil, G., and Tuppy, H. (1961). *Nature (London)* **192**, 1121–1127.
Margulis, L. (1970). "Origin of Eukaryotic Cells." Yale Univ. Press, New Haven, Connecticut.
Meyer, T. E., and Cusanovich, M. A. (1972). *Biochim. Biophys. Acta* **267**, 383–387.
Miller, M. J., and Rapoport, H. (1977). *J. Am. Chem. Soc.* **99**, 3479–3487.
Newton, B. A. (1968). *Annu. Rev. Microbiol.* **22**, 109–130.
Perini, F., Kamen, M. D., and Schiff, J. A. (1964). *Biochim. Biophys. Acta* **88**, 74–90.
Pettigrew, G. W. (1973). *Nature (London)* **241**, 531–533.
Pettigrew, G. W. (1974). *Biochem. J.* **139**, 449–459.
Pettigrew, G. W. (1978). *FEBS Lett.* **86**, 14–18.
Pettigrew, G. W., and Smith, G. M. (1977). *Nature (London)* **265**, 661–662.
Pettigrew, G. W., Aviram, I., and Schejter, A. (1975a). *Biochem. J.* **149**, 155–167.
Pettigrew, G. W., Leaver, J. L., Meyer, T. E., and Ryle, A. P. (1975b). *Biochem. J.* **147**, 291–302.
Pettigrew, G. W., Meyer, T. E., Bartsch, R. G., and Kamen, M. D. (1975c). *Biochim. Biophys. Acta* **430**, 197–208.
Redfield, A. G., and Gupta, R. K. (1971). *Cold Spring Harbor Symp. Quant. Biol.* **36**, 405–411.
Rieder, R., and Bosshard, H. R. (1977). *FEBS Meet., 11th, Copenhagen.*
Ryley, J. F. (1952). *Biochem. J.* **52**, 483–492.
Salemme, F. R., Freer, S. T., Xuong, Ng, H., Alden, R. A., and Kraut, J. (1973). *J. Biol. Chem.* **248**, 3910–3921.
Schechter, E., and Saludjian, P. (1967). *Biopolymers* **5**, 788–790.
Smith, H. T., Staudenmeyer, N., and Millet, F. (1977). *Biochemistry* **16**, 4971–4974.
Smith, L., Nava, M. E., and Margoliash, E. (1973a). *In* "Oxidases and Related Redox

Systems" (T. E. King, H. S. Mason, and M. Morrison, eds.), Vol. 2, 629–638. Univ. Park Press, Baltimore, Maryland.

Smith, L., Davies, H. C., Reichlin, M., and Margoliash, E. (1973b). *J. Biol. Chem.* **248**, 237–243.

Sneath, P. H. A. (1974). *Symp. Soc. Gen. Microbiol.* No. 24, 1–39.

Stevens, F. C., Glazer, A. N., and Smith, E. L. (1967). *J. Biol. Chem.* **242**, 2764–2779.

Takano, T., Kallai, O. B., Swanson, R., and Dickerson, R. E. (1973). *J. Biol. Chem.* **248**, 5234–5255.

Tarr, G., and Fitch, W. M. (1976). *Biochem. J.* **159**, 193–199.

Thatcher, D. R. (1976). *Nature (London)* **259**, 624.

Timkovich, R., Dickerson, R. E., and Margoliash, E. (1976). *J. Biol. Chem.* **251**, 2197–2206.

Van Beeumen, J., Ambler, R. P., Meyer, T. E., Kamen, M. D., Olson, J. M., and Shaw, E. K. (1976). *Biochem. J.* **159**, 757–774.

Wood, P. M. (1977). *Eur. J. Biochem.* **72**, 605–612.

Wuthrich, K. (1969). *Proc. Natl. Acad. Sci. U.S.A.* **63**, 1071–1078.

Yakushiji, E. (1971). *In* "Photosynthesis and Nitrogen Fixation," Part A (A. San Pietro, ed.), Methods in Enzymology, Vol. 23, pp. 364–368. Academic Press, New York.

Yamanaka, T., and Okunuki, K. (1968). *In* "Structure and Function of Cytochromes" (K. Okunuki, M. D. Kamen, and I. Sekuzu, eds.), pp. 390–403. Univ. of Tokyo Press, Tokyo.

Yamanaka, T., and Okunuki, K. (1974). *In* "Microbial Iron Metabolism" (J. B. Neilands, ed.), pp. 349–400. Academic Press, New York.

Yamanaka, T., Nagata, Y., and Okunuki, K. (1968). *J. Biochem. (Tokyo)* **63**, 753–760.

Zablen, L. B., Kissil, M. S., Woese, C. R., and Buetow, D. E. (1975). *Proc. Natl. Acad. Sci. U.S.A.* **72**, 2418–2422.

Isoprenoid Distribution and Biosynthesis in Flagellates

4

T. W. GOODWIN

I. INTRODUCTION

Isoprenoids, compounds built up from branched C_5 $(C-C-C-C)$ C

units, are probably the most widespread group of natural products known. In addition, the ingenious variations played during evolution on the biosynthetic isoprene rule have yielded a multiplicity of compounds varying in size from 5 to 100 carbon atoms. This profligacy reaches its apex in the higher plants, and a glance at any appropriate list of natural products reveals the very large number of, for example, sesquiterpenes (C_{15}) and triterpenes (C_{30}) in higher plants; phytoflagellates, on the other hand, have not accomplished such diverse biosynthetic feats and confine themselves mainly to synthesizing sterols (triterpene derivatives) and C_{40} carotenoids (tetraterpenes); this chapter concentrates on these two groups.

BIOCHEMISTRY AND PHYSIOLOGY OF PROTOZOA
SECOND EDITION, VOL. 1

Table I Structure of Sterols

Ring system	Side chain R				
	1	12	13	2	26
	9	3		13A	6
	10		4	5	
		24			

8   22   28

7

23   61   11

                                   15   14

                                   16   57

         60   19   27

                                   20

II. NATURE AND DISTRIBUTION

A. Sterols

1. General

The best known sterol in nature is probably cholesterol (1), which is the characteristic mammalian sterol; it is found in traces in many higher plants and in some phytoflagellates, but the major sterols in these groups are characterized by additional C_1 or C_2 groups in the side chain at C-24. (The structures of the various sterols considered in this chapter are collected in Table I, unless indicated otherwise.) Many also have a double bond between C-22 and C-23. Both these characteristics are illustrated in the structure of poriferasterol (2); the numbering of the carbon atoms in sterols is also indicated in the case of poriferasterol:

(2)

When a hydrogen at C-24 is replaced by a methyl or ethyl group, a chiral centre is produced; it was assumed until recently that the chirality of sterols from phytoflagellates was always β as indicated in (2), and that sterols with α chirality at C-24 were synthesized only by higher plants. However, 24α-sterols are now known to occur in some diatoms (see Section II,A,4). The prefixes α and β indicate whether the substituent is above or below the plane of the paper, respectively. The more formal R, S convention for defining a chiral centre when applied to sterols with a side chain at C-24 can lead to ambiguities (Goodwin, 1973a), and the simpler α, β convention is adequate for the present purpose.

2. Distribution

Euglenophyceae. Ergosterol (3) was at one time considered the major sterol of *Euglena gracilis* strain Z (Stern *et al.*, 1960; Avivi *et al.*, 1967), but a recent investigation showed that a complex mixture existed in which ergosterol (3), ergost-7-enol (4), chondrillasterol (5), and stigmasta-5,7,22-trienol (6) were the major components (Brandt *et al.*, 1970; Anding and Ourisson, 1973). The chirality at C-24 of the compound identified as (6) needs careful assessment with modern NMR techniques, because the presence of a 24α-sterol is unexpected. *Euglena* can grow heterotrophically in the dark as well as photosynthetically. It is then essentially colorless and contains plastids but no photosynthetic pigments. Table II shows that in such "bleached" cultures there is some change in the nature of the sterols synthesized compared with those from green cultures. Chondrillasterol (5) disappears whereas chondrillasta-5,7-dienol (7), clionasterol (8), and poriferasterol (2) appear. Ergosterol (3) dominates the dark-grown cultures, representing 73% of the total, whereas it represents only 30% of the total in green cultures; indeed, on occasion it is absent from green cultures (Brandt *et al.*, 1970). Esterified sterols are found in both types of cells.

Triterpene precursors of sterols which have one and two methyl groups at C-4 are also present in *Euglena*. The 4,4-dimethyl compounds demonstrated are cycloartenol (14) (80% of total), 24-methylenecycloartanol

Table II   Sterols in *Euglena gracilis*[a]

Sterol	Green cells	Bleached cells
Cholesta-5,7-dienol (9)	Trace	Trace
Cholest-7-enol (10)	Trace	Trace
Cholesterol (1)	+	+ +
Episterol (11)	−	+
Ergost-7-enol (4)	+ +	+ +
Ergosterol (3)	+ +	+ + +
Ergosta-5,22-dienol (12)	+ +	+
Ergost-5-enol (13)	−	Trace
Clionasterol (8)	−	+ +
Poriferasterol (2)	−	+
Chondrillasterol (5)	+ +	−
Stigmasta-5,7,22-trienol (6)	+ +	+ +
Chondrillasta-5,7-dienol (7)	Trace	Trace
4,4-Dimethyl sterols		
Cycloartenol (14)	+	+ +
24-Methylenecycloartanol (15)	+ + +	+ +
24-Methylenelanosterol (16)	+	−
24-Methyleneagnosterol (17)	+	−
4-Methyl sterols		
Obtusifoliol (18)	+ + +	+ + +
24-Methylenelophenol (19)	+ +	+
4α-Methylzymosterol (20)	+	−

[a] From Brandt *et al.* (1970) and Anding and Ourisson (1973). −, Not detected; +, present (number of +'s indicates quantitative distribution).

(15), 24-methylenelanosterol (16), and 24-methyleneagnosterol (17). In bleached cultures only (14) and (15) are present in about equal amounts. Obtusifoliol (18), 24-methylenelophenol (19), and 4α-methylzymosterol (20) are among the 4α-methyl compounds present in both green and bleached strains (Anding *et al.*, 1971). Rather unexpectedly, the pentacyclic triterpene β-amyrin (21) is present in both types of *Euglena* cells (And-

ing *et al.*, 1971). The colourless phytoflagellate *Astasia longa,* which is considered a natural mutant of *Euglena,* also contains a complex mixture of sterols (Table III) (Rohmer and Brandt, 1973).

3. Chrysophyceae

The only organism in this class which has been examined in detail for sterols is *Ochromonas*. The only sterol present to any detectable extent in *Ochromonas malhamensis* is poriferasterol (2) (Williams *et al.*, 1966; Gershengorn *et al.*, 1968), and not its C-24 epimer stigmasterol (26) as reported earlier (Bazzano, 1965; Avivi *et al.*, 1967). *Ochromonas danica* produces, in addition to poriferasterol (2) brassicasterol (ergosta-5,22-dienol) (12), 22-dihydrobrassicasterol (ergosta-5-enol) (13), clionasterol (8), and probably 7-dehydroporiferasterol (13A) (Gershengorn *et al.*, 1968) as well as ergosterol (3) (Stern *et al.*, 1960; Aaronson and Baker, 1961; Halevy *et al.*, 1966). Furthermore, stigmasterol (26) could not be observed in this species (Gershengorn *et al.*, 1968) or in *Ochromonas sociabilis* (W.

Table III Sterols in *Astasia longa*

Sterols
  Cholest-7-enol (**10**)
  Cholesterol (**1**)
  Ergost-7-enol (**4**)
  Ergost-5-enol (**13**)
  Isofucosterol (**22**)
  Chondrillast-7-enol (**23**)
  Clionasterol (**8**)
  Poriferasterol (**2**)

4-Methyl sterols
  4α-Methyl ergost-8-enol (**24**)

4,4-Dimethyl sterols
  Cycloartenol (**14**)
  24-Methylenecycloartanol (**15**)
  Cycloeucalenol (**25**)
  Obtusifoliol (**18**)

a From Rohmer and Brandt (1973).

Sach and L. J. Goad, unpublished observations) although reported earlier (Halevy *et al.*, 1966). *Synura petersenii* is said to produce cholesterol (**1**) and stigmasterol (**26**) (Collins and Kalnins, 1969), but this needs modern confirmation.

4. Bacillariophyceae (Diatoms)

Nitzschia closterium was early reported to contain fucosterol (**27**) (Heilbron, 1942), and more recently brassicasterol (**12**) (24β) was observed in *N. closterium* and *Cyclotella nana* (Kanazawa *et al.*, 1971). However, 250-MHz NMR studies have now revealed that in *Phaeodactylum tricornutum* (=*N. closterium* f. *minutissima*) the methyl sterol present (**28**) has the α configuration (Rubinstein and Goad, 1974). This was unexpected and is the first fully authenticated case of a 24α-methyl sterol in an alga. *Navicula pelliculosa* may contain chondrillasterol (**5**) (Low, 1955).

5. Pyrrophyta (Dinoflagellates)

Recently a sterol with a unique side chain (dinosterol) (**29**) was isolated from *Gonyaulax tamarensis* (Shimizu *et al.*, 1975). It has also been found in *Crypthecodinium cohnii* (Withers *et al.*, 1978) together with Δ^5-dinosterol and 3-ketodinosterol (dinosterone); other sterols present included cholesterol (**1**).

(29)

B. Carotenoids

1. General

The C_{40} carotenoids are characterized by a system of conjugated double bonds, which is responsible for their colour. All the carotenoids considered in this chapter can be related to the acyclic hydrocarbon pigment lycopene (30). Individual carotenes (hydrocarbons) are formally named by reference to the two C_9 end groups which they contain. For the present purpose three end groups, β, ϵ, and ψ, are involved. Lycopene has two ψ (i.e., acyclic) end groups and is formally ψ, ψ-carotene; γ-carotene (31) contains one ψ end group and one β (cyclic) end group and is β,ψ-

(30)

(31)

(32)

(33)

carotene, whilst α-carotene (**32**) contains two cyclic end groups, one β and one ϵ, and formally is β,ϵ-carotene; β-carotene (**33**) has two β end groups and is β,β-carotene. In all these cases the C_{22} portion of the molecule is the same. It can, however, vary as pointed out in Section II,B,3. Full details of carotenoid nomenclature can be found in Isler (1971).

The numbering of carotenoids, based on separating the molecule into two halves, is illustrated for α-carotene (**32**); the main rule is that, if the carotenoid under consideration is unsymmetrical, unprimed numbers are allotted to the half of the molecule associated with the Greek letter cited first in its systematic name. Carotenoids are divided into two main groups, carotenes, which are hydrocarbons, and xanthophylls, which are oxygenated carotenoids; numerous examples of xanthophylls will appear as the discussion proceeds.

2. Chlorophyta

It is only the Volvocales among the Chlorophyta which possess flagellae, but the carotenoid distribution in the members of this order (Table IV) is very similar to that found in other orders (Goodwin, 1978); that is, they produce the same major pigments as higher plants, namely, β-carotene

Table IV Carotenoids in Volvocales (Chlorophyta)

Organism	Additional pigments[a]	References
Chlamydomonadaceae		
Chlamydomonas agloeformis	α-Carotene	Strain (1958, 1966)
Chlamydomonas reinhardi	Loroxanthin	Sager and Zalokar (1958); Krinsky and Levine (1964)
Chlorogonium elongatum	α-Carotene	Francis *et al.* (1975)
Haematococcus pluvalis	α-Carotene	Hager and Stransky (1970a,b)
Pedinomonadaceae		
Pedinomonas tuberculata	α-Carotene	Hager and Stransky (1970a,b)
Polyblepharidaceae		
Dunaliella primolecta		Riley and Wilson (1965)
Dunaliella salina	α-Carotene	Strain (1958, 1966)
Dunaliella sp.	α-Carotene	Hager and Stransky (1970a,b)
Dunaliella tertiolecta	α-Carotene	Masyuk and Radchenko (1970); de Nicola (1961)
Volvocaceae		
Eudorina elegans		Bunt (1964) Hager and Stransky (1970a,b)

[a] All species contain β-carotene (**33**), lutein (3,3'-dihydroxy-α-carotene), violaxanthin (**34**), and neoxanthin (**35**). Additional pigments are indicated in this column.

(34)

(35)

(33), lutein (3,3'-dihydroxy-α-carotene), violaxanthin [3,3'-dihydroxy-β-carotene 5,6,5',6'-epoxide (34)], and neoxanthin (35); the last-named contains an allene (C=C=C) group, and thus the bridging part of the molecule is different from the common C_{22} portion in carotenes. In the present context this distribution generally holds for *Chlamydomonas agloeformis* (Strain, 1958, 1966), *C. reinhardi* (Sager and Zalokar, 1958; Krinsky and Levine, 1964), *Haematococcus pluvialis* (Hager and Stransky, 1970a), and various species of *Dunaliella* (Hager and Stransky, 1970a; Riley and Wilson, 1965). Traces of α-carotene (32) generally occur alongside β-carotene, but *C. agloeformis* is said to produce "much" α-carotene (Strain, 1958, 1966). Loroxanthin (lutein with the methyl group at C-19 oxidized to CH_2OH) is also present (Francis *et al.*, 1975), and this is a pigment not produced by higher plants.

Under adverse nutritional conditions, usually nitrogen deficiency, considerable amounts of carotenoids accumulate outside the chloroplast, so much so that the cells appear red. This phenomenon has been observed in *Chlamydomonas* spp. (Czygan, 1968) and in *H. pluvalis* (Tischer, 1941; Goodwin and Jamkorn, 1954a; Czygan, 1968), where the accumulating pigments are mainly keto carotenoids, echinenone (4-keto-β-carotene), canthaxanthin (4,4'-diketo-β-carotene), and astaxanthin (36), whereas in *Dunaliella salina* it is β-carotene (Fox and Sargent, 1938). In the eyespot

(36)

within the chloroplast of *C. reinhardi* there is a specific accumulation of β-carotene (Ohad *et al.*, 1969).

A series of mutants of *C. reinhardi* having impaired ability to photosynthesize contain essentially the normal carotenoid group, although there are some quantitative differences (Krinsky and Levine, 1964). Another mutant, which is pale green, synthesizes only about 0.2–0.5% of the usual amount of carotenoids, which are entirely carotenes, no xanthophylls being detected (Sager and Zalokar, 1958).

A start has recently been made on studying the changes of pigments during the life cycle of *C. reinhardi* in a 12-hr light–4-hr dark cycle. During the light period all pigments increased at a similar rate, but the synthesis of lutein and violaxanthin preceded that of β-carotene, neoxanthin, and loroxanthin. A marked drop in all pigments occurred after 9 hr, which corresponded to the known loss of RNA and the breakup of the nucleus which occur at this time. Only relatively small changes in the pigment pattern occurred in the dark (Francis *et al.*, 1975).

3. Euglenophyta

Chloroplast pigments in *Euglena* spp. include β-carotene (**33**), zeaxanthin (3,3'-dihydroxy-β-carotene), and neoxanthin (**34**) (Goodwin and Jamikorn, 1954b; Krinsky and Goldsmith, 1960); however, the major component is diadinoxanthin (**37**) (Aitzetmüller *et al.*, 1968). This pigment, which contains an acetylenic bond within the C_{22} portion of the molecule, was earlier confused with lutein and antheraxanthin (5,6-epoxyzeaxanthin).

(**37**)

(**38**)

(**39**)

Diatoxanthin (38) and heteroxanthin (39) have also been detected in small amounts (Aitzetmüller et al., 1968; Stransky and Hager, 1970). Essentially the same pigments are present in Trachelomonas hispida var. coronata (Johannes et al., 1971). Traces of keto carotenoids are usually present (Goodwin and Gross, 1958; Krinsky and Goldsmith, 1960; Stransky and Hager, 1970), and it is thought that they may be concentrated in the eyespot, although this has not yet been clearly demonstrated.

When grown heterotrophically in the dark, Euglena spp. produce not chloroplasts but plastids (etioplasts). Such cells form the usual carotenoids but in greatly reduced amounts (Helmy et al., 1967).

Mutants of Euglena can be obtained by treatment with either pyribenzamine or streptomycin or by exposure to high temperatures or high pressures; the main block is frequently in the ability to desaturate phytoene (40), the first C_{40} compound formed in the biosynthetic sequence to coloured carotenoids, e.g., lycopene (30) (Goodwin and Gross, 1958; Gross and Stroz, 1974; Gross et al., 1975). In mutants where the block is before phytoene, dark-grown cells do not produce plastids, whereas dark-grown cells of phytoene-synthesizing mutants do (Kivic and Vesk, 1974).

(40)

The heterotrophic euglenophyte Astasia ocellata synthesizes α-carotene (32) as its major pigment, together with a number of ketonic xanthophylls including echinenone, canthaxanthin, and possibly phoenicopterone (41)

(41)

(Thomas et al., 1967). Astasia longa does not accumulate either colourless polyenes or coloured carotenoids (Gross and Stroz, 1974). The heterotrophic chlorophycean phytoflagellate Polytoma uvella synthesizes a weakly acidic, incompletely characterized carotenoid, polytomaxanthin; the xanthophyll: carotene ratio in this organism is 4 : 1 (Links et al., 1960). The apochlorotic dinoflagellate C. cohnii produces β- and γ-carotenes as its major pigments (Tuttle et al., 1973).

Table V Xanthophyceae Recently Examined for Carotenoids

Organism	Pigments[a]	Reference
Vaucheriales		
Botrydium granulatum	1, 2, 3, 4, 5, 6, 7, 8, 9	Egger *et al.* (1969); Stransky and Hager (1970)
Vaucheria sp.	4, 6, 8	Strain *et al.* (1968)
Vaucheria sessilis	1, 4, 6, 7, 8	Egger *et al* (1969); Norgärd *et al.* (1974)
Vaucheria tenestris	6, 7	Egger *et al.* (1969)
Bumilleriopsis filiformis	1, 2, 3, 4, 5, 6, 7, 8, 10	Stransky and Hager (1970)
Tribonemetales		
Bumilleria sicula	1, 2, 3, 4, 5, 6, 7, 8, 10	Stransky and Hager (1970)
Heterothrix debilis	1, 2, 3, 4, 5, 6, 7, 8, 10	Falk and Kleinig (1968)
Tribonaema aequale	1, 2, 3, 4, 5, 6, 7, 8, 10, 11	Strain *et al.* (1970)
Heterococcus caespitosus	1, 2, 3, 4, 5, 6, 7, 10	Stransky and Hager (1970)

[a] 1, β-Carotene (**33**); 2, β-cryptoxanthin 5,6,5′,6′-diepoxide; 3, β-cryptoxanthin 5,6-epoxide; 4, vaucheriaxanthin ester (**43**); 5, neoxanthin (**35**); 6, diadinoxanthin (**37**); 7, diatoxanthin (**38**); 8, heteroxanthin (**39**); 9, zeaxanthin; 10, β-carotene 5,6-epoxide; 11, unidentified carotene.

4. Chrysophyta

a. Xanthophyceae. A relatively large number of pure cultures of Xanthophyceae have been examined recently (Goodwin, 1979a), and it is clear (Table V) that the constant components are the xanthophylls diadinoxanthin (**37**), diatoxanthin (**38**), and heteroxanthin (**39**) (Nörgard *et al.*, 1974; Stransky and Hager, 1970; Strain *et al.*, 1968, 1970; Egger *et al.*, 1969; Buchecker and Liaaen-Jensen, 1977). Table V is limited to species in which the pigments have been fully characterized. Other members examined earlier, but in less detail, have been listed by Goodwin (1979a). The absence of fucoxanthin (**42**), the characteristic pigment of other Chrysophyta, should be noted.

(**42**)

(**43**)

Table VI Carotenoids in the Haptophyceae

Organism	Pigment[a]	Reference
Chrysochromulina ericina	1	Dales (1960)
Dictaeria inomata	1	Dales (1960)
Emiliana huxleyi[b]	2, 3	Norgärd et al. (1974); Hertzberg et al. (1977); Arpin et al. (1976); Berger et al. (1977)
Isochrysis galbana	1, 3, 4, 5, 7[c], 8	Dales (1960); Hager and Stransky (1970b); Hertzberg et al. (1977)
Hymenomonas carteae	1, 3, 4, 5, 9, 10	Arpin et al. (1976)
Pavlova gyrans	1	Dales (1960)
Pavlova lutheri[d]	1, 3, 4, 5, 9, 11, 12[c], 13	Hertzberg et al. (1977); Hager and Stransky (1970b)
Phaeocystis puchetti	1	Dales (1960)
Prymnesium parvum	1, 3, 4, 5, 7, 11	Hertzberg et al. (1977)
Sphaleromantis sp.	1	Dales (1960)
Syracophera carterae	1	Hager and Stransky (1970b)

[a] 1, Fucoxanthin (**42**); 2, 19′-*n*-hexanoyloxyfucoxanthin; 3, diadinoxanthin (**37**); 4, β-carotene (**33**); 5, diatoxanthin (**38**); 6, fucoxanthol (**44**); 7, α-carotene (**32**); 8, echinenone; 9, β-carotene 5,6-epoxide; 10, β-cryptoxanthin; 11, β-cryptoxanthin 5,6,5′,6′-epoxide; 12, canthaxanthin; 13, dinoxanthin (**45**).

[b] Formerly *Coccolithus huxleyi*.

[c] Traces.

[d] Formerly *Monochrysis lutheri*.

Table VII Carotenoid Distribution in the Eustigmatophyceae

Organism	Pigment[a]	Reference
Botyridiopsis alpina	1, 2, 3, 4, 5, 6	Thomas and Goodwin (1965)
Chlorobotrys regularis	3[b], (4?), 8	Whittle (1977)
Pleurochloris commutata	3, 4, 7, 8, 9	Whittle and Castleton (1968)
Pleurochloris magna	1, 3[c], 4[d], 8[c] (?11)	Nörgard et al. (1974)
Polyedrilla helvetica	1, 4, 8	Whittle and Castleton (1968)
Vischeria sp.	1, 9[e]	Thomas and Goodwin (1965)
Vischeria stellata	5, 6, 10	Strain (1958)
GBS-Sticho[f]	1, 3, 8[c], 11	Whittle and Castleton (1968)
Clone Tunis[f]	1, 3[b,c], 8[c], 11	Whittle and Castleton (1968)

[a] 1, β-Carotene (**33**); 2, β-cryptoxanthin 5,6-epoxide; 3, vaucheriaxanthin (**43**); 4, neoxanthin (**35**); 5, diadinoxanthin (**37**); 6, diatoxanthin (**38**); 7, zeaxanthin; 8, violaxanthin (**34**); 9, antheraxanthin; 10, heteroxanthin (**39**); 11, canthaxanthin.

[b] Free and esterified.

[c] Furanoid derivatives also present.

[d] Not reported in later investigations.

[e] Diadinoxanthin probably also present.

[f] Unnamed but probably Eustigmatophytes.

b. Chrysophyceae. *Ochromonas danica* (Allen *et al.,* 1960) and *O. stipitata* (Hager and Stransky, 1970b) synthesize fucoxanthin (42) as their major pigment, together with β-carotene (33), violaxanthin (35), antheraxanthin, and zeaxanthin. β-Cryptoxanthin (3-hydroxy-β-carotene) and its 5,6-epoxide are present in traces in *O. stipitata. Pseudopedinella* also synthesizes fucoxanthin (42) (Dales, 1960).

(44)

(45)

c. Haptophyceae. The major carotenoid in all Haptophyceae so far examined is fucoxanthin (42), although in *Emiliana huxleyi* it is esterified with *n*-hexanoic acid at C-19′ (Table VI). The appearance of 4-keto carotenoids in *Isochrysis galbana* and *Pavlova lutheri* is unexpected (Berger *et al.,* 1977).

d. Eustigmatophyceae. A number of members of this relatively recently defined class have been examined. The most important aspect (Table VII) is the absence of fucoxanthin (42) and the appearance of either vaucheriaxanthin (43) or heteroxanthin (39) in all species.

e. Bacillariophyceae (Diatoms). The pigment distribution in diatoms examined in pure culture (Table VIII) is constant and consists of β-carotene (33), diatoxanthin (38), and fucoxanthin (42) (Strain, 1958, 1966). The rare ε-carotene (46) is present in *N. closterium* (Strain *et al.,* 1944) and *N. pelliculosa* (Hager and Stransky, 1970b).

5. Pyrrophyta

The characteristic pigment of this phylum is peridinin (47), first isolated in 1927 (Carter *et al.,* 1939) but only fully characterized within the last few

(46)

Table VIII   Diatoms Which Contain β-Carotene (**33**), Diatoxanthin (**38**), Diadinoxanthin (**37**), and Fucoxanthin (**42**)

Organism	Reference
Cymbella cymbiformis[a]	Illyes *et al.* (1975)
Fragilaria sublinearis	Jeffrey and Haxo (1968)
Isthmia nervosa	Strain *et al.* (1944)
Melosira sp.	Hager and Stransky (1970a,b)
Navicula spp.	Strain *et al.* (1944); Geneves *et al.* (1976); Hager and Stransky (1970a,b)
Nitzschia closterium[a,b]	Strain *et al.* (1944); Jeffrey (1961); Strain (1966); Carreto and Catoggio (1976)
Nitzschia dissipata	Wassink and Kersten (1946)
Nitzschia palea	Strain *et al.* (1944)
Phaeodactylum tricornutum[c]	Hager and Stransky (1970a,b); Strain *et al.* (1944)
Stephanodyscis turris	Strain *et al.* (1944)
Thalassioseira gravida	Strain *et al.* (1944)

[a] Also possibly lutein.
[b] The furanoid isomer of diadinoxanthin appears in old cultures.
[c] Formerly *N. closterium* f. *minutissima*.

years (Strain *et al.*, 1971, 1976). Its similarity to fucoxanthin (**42**) is clear; the oxidation of an in-chain methyl group, first seen in green algae as a C-19 hydroxymethyl group in loroxanthin, has been carried further to yield a carboxylic acid which then forms a lactone. There are, however, reports of fucoxanthin in some Pyrrophyta (Table IX). Its presence in *Peridinium foliaceum* and *P. balticum* is interesting, because both algae contain two nuclei, one dinokaryotic and one eukaryotic (Dodge, 1971;

(**47**)

(**48**)

(49)

(50)

(51)

(52)

(53)

Tomas *et al.,* 1973). It is suggested that fucoxanthin (42) and diadinoxan-thin (37) are produced by an endosymbiont chrysophyte (Tomas *et al.,* 1973; Jeffrey, 1976), whilst phytoene (40), phytofluene (51), ζ-carotene (52), β-zeacarotene (53), β-carotene (33), and γ-carotene (31), which in *P. foliaceum* appear in oil droplets (Withers and Haxo, 1975), are formed by an apochlorotic host dinoflagellate. It may be significant that, as noted earlier, the apochlorotic dinoflagellate *C. cohnii* contains β-carotene and γ-carotene as its major carotenoids (Tuttle *et al.,* 1973).

Peridinin represents 38–84% of the total pigments in the Pyrrophyta

which synthesize it (Johansen *et al.*, 1974) and is accompanied by
β-carotene **(33)**, diatoxanthin **(38)**, pyrroxanthin **(48)**, dinoxanthin **(45)**,
diadinoaxanthin **(37)**, pyrroxanthinol **(49)**, and peridinol **(50)** (Johansen *et al.*, 1974; Loeblich and Smith, 1968). Characteristic of many of these

Table IX Carotenoid Distribution in the Pyrrophyta

Organism	Pigments[a]	Reference
Class Desmophyceae		
Order Prorocentriales		
Exuviella cassubica	1, 2, 3, 4	Jeffrey *et al.* (1975)
Exuviella sp.	1, 2, 3, 5[b]	Tomas and Cox (1973); Riley and Segar (1969)
Prorocentrum micans	1, 2, 3, 4	Pinckard *et al.* (1953); Scheer (1940)
Class Dinophyceae		
Order Gymnodiniales		
Family Gymnodiniaceae		
Amphidinium carterae	1, 2, 3, 4, 5?, 6, 7	Jeffrey *et al.* (1975); Johansen *et al.* (1974); Parsons and Strickland (1963); Heilbron *et al.* (1935)
Amphidinium corpulentum	1, 2, 3, 4	Jeffrey *et al.* (1975)
Amphidinium hoeffleri	1, 2, 3, 4	Jeffrey *et al.* (1975)
Amphidinium klebsii	1, 2, 3, 4	Mandelli (1968); Jeffrey *et al.* (1975); Bunt (1964)
Amphidinium rhyncocephalum	1, 2, 3, 4	Jeffrey *et al.* (1975)
Amphidinium spp.	1, 2, 3, 4	Stransky and Hager (1970); Johannes *et al.* (1971); Jeffrey and Haxo (1968)
Gymnodinium nelsoni	1, 2, 3, 4, 5? 6?, 7	Johansen *et al.* (1974)
Gymnodinium punctatum	1, 2, 3, 4	Jeffrey *et al.* (1975)
Gymnodinium simplex	1, 2, 3, 4	Jeffrey *et al.* (1975)
Gymnodinium sp.	1, 2, 3, 4	Jeffrey *et al.* (1975); Bunt (1964); Jeffrey (1961)
Gymnodinium splendens	1, 2, 3, 4, 7?	Jeffrey *et al.* (1975); Johansen *et al.* (1974)
Gymnodinium veneficum	1, 4, 8	Riley and Wilson (1965)
Family Glenodiniaceae		
Glenodinium sp.[c]	1, 2, 3, 4, 5, 6, 7, 9	Jeffrey *et al.* (1975); Johansen *et al.* (1974)
Gyrodinium dorsum[c]	1, 2, 3, 4, 5, 7, 10	Jeffrey *et al.* (1975); Johansen *et al.* (1974)

Table IX *(Continued)*

Organism	Pigments[a]	Reference
Family Glenodiniaceae *(continued)*		
Gyrodinium resplendens[c]	1, 2, 3, 4, 6	Jeffrey *et al.* (1975); Loeblich and Smith (1968)
Family Peridinaceae		
Aureodinium pigmentosum	3	Jeffrey (1976); Whittle and Castleton (1968)
Cachonina niei[c]	1, 2, 3, 4	Jeffrey *et al.* (1975); Strain *et al.* (1944)
Ensiculifera loeblichii[c]	1, 2, 3, 4	Jeffrey *et al.* (1975)
Peridinium balticum	1, 4, 5, 8, 11	Jeffrey *et al.* (1975); Tomas and Cox (1973)
Peridinium cinctum	1, 2, 3, 4	Strain *et al.* (1944)
Peridinium foliaceum[c]	1, 4, 8, 11	Mandelli (1968); Jeffrey *et al.* (1975); Withers and Haxo (1975)
Peridinium trochoideum	1, 2, 3, 4	Riley and Wilson (1965); Jeffrey *et al.* (1975); Whittle and Castleton (1968); Riley and Sager (1969)
Woloszynskia micra	8	Whittle and Castleton (1968)
Woloszynskia sp.	8	Whittle and Castleton (1968)
Family Gonyaulacaceae		
Gonyaulax polyedra	1, 2, 3, 4	Jeffrey *et al.* (1975); Sweeney *et al.* (1959)
Zooxanthellae[d]		
Anemonia sulcata (sea anemone)	3[e]	Heilbron *et al.* (1935)
Anthopleura xanthogrammica[f] (sea anemone)	1, 2, 3, 4	Strain *et al.* (1944)
Pocillipora sp. (coral)[f]	1, 2, 3, 4	Jeffrey and Haxo (1968)
Tridacna crocea (clam)[f]	1, 2, 3, 4	Jeffrey and Haxo (1968)
Family Protoceratiaceae		
Protoceratium reticulatum[c]	1, 2, 3, 4	Jeffrey *et al.* (1975)
Family Phytodinaceae		
Pyrocystis lunula	1, 2, 3, 4	Jeffrey *et al.* (1975)

[a] 1, β-Carotene (**35**); 2, dinoxanthin (**45**); 3, peridinin (**47**); 4, diadinoxanthin (**37**); 5, diatoxanthin (**38**); 6, pyroxanthin (**49**); 7, peridinol (**50**); 8, fucoxanthin (**42**); 9, astaxanthin (**36**); 10, pyrroxanthinol (**49**); 11, γ-carotene (**31**).

[b] Both fucoxanthin and peridinin were reported in different investigations.

[c] Also unidentified pigments.

[d] The carotenoids are in the symbiotic Dinophyceae living in the animals listed.

[e] First isolated as sulcatoxanthin (Heilbron *et al.*, 1935).

[f] The isolated organism is *Gymnodinium microadriaticum* (Jeffrey *et al.*, 1975).

pigments is the presence of acetylenic linkages (diadinoxanthin and pyrroxanthin) and acetoxy groups (peridinin, pyrroxanthin, and dinoxanthin). The detection for the first time of deacylated derivatives such as peridinol and pyrroxanthin is significant; no such derivatives of fucoxanthin have been reported. The known distribution of carotenoids in the Pyrrophyta is given in Table IX.

III. BIOSYNTHESIS

A. General

This is not the place to elaborate on the subtle details of the biosynthesis of sterols (Goad, 1970; Goad *et al.,* 1974) and carotenoids (Britton, 1976), but there are certain general aspects which are appropriate to consider. Many of the early biosynthetic steps are common to both pathways: both involve the formation of the first committed precursor mevalonic acid from acetyl-CoA and its conversion into the universal C_5 intermediate isopentenyl pyrophosphate (Figure 1). The chain elongation steps to farnesyl pyrophosphate are also common (Figure 2). Here the pathway divides, farnesyl pyrophosphate dimerizing to form the C_{30} hydrocarbon squalene (54) via presqualene pyrophosphate (55) (Figure 3). In the case of carotenoid biosynthesis the chain elongation continues to C_{20}, that is, to geranylgeranyl pyrophosphate which then dimerizes, in a reaction analogous to but not identical with that producing squalene, to form phytoene (40) via prephytoene pyrophosphate (56) (Figure 3). It is noted that two mechanisms are involved, one producing all-*trans*-phytoene and the other 15-*cis*-phytoene. Both are produced in nature, and the biosynthetic significance of this is not yet clear (Goodwin, 1979a,b). The difference is that squalene production involves NADPH in the reaction, whereas it is not required in the production of phytoene.

B. Sterols

In animals and nonphotosynthetic fungi the first cyclic product in sterol biosynthesis is lanosterol (57) formed via squalene 2,3-oxide (58) (Figure 4). In photosynthetic tissues the first product is not lanosterol but its isomer cycloartenol (14). The internal rearrangement during cyclization is stabilized in animals and fungi by the elongation of a proton from C-9 to yield lanosterol, whilst in photosynthetic tissues a proton is lost from C-19 with the concomitant formation of a cyclopropane ring in cycloartenol (Goad, 1970). Experiments with *O. danica* were instrumental in demonstrating the cycloartenol pathway in plants; thus the formation of squalene

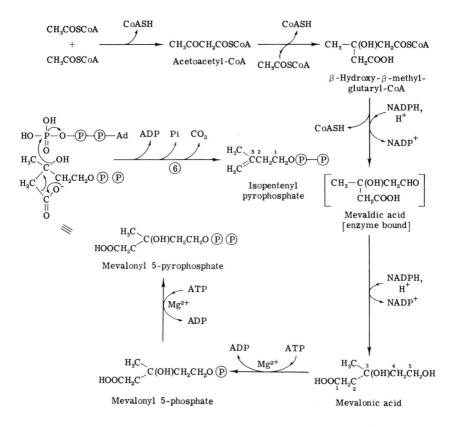

Figure 1. Conversion of acetyl-CoA into mevalonic acid and isopentenyl pyrophosphate. (Redrawn from Mercer and Goodwin, 1972.)

2,3-oxide and its conversion to sterols in this organism was the first report in a photosynthetic organism, as was the demonstration that the formation of cycloartenol was a primary result of cyclization and not merely a consequence of the isomerization of lanosterol (Rees *et al.*, 1969).

As indicated earlier, a unique feature of sterol synthesis in higher plants, algae, and fungi is the appearance of one or two supernumerary carbon atoms at C-24. So, superimposed on the question of the sequence of steps involved in converting the ring system of cycloartenol (14) into that of poriferasterol (2) is the question of when and how the two carbons at C-24 appear in the molecule. The proposed pathway outlined in Figure 5 has been developed from experiments with *Ochromonas* spp. The sequence has been made likely by feeding experiments with labelled precur-

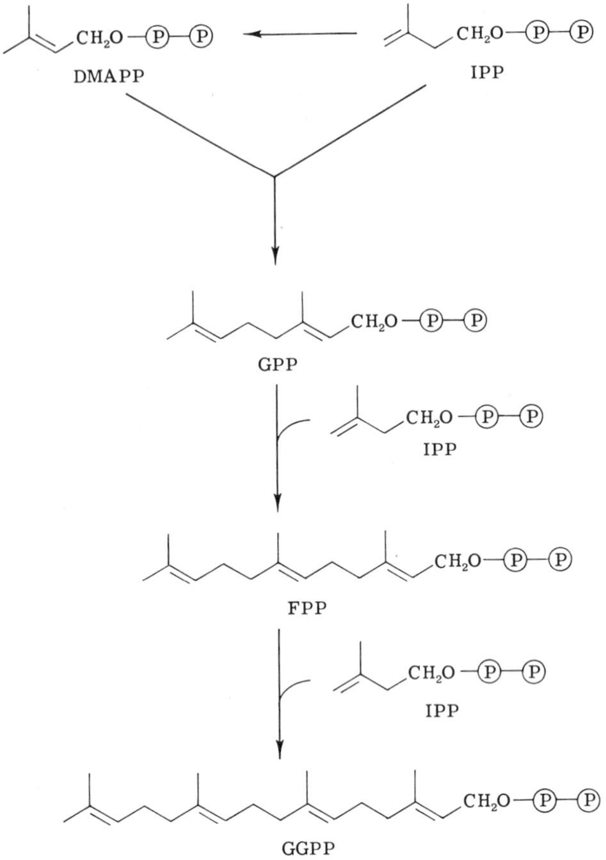

Figure 2. Chain elongation from C_5 (isopentenyl pyrophosphate) to C_{15} (farnesyl pyrophosphate) and C_{20} (geranylgeranyl pyrophosphate). (Redrawn from Goodwin, 1976.)

sors (Hall *et al.*, 1969; Lenton *et al.*, 1971; Knapp *et al.*, 1977). This approach can never be as definitive as enzyme studies, and so far most enzyme experiments have been on tissue cultures of higher plants and chlorophyte algae. They have confirmed the proposed points of methylation (Figure 5) (Goodwin, 1979b).

The alkylation mechanism indicated in Figure 6 was clearly demonstrated in *Ochromonas* spp. The basic evidence was as follows: (1) [CD$_3$]methionine is incorporated into dihydrobrassicasterol (**13**) and poriferasterol (**2**) with the maximum incorporation of two and four deuterium atoms into the methyl and ethyl side chains, respectively (Smith *et al.*, 1967; Knapp *et al.*, 1977); (2) labelled isofucosterol (**58**) is

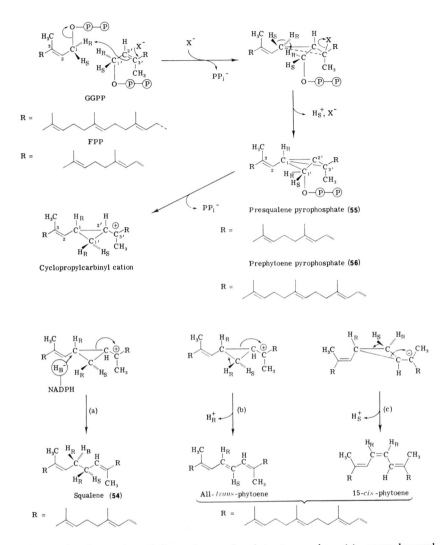

Figure 3. Conversion of farnesyl pyrophosphate to squalene (a), geranylgeranyl pyrophosphate to all-*trans*-phytoene (b), and geranylgeranyl pyrophosphate into 15-*cis*-phytoene (c). (Redrawn from Davies, 1976.)

very effectively incorporated into poriferasterol (Lenton *et al.*, 1971); and (3) the hydrogen originally at C-24 in cycloartenol migrates to C-25 in isofucosterol and is retained there in poriferasterol. It is ironic that this, the first alkylation mechanism elucidated in photosynthetic tissues, is at the time of writing of limited distribution. Of course, only a relatively few

Squalene 2, 3-oxide
(58)
(a)
b
b
19
10
9
8
HO
H
(b)
H^+ (C-19)
Cycloartenol

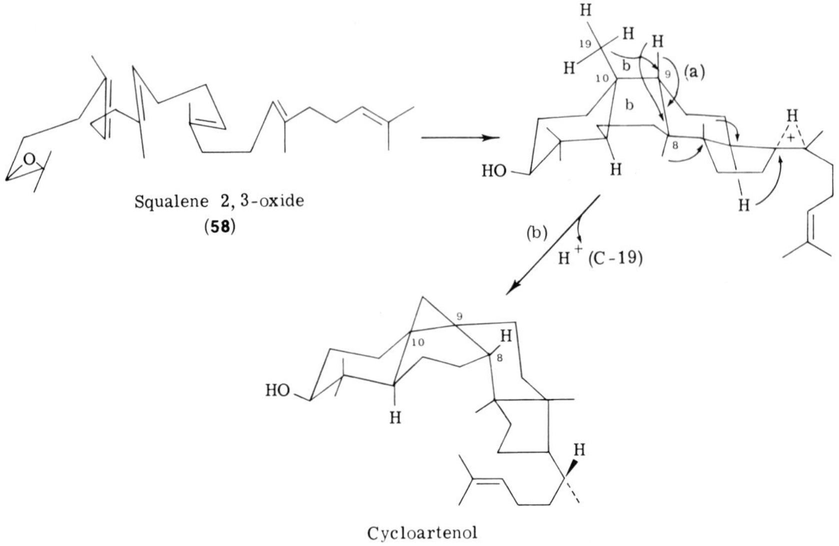

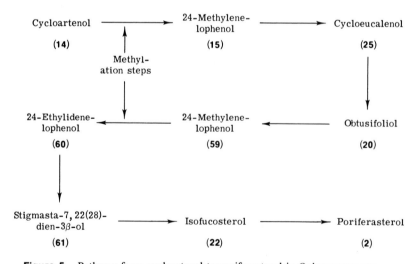

Figure 4. Cyclization of squalene 2,3-oxide to cycloartenol. (Redrawn from Goodwin, 1971.)

Cycloartenol
(14)
24-Methylene-
lophenol
(15)
Cycloeucalenol
(25)
Methyl-
ation steps
24-Ethylidene-
lophenol
(60)
24-Methylene-
lophenol
(59)
Obtusifoliol
(20)
Stigmasta-7, 22(28)-
dien-3β-ol
(61)
Isofucosterol
(22)
Poriferasterol
(2)

Figure 5. Pathway from cycloartenol to poriferasterol in *Ochromonas* spp.

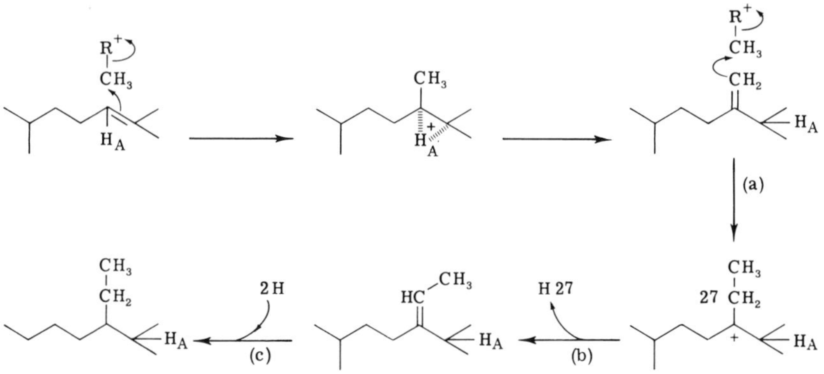

Figure 6. Alkylation mechanism of sterols in *Ochromonas* spp. (Redrawn from Goodwin, 1973b.)

organisms have been examined, but there is no doubt that the alkylation mechanisms in nonmotile algae (*Trebouxia,* etc.) (Goad *et al.,* 1974) and higher plants (Lenton *et al.,* 1975; Goodwin, 1974, 1977) are different.

C. Carotenoids

The general pathway to the "basic" carotenoids such as α- and β-carotenes, via lycopene, itself formed by a sequential desaturation of phytoene (Figure 7), is now well known (Figure 8) (Britton, 1976). Little or no evidence for this pathway has come from studies on phytoflagellates, except for the early demonstration in *Euglena* mutants that phytoene accumulated when β-carotene synthesis was blocked (Goodwin and Gross, 1958). Suggestions for pathways to the more exotic carotenoids found in these microorganisms have not been lacking (see, e.g., Goodwin, 1979a). However, no experimental evidence for such pathways has been forthcoming, and there is no need here to elaborate on the proposals.

IV. CONCLUSIONS

From an overview of sterols and carotenoids in phytoflagellates one comes to almost opposite conclusions about developments in the study of these two main groups of terpenoids. Sterol distribution has been studied much less exhaustively than carotenoid distribution. The latter investigations have been extremely rewarding to the organic chemist in revealing

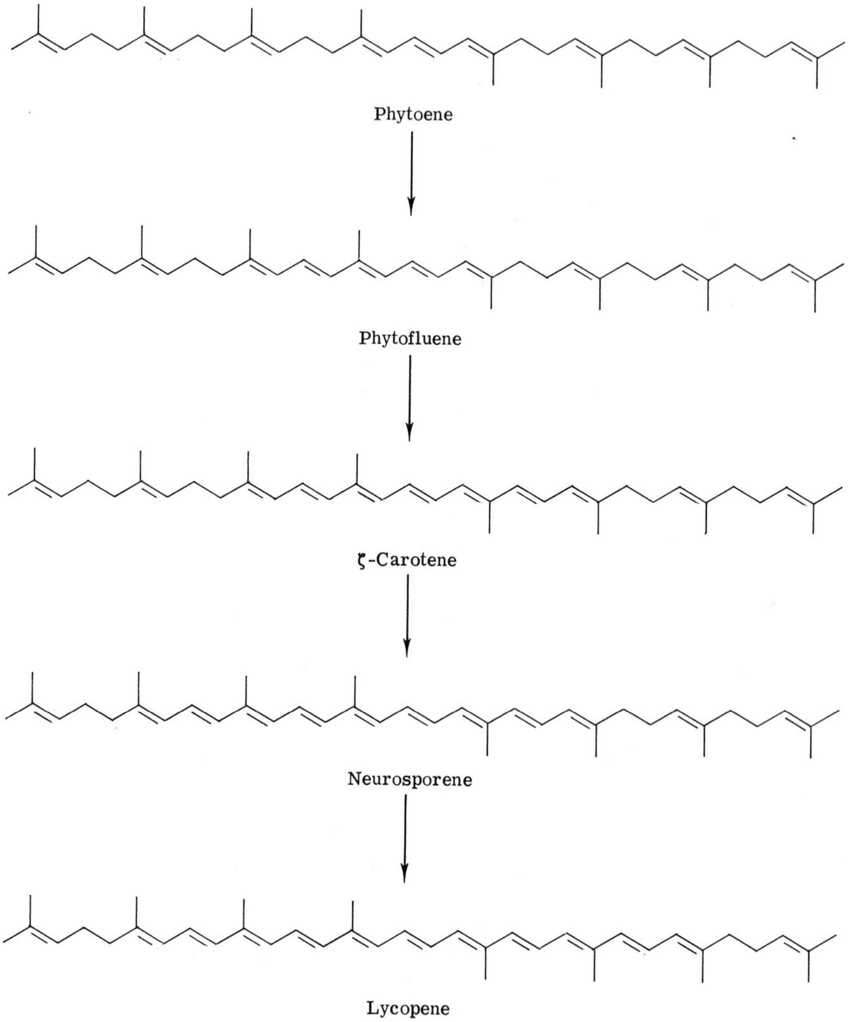

Phytoene

Phytofluene

ζ-Carotene

Neurosporene

Lycopene

Figure 7. Pathway for the conversion of phytoene to lycopene.

numerous unexpected and fascinating new structures. There is no reason why such an attack on the problem of sterol distribution should not equally yield new compounds; indeed, these are already beginning to appear as indicated by the studies now going on into the nature of the sterols of dinoflagellates. Additionally, the increasing knowledge of sterol distribution in phytoflagellates and algae in general will help the study of algal evolution at the biochemical level. Carotenoid studies have already

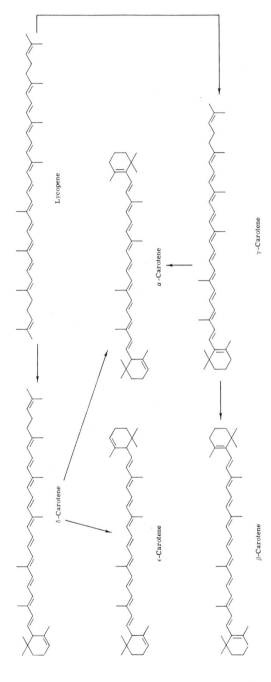

Figure 8. Pathway for the conversion of lycopene to cyclic carotenes. (Redrawn from Goodwin, 1976.)

made some progress in this direction (Goodwin, 1974, 1979a). In contrast, sterol biosynthesis owes a great deal to studies with phytoflagellates, whilst carotenoid biosynthesis has profited hardly at all. There is great scope here for enzymological studies on the later stages of carotenoid biosynthesis which lead to pigments characteristic of phytoflagellates.

REFERENCES

Aaronson, S., and Baker, H. (1961). *J. Protozool.* **8,** 274.
Aitzetmüller, K., Svec, W. A., Katz, J. J., and Strain, H. H. (1968). *Chem. Commun.* p. 32.
Allen, M. B., Goodwin, T. W., and Phagpolngarm, S. (1960). *J. Gen. Microbiol.* **23,** 93.
Anding, C., and Ourisson, G. (1973). *Eur. J. Biochem.* **36,** 446.
Anding, C., Brandt, R. D., and Ourisson, G. (1971). *Eur. J. Biochem.* **24,** 259.
Arpin, N., Svec, W. A., and Liaaen-Jensen, S. (1976). *Phytochemistry* **15,** 529.
Avivi, L., Iaron, O., and Halevy, S. (1967). *Comp. Biochem. Physiol.* **21,** 321.
Bazzano, G. (1965). Univ. Microfilm 65-4100.
Berger, R., Liaaen-Jensen, S. L., McAlister, V., and Guillard, R. R. L. (1977). *Biochem. Syst. Ecol.* **5,** 71.
Brandt, R. D., Pryce, R. J., Anding, C., and Ourisson, G. (1970). *Eur. J. Biochem.* **17,** 344.
Britton, G. (1976). *Pure Appl. Chem.* **47,** 223.
Buchecker, R., and Liaaen-Jensen, S. (1977). *Phytochemistry* **16,** 729.
Bunt, J. S. (1964). *Nature (London)* **203,** 1261.
Carreto, J. I., and Catoggio, J. A. (1976). *Mar. Biol.* **36,** 105.
Carter, P. W., Heilbron, I. M., and Lythgoe, B. (1939). *Proc. R. Soc. London, Ser. B* **128,** 82.
Collins, R. P., and Kalnins, K. (1969). *Comp. Biochem. Physiol.* **30,** 779.
Czygan, F. C. (1968). *Arch. Mikrobiol.* **61,** 81; **62,** 209.
Dales, R. P. (1960). *J. Mar. Biol. Assoc. U.K.* **39,** 693.
Davies, B. H. (1976). *Pure Appl. Chem.* **47,** 211.
de Nicola, M. G. (1961). *Boll. Ist Bot. Univ. Catania* **6,** 51.
Dodge, J. D. (1971). *Protoplasma* **73,** 145.
Egger, K., Nitsche, H., and Kleinig, H. (1969). *Phytochemistry* **8,** 1583.
Falk, H., and Kleinig, H. (1968). *Arch. Mikrobiol.* **61,** 347.
Fox, D. L., and Sargent, M. C. (1938). *Chem. Ind (London)* **57,** 1111.
Francis, G. W., Strand, L. P., Lien, T., and Knutsen, G. (1975). *Arch. Microbiol.* **73,** 216.
Geneves, L., Choussy, M., Barbier, M., Neuville, D., and Daste, P. (1976). *C. R. Acad. Sci., Ser. D* **282,** 449.
Gershengorn, M. C., Smith, A. R. H., Goulston, G., Goad, L. J., Goodwin, T. W., and Haines, T. H. (1968). *Biochemistry* **7,** 1698.
Goad, L. J. (1970). *In* "Naturally Occurring Compounds Formed Biologically from Mevalonic Acid" (T. W. Goodwin, ed.), p. 45. Academic Press, New York.
Goad, L. J., Lenton, J. R., Knapp, F. F., and Goodwin, T. W. (1974). *Lipids* **9,** 582.
Goodwin, T. W. (1971). *Biochem. J.* **123,** 293.
Goodwin, T. W. (1973a). *Essays Biochem.* **9,** 103.
Goodwin, T. W. (1973b). *In* "Lipids and Biomembranes of Eukaryotic Microorganisms" (J. A. Erwin, ed.). Academic Press, New York.
Goodwin, T. W. (1974). *In* "Algal Biochemistry and Physiology" (W. D. P. Stewart, ed.), p. 176. Blackwell, Oxford.
Goodwin, T. W. (1977). *Trans. Biochem. Soc.* **5,** 1252.

Goodwin, T. W., ed. (1976). *In* "Chemistry and Physiology of Plant Pigments." Academic Press, New York.

Goodwin, T. W. (1979a). "Comparative Biochemistry of Carotenoids," 2nd Ed., Vol. 1. Chapman & Hall, London.

Goodwin, T. W. (1979b). *Annu. Rev. Plant Physiol.* (in press).

Goodwin, T. W., and Gross, J. A. (1958). *J. Protozool.* **5,** 292.

Goodwin, T. W., and Jamikorn, M. (1954a). *Biochem. J.* **47,** 513.

Goodwin, T. W., and Jamikorn, M. (1954b). *J. Protozool.* **1,** 216.

Gross, J. A., and Stroz, R. J. (1974). *Plant Sci. Lett.* **3,** 67.

Gross, J. A., Stroz, R. J., and Britton, G. (1975). *Plant Physiol.* **55,** 175.

Hager, A., and Stransky, H. (1970a). *Arch. Mikrobiol.* **72,** 68.

Hager, A., and Stransky, H. (1970b). *Arch. Microbiol.* **73,** 77.

Halevy, S., Avivi, L., and Katan, H. (1966). *J. Protozool.* **13,** 480.

Hall, J., Smith, A. R. H., Goad, L. J., and Goodwin, T. W. (1969). *Biochem. J.* **112,** 129.

Heilbron, I. M. (1942). *J. Chem. Soc.,* 79.

Heilbron, I. M., Jackson, H., and Jones, E. R. H. (1935). *Biochem. J.* **29,** 1384.

Helmy, F. M., Hack, M. H., and Yaeger, R. G. (1967). *Comp. Biochem. Physiol.* **23,** 565.

Hertzberg, S., Mortensen, T., Borch, G., Siegelman, H. W., and Liaaen-Jensen, S. (1977). *Phytochemistry* **16,** 587.

Illyes, G., Neamtu, G., and Bodea, C. (1975). *Stud. Cercet. Biochim.* **18,** 109.

Isler, O., ed. (1971). "Carotenoids." Birkhaeuser, Basel.

Jeffrey, S. W. (1961). *Biochem. J.* **80,** 336.

Jeffrey, S. W. (1976). *CSIRO Mar. Biochem. Unit Annu. Rep.* 1974–1975.

Jeffrey, S. W., and Haxo, F. J. (1968). *Biol. Bull. (Woods Hole, Mass.)* **135,** 149.

Jeffrey, S. W., Sielicki, M., and Haxo, F. T. (1975). *J. Phycol.* **11,** 374.

Johannes, B., Brezinka, H., and Budzickiewicz, H. (1971). *Z. Naturforsch., Teil B* **24,** 377.

Johansen, J. E., Svec, W. A., Liaaen-Jensen, S., and Haxo, F. T. (1974). *Phytochemistry* **13,** 2261.

Kanazawa, A., Yoshioka, M., and Teshima, S. (1971). *Nippon Suisan Gakkaishi* **37,** 899.

Kivic, P. A., and Vesk, M. (1974). *Can. J. Bot.* **52,** 695.

Knapp, F. F., Goad, L. J., and Goodwin, T. W. (1977). *Phytochemistry* **16,** 1677.

Krinsky, N. I., and Goldsmith, T. H. (1960). *Arch. Biochem. Biophys.* **91,** 271.

Krinsky, N. I., and Levine, R. P. (1964). *Plant Physiol.* **39,** 680.

Lenton, J. L., Hall, J., Smith, A. R. H., Ghisalberti, E. L., Rees, H. R., Goad, L. J., and Goodwin, T. W. (1971). *Arch. Biochem. Biophys.* **143,** 664.

Lenton, J. L., Goad, L. J., and Goodwin, T. W. (1975). *Phytochemistry* **16,** 1925.

Links, J., Verloop, A., and Havinga, E. (1960). *Arch. Mikrobiol.* **36,** 306.

Loeblich, A. R., and Smith, V. E. (1968). *Lipids* **3,** 5.

Low, E. M. (1955). *J. Mar. Res.* **14,** 199.

Mandelli, E. F. (1968). *J. Phycol.* **4,** 347.

Masyuk, N. P., and Radchenko, M. I. (1970). *Gidrobiol. Zh.* **6,** 51.

Mercer, E. I., and Goodwin, T. W. (1972). *In* "Plant Biochemistry." Pergamon, Oxford.

Norgärd, S., Svec, W. A., Liaaen-Jensen, S., Jensen, A., and Guillard, R. R. L. (1974). *Biochem. Syst. Ecol.* **2,** 37.

Ohad, I., Goldberg, I., Broza, R., Schuldiner, S., and Gan-Zvi, E. (1969). *Prog. Photosynth. Res.* **1,** 284.

Parsons, P. R., and Strickland, J. D. H. (1963). *J. Mar. Res.* **21,** 155.

Pinckard, J. H., Kittredge, J. S., Fox, D. L., Haxo, F. J., and Zechmeister, L. (1953). *Arch. Biochem. Biophys.* **44,** 189.

Rees, H. R., Goad, L. J., and Goodwin, T. W. (1969). *Biochim. Biophys. Acta* **176,** 894.

Riley, J. P., and Segar, D. A. (1969). *J. Mar. Biol. Assoc. U.K.* **49,** 1047.

Riley, J. P., and Wilson, T. R. S. (1965). *J. Mar. Biol. Assoc. U.K.* **45**, 583.
Rohmer, M., and Brandt, R. D. (1973). *Eur. J. Biochem.* **34**, 345.
Rubinstein, I., and Goad, L. J. (1974). *Phytochemistry* **13**, 485.
Sager, R., and Zalokar, M. (1958). *Nature (London)* **182**, 98.
Scheer, B. T. (1940). *J. Biol. Chem.* **136**, 275.
Shimizu, Y., Alam, M., and Kobayashi, A. (1975). *J. Am. Chem. Soc.* **98**, 1059.
Smith, A. R. H., Goad, L. J., Goodwin, T. W., and Lederer, E. (1967). *Biochem. J.* **104**, 56C.
Stern, A. I., Schiff, J. A., and Klein, H. P. (1960). *J. Protozool.* **7**, 52.
Strain, H. H. (1958). "32nd Annual Priestley Lectures." Pennsylvania State University.
Strain, H. H. (1966). *In* "The Biochemistry of Chloroplasts" (T. W. Goodwin, ed.), Vol. 1, p. 387. Academic Press, New York.
Strain, H. H., Manning, W. M., and Hardin, G. J. (1944). *Biol. Bull. (Woods Hole, Mass.)* **86**, 169.
Strain, H. H., Svec, W. A., Aitzetmüller, K., Grandolfo, M., and Katz, J. J. (1968). *Phytochemistry* **7**, 1417.
Strain, H. H., Benton, F. L., Grandolfo, M. C., Aitzetmüller, K., Svec, W. A., and Katz, J. J. (1970). *Phytochemistry* **9**, 2561.
Strain, H. H., Svec, W. A., Aizetmüller, G., Grandolfo, M. C., Katz, J. J., Kjösen, H., Norgärd, S., Liaaen-Jensen, S., Haxo, F. T., Wegfahrt, P., and Rapoport, H. (1971). *J. Am. Chem. Soc.* **93**, 1823.
Strain, H. H., Svec, W. A., Wegfahrt, P., Rapoport, H., Haxo, F. T., Norgärd, S., Kjosen, H., and Liaaen-Jensen, S. (1976). *Acta Chem. Scand., Ser. B* **30B**, 109.
Stransky, H., and Hager, A. (1970). *Arch. Mikrobiol.* **72**, 84.
Sweeney, B. M., Haxo, F. T., and Hastings, J. W. (1959). *J. Gen. Physiol.* **43**, 285.
Thomas, D. M., and Goodwin, T. W. (1965). *J. Phycol.* **1**, 118.
Thomas, D. M., Goodwin, T. W., and Ryley, J. F. (1967). *J. Protozool.* **14**, 654.
Tischer, J. (1941). *Hoppe-Seyler's Z. Physiol. Chem.* **267**, 281.
Tomas, R. N., and Cox, E. R. (1973). *J. Phycol.* **9**, 304.
Tomas, R. N., Cox, E. R., and Steidinger, K. A. (1973). *J. Phycol.* **9**, 91.
Tuttle, R. C., Loeblich, A. R., III, and Smith, V. E. (1973). *J. Protozool.* **20**, 521.
Wassink, E. C., and Kersten, J. A. H. (1946). *Enzymologia* **12**, 1.
Whittle, S. J. (1977). *Br. J. Phycol.* **3**, 602.
Whittle, S. J., and Castleton, J. P. (1968). *Br. Phycol. Bull.* **3**, 602.
Williams, B. L., Goodwin, T. W., and Ryley, J. F. (1966). *J. Protozool.* **13**, 227.
Withers, N., and Haxo, F. T. (1975). *Plant Sci. Lett.* **5**, 7.
Withers, N. W., Tuttle, R. C., Holtz, G. G., Beach, D. H., Goad, L. J., and Goodwin, T. W. (1978). *Phytochemistry* **17**, 1987.

Phycobiliproteins of Cryptophyceae

5

ELISABETH GANTT

I. INTRODUCTION

The photosynthetic members of the Cryptophyceae have long been recognized as a distinct algal group, based primarily on their pigment characteristics (Dougherty and Allen, 1960) and certain ultrastructural features. Phycobiliproteins are their major accessory pigments, and the principal photoreceptors in photosynthesis. Energy absorbed by phycobiliproteins is passed to chlorophyll a, presumably to a reaction center within the photosynthetic lamella. The Cryptophyceae also contain chlorophyll c_2 (Jeffrey, 1976). However, chlorophyll c_2 and carotenoids, which are also regarded as accessory pigments, are not known to have any direct involvement in the phycobiliprotein energy transfer.

The phycobiliproteins of red and blue-green algae have been much more extensively studied than those of cryptophytes. The similarities of the three groups are greater than their differences. Basic similarities common to all groups are an open-chain tetrapyrrole type of chromophore, the covalent binding of the chromophore to the apoprotein, the existence of α and β subunits in a 1 : 1 relationship, the size of these subunits (molecular

BIOCHEMISTRY AND PHYSIOLOGY OF PROTOZOA
SECOND EDITION, VOL. 1

weight range of about 10,000–20,000), and their amino acid sequence from the NH_2-terminus.

Red phycoerythrin (PE) and blue phycocyanin (PC) are the two major types, and range in absorption from about 542 to 645 nm. However, further subdivision of types has become necessary because of the great spectral variability. From those examined, it appears that the variability may become greater in cryptophytan than in rhodophytan and cyanophytan phycobiliproteins. One remarkable peculiarity of cryptophytes is the existence of only one phycobiliprotein in any species—PE or PC, but never both. In the red and blue-green algae at least two, and often three, types are present. Most notable is the lack of a pigment equivalent to allophycocyanin, the terminal pigment before chlorophyll in the energy transfer chain of red and blue-green algae. Its existence may not be as critical as in red and blue-green algae, because the spectral overlap of the cryptophytan phycobiliproteins and chlorophyll a forms may be more advantageous to maximum energy transfer (at least the cryptophytan PC absorption and emission range is wider than in the other two algal groups).

The emphasis in this chapter is on the functional involvement of cryptophytan phycobiliproteins in photosynthesis, their photophysical characteristics, their biochemical properties, and their possible evolutionary relationships with other phycobiliproteins. This is the first time they have been treated separately; hence an attempt is made to present the essential characteristics of cryptophytan phycobiliproteins which have become known since the pioneering work by ÓhEocha and Raftery (1959), Haxo and Fork (1959), and Allen et al. (1959) and the extension by ÓhEocha (1960) and Haxo (1960) and colleagues. For additional information on phycobiliproteins, particularly those of red and blue-green algae, several excellent recent reviews are available by Bogorad (1975), Rüdiger (1975), Glazer (1976), ÓCarra and ÓhEocha (1976), and Troxler (1977), as is a thesis by Bryant (1977).

II. FUNCTION AND LOCALIZATION

Phycobiliproteins as accessory pigments fill an optical gap where chlorophyll absorption is minimal. Action spectra determined by Haxo and Fork (1959) and Haxo (1960) provide clear evidence of their function as major contributors to the photosynthetic process in these algae. An example of a typical action spectrum is shown for *Rhodomonas lens* in Figure 1. Activity is particularly prominent in the green region, which is due to PE, and in the orange-red region, for which chlorophylls a and c_2 are responsible. In contrast to that of red and blue-green algae, the acces-

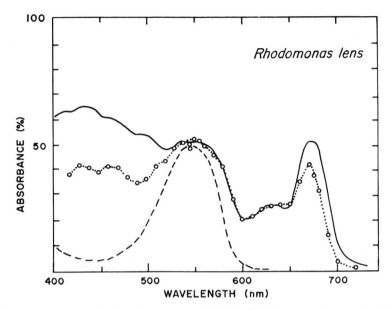

Figure 1. Photosynthetic action spectrum (circles) of *Rhodomonas lens* compared with the absorption spectrum of whole cells (solid line) and the aqueous extract (dashed line) containing PE I. (After Haxo and Fork, 1959.)

sory pigment system of cryptophytes is unusual in that it contains both phycobiliproteins and chlorophyll c_2, which may allow greater photosynthetic activity.

Cells grown photoautotrophically under nonsaturating light conditions appear to be largely dependent on phycobiliproteins, as indicated by their increased phycobiliprotein production. The phycobiliprotein content is lowered under heterotrophic conditions (Antia *et al.*, 1973). In *Chroomonas* sp., the PC content was substantially greater in cells grown in low light (100 μW/cm^2) than in high light (2700 μW/cm^2) (Faust and Gantt, 1973). The chlorophyll c_2 content, under the same conditions, was relatively little affected, which is uncharacteristic of an accessory pigment. In addition to an intensity response, a change in the phycobiliprotein content has also been elicited by a change in light quality. Exposure of cells at equal energies to white and blue-green light enhanced the phycobiliprotein content in blue-green light-grown cells (Vesk and Jeffrey, 1977), again without appreciably affecting chlorophyll c_2.

For maximum energy transfer, which is assumed to occur by resonance transfer (MacColl and Berns, 1978), a prime requirement is close proximity of the phycobiliproteins and chlorophyll. This seems to have been satisfied by enclosure of the phycobiliproteins in intrathylakoid membrane sacs. It has been possible to show indirectly that phycobiliproteins are located within the intrathylakoid spaces (Figure 2) of the chloroplast. Dodge (1969), followed by Wehrmeyer (1970), suggested this possibility in examining the chloroplast ultrastructure of *Chroomonas mesostigmatica*. Evidence for an intrathylakoid phycobiliprotein location has been presented for three species: *Chroomonas* sp., *Cryptomonas ovata* var. *palustris*, and *R. lens* (Gantt *et al.*, 1971). When cells, fixed with glutaraldehyde, were treated with protease, the electron-dense material within the intrathylakoid space was removed as the phycobiliproteins were released into the medium. Neither the electron-dense intrathylakoidal material nor the phycobiliprotein color was removable by acetone extraction. Location in the intrathylakoidal space is further suggested by both the fact that an increase in the width of the dense intrathylakoidal material occurs concomitantly with an increase in the PC content (Faust and Gantt, 1973), and by the lack of phycobilisomes which occur in other phycobiliprotein-containing algae (Gantt, 1975). Location of the phycobiliproteins in the intrathylakoid space raises some interesting points about the lamellae. It implies that the thylakoid membranes may be inverted when compared to the situation in red and blue-green algae, where the phycobilisomes are on the stroma side.

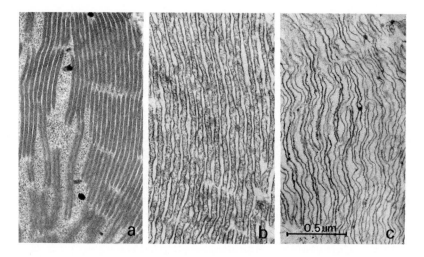

Figure 2. Chloroplast sections of *Rhodomonas lens* with electron-dense material in between thylakoids in which PE is assumed to be localized. Protease treatment at 0 hr (a), 1 hr (b), and 2 hr (c) resulted in extraction of the pigment (0, 75, and 95%, respectively) and concomitant removal of the electron-dense material. Bar, 0.5 μm.

III. PHOTOPHYSICAL PROPERTIES AND BIOCHEMICAL CHARACTERISTICS

A. Absorption and Fluorescence Characteristics

Several classes of phycobiliproteins can be spectrally distinguished. Based on their absorption characteristics, ÓhEocha *et al.* (1964) proposed a classification scheme which has been adhered to in this chapter. The principal types of PE and PC, shown in Figure 3, reveal a broad absorption range in the visible portion of the spectrum, suggesting the presence of chromophores in different environments, or the presence of multiple chromophores. As seen in Table I, PC I has two peaks at ca. 581–585 nm and at 641–645 nm and often a shoulder at 620 nm. The second spectral class, PC II, is somewhat less distinct, because the absorption peaks at 583–588 and 615–630 nm fall within the PC I class. A further complication is evident in *Hemiselmis virescens,* where the Millport 64 strain has peaks at 581 and 641 nm, and the Plymouth 157 strain has peaks at 577 and 612

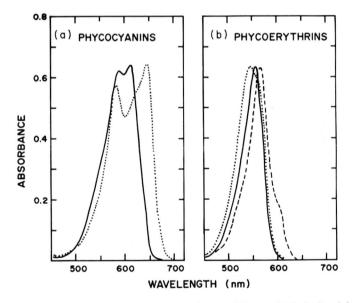

Figure 3. Absorption spectra in aqueous solution of PCs and PEs isolated from cryptophycean algae. (a) PC I (dotted line) from *Chroomonas* sp. and PC II (solid line) from *Hemiselmis virescens* (Plymouth 157). (b) PE I (dotted line) from *Rhodomonas lens,* PE II (solid line) from *Hemiselmis rufescens,* and PE III (dashed line) from *Cryptomonas ovata.* [Spectra adapted from: PC I, Mörschel and Wehrmeyer (1975); PC II, ÓhEocha and Raftery (1959); PE I and III, MacColl *et al.* (1976); PE II, ÓhEocha *et al.* (1964).]

Table I Photophysical Properties of Isolated Phycobiliproteins

Type	Species	Absorption maxima and shoulders (nm)[a]	Fluorescence emission maxima and shoulders (nm)[a]	Extinction coefficient $(A_{1\,cm}^{1\%})$	References
PE I	*Cryptomonas* sp.	545	—	—	ÓhEocha and Raftery (1959)
	Cryptomonas atrorosea	542, 560–570	—	—	ÓhEocha et al. (1964)
	Cryptomonas maculata	544, 555–560	584, 620	—	Mörschel and Wehrmeyer (1977)
	Cryptomonas suberosa	544	—	—	ÓhEocha et al. (1964)
	Cryptomonas suberosa[b]	542	—	—	Barber et al. (1969)
	Plagioselmis prolonga	544, 560–570	—	—	ÓhEocha et al. (1964)
	Rhodomonas sp.	545	—	—	ÓhEocha and Raftery (1959)
	Rhodomonas sp. 3-C	542, 562	—	—	Brooks and Gantt (1973)
	Rhodomonas lens	545, 560	585, 625	126	MacColl et al. (1976)
PE II	*Chroomonas diplococcus*	553	—	—	ÓhEocha et al. (1964)
	Cryptochrysis sp.	554	—	—	ÓhEocha and Raftery (1959)
	Hemiselmis cyclopaea	554	—	—	ÓhEocha et al. (1974)
	Hemiselmis rufescens	556	580	—	Nolan and ÓhEocha (1967)
					ÓhEocha and Raftery (1959)

PE III	*Cryptomonas ovata*	*566, 620*	*617, 650*	—	MacColl *et al.* (1976)
	Cryptomonas sp.	*567*	—	—	Glazer *et al.* (1971)
	Cryptomonas ovata var. *palustris*	*566, 600*	—	—	Allen *et al.* (1959); Brooks and Gantt (1973)
PC I	*Chroomonas* sp.	*585, 620, 645*	*655 (661, 715)*[c]	114	ÓhEocha *et al.* (1964); MacColl *et al.* (1973, 1976); Mörschel and Wehrmeyer (1975)
	Hemiselmis virescens (Millport 64)	*581, 641*	—	—	ÓhEocha and Raftery (1959); Glazer and Cohen-Bazire (1975)
PC II	*Chroomonas mesostigmatica*	*583, 643*	—	—	ÓhEocha *et al.* (1964)
	Cryptomonas cyanomagna	*583, 625–630*	—	—	ÓhEocha *et al.* (1964)
	Hemiselmis virescens	*588, 615*	637	—	Allen *et al.* (1959); ÓhEocha and Raftery (1959)
	Hemiselmis virescens (Plymouth 157)	*577,*[d] *612*	—	—	Glazer and Cohen-Bazire (1975)

[a] Major peaks are italicized.
[b] Tentative identification as a symbiont in the ciliate *Cyclotrichium meunieri*.
[c] Reevaluated by MacColl *et al.* (1976).
[d] For consistency with PE, this designation should be PC I and others PC II.

nm (Glazer and Cohen-Bazire, 1975) or 588 and 615 nm (Allen *et al.*, 1959; ÓhEocha and Raftery, 1959). In addition, according to ÓhEocha *et al.* (1964), *Cryptomonas cyanomagna* does not fit into either class. In the future it may be more meaningful to separate them according to fluorescence emission spectra rather than absorption spectra. Of the two species studied by fluorescence, *Chroomonas* sp. has a peak at 661 nm (Mörschel and Wehrmeyer, 1975; O'hEocha *et al.*, 1964; MacColl *et al.*, 1976), and *H. virescens* (Plymouth 157) has a peak at 637 nm (ÓhEocha *et al.*, 1964). As more corrected fluorescence spectra became available, the pigment classification may resolve itself.

By absorption the PE classes are more distinct, with PE I having a peak at 542–545 nm, PE II at 553–554 nm, and PE III at 566–567 nm (Table I and Figure 3b). Fluorescence emission data are also limited, but peaks are at ca. 580 nm for types I and II, and at 617 nm for type III. Additional far-emitting shoulders have been observed in PE (620–650 nm) and in PC (715 nm), but the nature of these additional shoulders is as yet unknown.

Fluorescence polarization studies (MacColl *et al.*, 1973, 1976) indicate that there are two "functional" types of chromophores identified as "sensitizing" and "fluorescing" chromophores (Teale and Dale, 1970). Only the latter, which occur on only one polypeptide (β-chain), can both absorb and emit light. The sensitizing absorption peak (544 nm) and the fluorescing peak (569 nm) of PE (*R. lens*) are well defined (MacColl *et al.*, 1976), but in PC (*Chroomonas* sp.) similar determinations are more complex.

The broad absorption and emission range in cryptophyte phycobiliproteins appears to be of considerable functional advantage. In this group, where only one phycobiliprotein type exists in each alga, the presence of multiple absorbing forms seems to be advantageous over the greater compartmentalization on separate proteins that generally occurs in red and blue-green algae. The wide absorption range is apparently not due to pigment mixtures, since highly purified phycobiliproteins prepared by several isolation techniques have been used in most studies. Future investigations on fluorescence emission spectra of whole cells and isolated pigments in different solvents may provide some indications of the effectiveness of the energetic coupling in living cells and in isolated pigments, and how it compares with that in red and blue-green algae, where tight energetic coupling exists because of phycobilisomes.

B. Chromophores and Possible Linkages

The chromophores of phycobiliproteins represent a highly conjugated system of tetrapyrroles (Figure 4), which are not combined with a metal as in chlorophyll. Phycoerythrin (and phycoerythrobilin) contains fewer

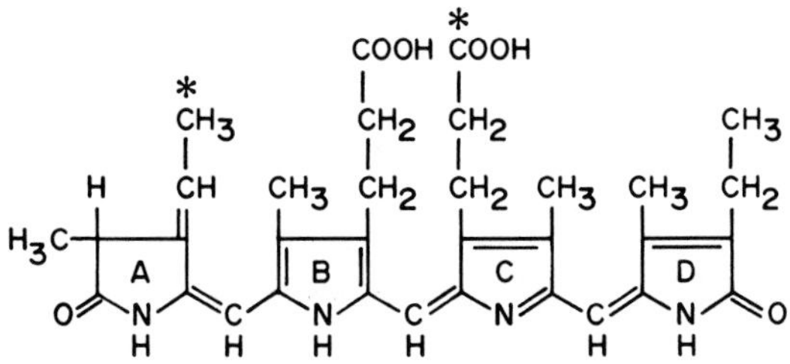

PHYCOCYANOBILIN

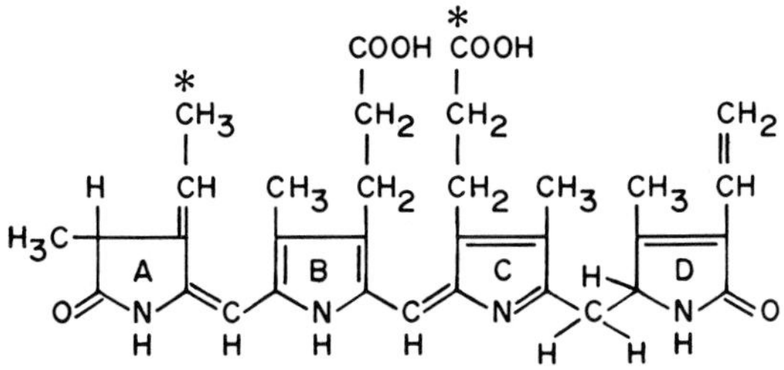

PHYCOERYTHROBILIN

Figure 4. Proposed chromophore structures of the two principal phycobiliproteins. Phycocyanobilin (after Cole *et al.*, 1967; Rüdiger *et al.*, 1967) and phycoerythrobilin (after Chapman *et al.*, 1967; Crespi *et al.*, 1967; Rüdiger *et al.*, 1967). The asterisks indicate the probable apoprotein attachment sites, which may be formed via cysteine on ring A, and via serine on ring C (ÓCarra and ÓhEocha, 1976).

conjugated double bonds than PC (and phycocyanobilin). The difference is reflected in their absorption; i.e., PC with a greater number of conjugated double bonds absorbs at a longer wavelength. When they exist as free chromophores, they contain extra double bonds, one of which occurs in the ethylidene side chain on ring A probably as a result of cleavage from the apoprotein.

The open-chain tetrapyrrole chromophores are covalently linked to the apoprotein. It can be assumed that one chromophore (ca. 590 MW) (Crespi *et al.*, 1967) comprises at least 5% of the total polypeptide weight. Very few studies have been carried out using purified chromophores from cryptophytes. From comparisons of cryptophytan chromophores and from spectral evidence of denatured pigments (ÓhEocha *et al.*, 1964; Chapman *et al.*, 1968; Siegelman *et al.*, 1968; Köst *et al.*, 1975), and those from blue-green and red algae, it can be concluded that they are basically similar.

Although the absorption characteristics of phycobiliproteins appear rather complex, only two chromophore types are generally acknowledged by those investigating chromophore structure directly (Siegelman *et al.*, 1968; Rüdiger, 1975). One is the blue phycocyanobilin and the other the red phycoerythrobilin (Figure 4). Since the structures proposed were derived from material exposed to mild chemical treatment (with methanol or hydrochloric acid), Rüdiger (1970) suggested that they be more accurately designated phycobiliviolin and phycobiliverdin.

A considerable amount of controversy has been engendered by the procedures used to characterize the chromophore, as to the types of native chromophore in existence (see references in Bogorad, 1975; Rüdiger, 1975; ÓCarra and ÓhEocha, 1976; Bryant, 1977). Different forms have been obtained depending on the cleavage conditions. Particularly severe is cleavage by acid hydrolysis, because of the ready isomerization and oxidation of the chromophores. Another factor is the method of analysis used for elucidating the structure, but recently Beuhler *et al.* (1976) examined PC from a blue-green alga by cleavage with either methanol, concentrated hydrochloric acid, or subtilisin and showed that the principal product obtained was phycocyanobilin by all three methods.

Additional chromophore types have been suggested from spectral evidence reported by ÓhEocha and co-workers (ÓCarra and ÓhEocha, 1976) and by Glazer and Cohen-Bazire (1975). One of these additional phycobilins in cryptophytes is phycoviolin (another is phycourobilin in red algae). Upon electrophoretic separation on acrylamide gels, Glazer and Cohen-Bazire (1975) obtained differently colored α and β polypeptides in *H. virescens*. In strains Millport 64 and Plymouth 157, they describe three chemically distinct forms with peaks at 600, 660, and 694 nm, which they identified as separate bilins. Of these the 694-nm form, occurring only on the α subunit of Millport 64, has not been previously described. Definite identity of this chromophore awaits its isolation and characterization in future investigations.

The diversity in the spectral forms is greatly influenced by the specific folding of the native polypeptide chains and by the sites of covalent bonding of the chromophores to the apoproteins. Covalent bonding between the apoprotein and the chromophore probably occurs at two sites involving rings A and C (Figure 4). Brooks and Chapman (1972) prepared a chromopeptide from PE of *Rhodomonas* sp. by proteolytic digestion, which contained four amino acids: alanine, cysteine, serine, and glutamic acid. Their analysis indicated that the chromophore was bound to glutamic acid via the propionic side chain of ring C, and to serine via the propionic side chain of ring B. This scheme was subsequently revised to a proposed seryl ether linkage to the ethylidene side chain of ring A, with ring B left free (Chapman, 1973). Köst *et al.* (1975) examined numerous

PEs and PCs, including those of *Chroomonas* sp., *Rhodomonas* sp., and *C. ovata*. From their analysis they present strong evidence that ring A is linked to the apoprotein, as is probably also ring C. In PE the chromophore–protein linkage on ring A is formed by a thioether linkage between the ethylidene side chain and cysteine of the apoprotein (Köst *et al.*, 1975; ÓCarra and ÓhEocha, 1976). More tentative is the ester linkage between serine and ring C. Direct proof for the cysteine linkage to ring A has been obtained by Köst-Reyes *et al.* (1975).

C. Chemical and Physical Properties

Phycobiliproteins are extremely soluble in water, and can be readily extracted from cells. In cryptophytes, unlike the situation in red and blue-green algae, only low molecular weight forms exist in solution. Their "native" molecular weights range from 28,000 to 57,000 (Table II), and they exhibit little tendency toward self-association *in vitro*. MacColl *et al.* (1973, 1976) explored separately various PE and PC concentrations (0.04–5 gm/liter) and found that only a single species was present with an $s_{20,w}$ of about 4.4 S.

Cryptophytan phycobiliproteins are rather labile and probably are easily altered by normal isolation procedures. Isoelectric focusing of PE and PC has revealed numerous bands ranging from pH 4.8 to 7.8 (Glazer *et al.*, 1971; MacColl *et al.*, 1973; Brooks and Gantt, 1973; Mörschel and Wehrmeyer, 1975, 1977; Glazer and Cohen-Bazire, 1975). Whether or not these are charge isomers or size isomers is not clear. However, in considering the sedimentation data, and molecular weight determinations made by gel filtration, it seems unlikely that they represent size isomers. Mörschel and Wehrmeyer (1975, 1977) have noted five PC charge isomers in *Chroomonas* sp. and three PE charge isomers in *C. maculata,* which they have designated as "multiple pigment forms" since they exhibited slight changes in their relative peak heights. These charge isomers, however, had the same basic polypeptide subunits.

Phycoerythrin and PC are each composed of two polypeptides which resolve into α and β types on sodium dodecyl sulfate (SDS) gel electrophoresis and generally occur with a 1:1 stoichiometry as evidenced from stained SDS gels and amino acid analysis (Table III). As seen in Table II, the α subunits are smaller (ca. 9000–11,000 MW) than the β subunits (ca. 15,000–19,000 MW). Generally, they exist as pairs, resulting in a quarternary protein structure consisting of $\alpha_2\beta_2$. This has been confirmed by cross-linking experiments on PE and PC with dimethyl suberimidate and SDS gel electrophoresis (MacColl *et al.,* 1976). The molecular weights reported from gel filtration suggest an $\alpha_1\beta_1$ structure (Glazer *et*

Table II Molecular weights of *in Vitro* Aggregates and Polypeptides of Cryptophytan Phycobiliproteins

Type	Species	MW of polypeptides	Polypeptide color or peak (nm)	MW of aggregate	Reference
PE I	*Rhodomonas* sp. 3-C	α: 11,000 β: 17,700	Blue (SDS) Red (SDS)	30,800 —	Brooks and Gantt (1973)
	Cryptomonas maculata	α: 9,900 β: 15,700	565 (urea) 555 (urea)	44,500 —	Mörschel and Wehrmeyer (1977)
	Rhodomonas lens	α: 9,800 β: 17,700	565 (SDS) 531 (SDS)	53,000 —	MacColl et al. (1976)
PE II	*Hemiselmis cyclopaea*	—	—	27,800	Nolan and ÓhEocha (1967)
	Cryptomonas sp.	α: 11,800 β: 19,000	Pink (SDS) Deep pink (SDS)	35,000 —	Glazer et al. (1971)
PE III	*Cryptomonas ovata* var. *palustris*	α: 11,000 β: 17,700	Blue (SDS) Red (SDS)	30,900 —	Brooks and Gantt (1973)
	Cryptomonas ovata	α: 9,700 β: 18,200	— —	53,000 —	MacColl et al. (1976)
PC I	*Chroomonas* sp.	α: 10,000 β: 16,000	644 (SDS) 566 (SDS)	54,300 —	MacColl et al. (1973, 1976)
	Chroomonas sp.	α: 9,200 α': 10,400 β: 15,500	645 (SDS) 645 (SDS) 560 (SDS)	45,000 — —	Mörschel and Wehrmeyer (1975)
PC II	*Hemiselmis virescens* (Millport 64)	α: 10,000 β: 19,000	600, 660, 694 (urea) 600, 660 (urea)	57,000 —	Glazer and Cohen-Bazire (1975)
	Hemiselmis virescens (Plymouth 157)	α: 10,000 β: 19,000	600, 660 (urea) 600, 660 (urea)	57,000 —	Glazer and Cohen-Bazire (1975)

al., 1971; Brooks and Gantt, 1973; Nolan and ÓhEocha, 1967) and may require reevaluation. An unusual combination was reported in *Chroomonas* sp., where Mörschel and Wehrmeyer (1975) found two light chains with a resulting combination of $\alpha\alpha'\beta_2$ and a combined molecular weight of 45,100.

On denaturation of phycobiliproteins by SDS treatment considerable fading of the chromophore ensues. Separation of the subunits on Bio-Rex 70 in a urea solution at pH 3.0 seems to preserve the natural chromophore environment much better. Regardless of the method of separation, it can be seen in Table II that the α subunits generally have absorption maxima at wavelengths longer than those of the β subunits. An interesting case exists in the two strains of *H. virescens* (Glazer and Cohen-Bazire, 1975), where the α and β polypeptides of the Plymouth 157 strain are purple as is also the β of the Millport 64 strain, but the α of the latter is green, presumably because of a newly designated chromophore.

A comparison of the amino acid composition (Table III) of PE from *R. lens* and PC from *Chroomonas* sp. shows that they are highly similar (MacColl *et al.,* 1973, 1976) and occur in a 1 : 1 stoichiometry. They are

Table III   Amino Acid Composition of Phycocyanin and Phycoerythrin

Amino acid	Phycoerythrin[a]	Phycocyanin[b]		
	Residues per 24,000 MW	Residues per 24,000 MW	α Chain	β Chain
Lysine	16.4	16.6	7.9	7.7
Histidine	1.3	0.96	0.61	Trace
Arginine	9.2	8.9	3.7	7.4
Aspartic acid	27.3	28.3	10.5	16.5
Threonine	11.4	10.1	4.0	5.8
Serine	31.0	23.7	5.3	18.2
Glutamic acid	17.2	16.6	8.6	8.4
Proline	7.0	6.4	3.3	4.2
Glycine	19.7	19.4	6.7	13.4
Alanine	30.3	33.4	10.0	20.2
Valine	17.0	16.4	5.3	9.5
Methionine	6.2	4.1	1.6	2.0
Isoleucine	10.7	9.8	4.2	6.0
Leucine	16.3	18.6	4.8	13.3
Tyrosine	6.1	6.7	2.4	4.9
Phenylalanine	6.3	5.3	1.7	3.0
Half-cystine[c]	9.4	11.1	3.2	5.9

[a] *Rhodomonas lens,* MacColl *et al.* (1976).
[b] *Chroomonas* sp., MacColl *et al.* (1973).
[c] Measured as cysteic acid.

more closely related to one another than to their spectral counterparts in the red and blue-green algae. In comparing cryptophytan PC with cyanophytan PC, MacColl *et al.* (1973) noted larger quantities of serine, half-cystine, and lysine in the cryptophytan PC. The α and β chains differ not only in size but also in composition. As seen in Table III, the α chain is richer in the polar amino acids lysine and glutamic acid, whereas the β chain is richer in serine and leucine.

IV. PHYLOGENETIC RELATIONSHIP WITH RHODOPHYTA AND CYANOPHYTA

Phylogenetically cryptophytes are distinct from all algae thus far studied, and taxonomically they have been placed in numerous groups. Because of the presence of phycobiliproteins, they have even been suggested as a transition form, between blue-green and red algae, to flagellated algae and higher plants. The greater the lack of pertinent information, the greater the speculations seem to be.

Until recently, cryptophytan phycobiliproteins seemed to be a class distinct from those in red and blue-green algae. They differed in aggregation properties, in location in the photosynthetic apparatus, and in that there was generally no immunological cross-reactivity. This was first pointed out by Berns (1967) and confirmed by various laboratories (Glazer *et al.*, 1971; Glazer and Cohen-Bazire, 1975; Gantt, unpublished data). Antisera prepared against PEs and PCs of red and blue-green algae only cross-reacted, respectively, with PEs and PCs of these two classes and did not cross-react with PE or PC of cryptophytes with one exception.

In his extensive screening of the three algal groups, Berns (1967) noted a weak interaction between antisera prepared against a cryptophytan PE (*R. lens*) and a rhodophytan PE (*P. cruentum*). This observation was recently confirmed and extended by MacColl *et al.* (1976), who showed a high degree of immunochemical relatedness within cryptophytes and with PE of *P. cruentum*. By Ouchterlony double-diffusion experiments (pH 6.0) they showed positive reactions, with spurring, between anticryptophytan PE (*R. lens* and *C. ovata*) and rhodophytan PE (*P. cruentum*), and the reverse. Positive reactions were also obtained with antirhodophytan PE and cryptophytan PC, but the reserve combination produced negative results. A reaction between rhodophytan PE and cryptophytan PC is particularly surprising, since antirhodophytan PE does not react with any rhodophytan or cyanophytan PC tested. These results lead one to speculate that an antigenic site common to several phycobiliproteins was preserved in the more stable rhodophytan PE, but not during antigen production by the more labile cryptophytan PC.

Glazer (1976), in his recently proposed hypothetical scheme for the evolutionary relationships of phycobiliproteins, suggests that cryptophytan biliproteins split off from the ancestral line very early in development. His scheme indicates that the α chain containing phycocyanobilin was modified by amino acid substitutions on the β chain, and that other chromophore types and binding sites probably arose by mutations. The most convincing evidence for one common ancestral gene is based on the nature of the chromophore structure and the similar nature of amino acid sequences of the polypeptides.

Sequence analyses of the NH_2-terminus of the separated α and β chains from cryptophytan, rhodophytan, and cyanophytan phycobiliproteins made by Harris and Berns (1975; see also Troxler et al., 1975) showed that methionine was the NH_2-terminal amino acid in most of the species studied. Furthermore, a comparison of partial amino acid sequences showed an invariant sequence in PEs and PCs. Recently, Glazer and Apell (1977) compared NH_2-terminal sequences of PC from H. virescens (Millport 64) with corresponding sequences (about 15 to 25 amino acids long) derived from red and blue-green algal phycobiliproteins and concluded that the three algal groups, on the basis of their phycobiliproteins, have a closer phylogenetic relationship than had previously been assumed.

The presence of chlorophyll c_2 in addition to phycobiliproteins, along with the absence of allophycocyanin, had previously allowed for the possibility that this accessory chlorophyll was the intermediate between phycobiliproteins and chlorophyll a. This, however, is not likely in view of recent calculations made by MacColl and Berns (1978). In applying Förster's energy transfer theory to the energy migration in cryptophyte pigments, they found that there was a large spectral overlap integral between PC and chlorophyll a, but a small one between PC and chlorophyll c_2. This suggests that PC transfers energy directly to chlorophyll a, without going through chlorophyll c_2. In PE-containing cryptophytes there is presently no evidence on whether or not chlorophyll c_2 is an intermediate in the transfer. Probably chlorophyll c_2 functions independently as an accessory pigment to chlorophyll a.

Even though convincing similarities exist between the phycobiliproteins of the three algal groups, the phylogenetic status of cryptophytes cannot be resolved on phycobiliprotein characteristics alone, in spite of the importance of these pigments in the energy conversion process of these cells. For the present, cryptophyte taxonomy remains, as the name of the group implies, a puzzling secret.

REFERENCES

Allen, M. B., Dougherty, E. C., and McLaughlin, J. J. A. (1959). *Nature (London)* **184**, 1047–1049.

Antia, N. J., Kalley, J. P., McDonald, J., and Bisalputra, T. (1973). *J. Protozool.* **20**, 379–385.

Barber, R. T., White, A. W., and Siegelman, H. W. (1969). *J. Phycol.* **5**, 86–88.

Berns, D. S. (1967). *Plant Physiol.* **42**, 1569–1586.

Beuhler, R. J., Pierce, R. C., Friedman, L., and Siegelman, H. W. (1976). *J. Biol. Chem.* **251**, 2405–2411.

Bogorad, L. (1975). *Annu. Rev. Plant Physiol.* **26**, 369–401.

Brooks, C., and Chapman, D. J. (1972). *Phytochemistry* **11**, 2663–2670.

Brooks, C., and Gantt, E. (1973). *Arch. Mikrobiol.* **88**, 193–204.

Bryant, D. A. (1977). Ph.D. Thesis, Univ. of California, Los Angeles. (Xerox Univ. Microfilms, Ann Arbor, Michigan.)

Chapman, D. J. (1973). *In* "Biology of Blue-Green Algae" (N. G. Carr and B. A. Whitton, eds.), pp. 162–185. Univ. of California Press, Berkeley.

Chapman, D. J., Cole, W. J., and Siegelman, H. W. (1967). *J. Am. Chem. Soc.* **89**, 5976–5977.

Chapman, D. J., Cole, W. J., and Siegelman, H. W. (1968). *Phytochemistry* **7**, 1831–1835.

Cole, W. J., Chapman, D. J., and Siegelman, H. W. (1967). *J. Am. Chem. Soc.* **89**, 3643–3645.

Crespi, H. L., Boucher, L. J., Norman, G. D., Katz, J. J., and Dougherty, R. C. (1967). *J. Am. Chem. Soc.* **89**, 3642–3643.

Dodge, J. (1969). *Arch. Mikrobiol.* **69**, 266–280.

Dougherty, E., and Allen, M. B. (1960). *In* "Comparative Biochemistry of Photoreactive Systems" (M. B. Allen, ed.), pp. 129–144. Academic Press, New York.

Faust, M. A., and Gantt, E. (1973). *J. Phycol.* **9**, 489–495.

Gantt, E. (1975). *BioScience* **25**, 781–788.

Gantt, E., Edwards, M. R., and Provasoli, L. (1971). *J. Cell Biol.* **48**, 280–290.

Glazer, A. N. (1976). *Photochem. Photobiol. Rev.* **1**, 71–115.

Glazer, A. N., and Apell, G. S. (1977). *FEMS Lett.* **1**, 113–116.

Glazer, A. N., and Cohen-Bazire, G. (1975). *Arch. Microbiol.* **104**, 29–32.

Glazer, A. N., Cohen-Bazire, G., and Stanier, R. Y. (1971). *Arch. Mikrobiol.* **80**, 1–18.

Harris, J. U., and Berns, D. S. (1975). *J. Mol. Evol.* **5**, 153–163.

Haxo, F. T. (1960). *In* "Comparative Biochemistry of Photoreactive Systems" (M. B. Allen, ed.), pp. 339–360. Academic Press, New York.

Haxo, F. T., and Fork, D. C. (1959). *Nature (London)* **184**, 1051–1052.

Jeffrey, S. W. (1976). *J. Phycol.* **12**, 349–354.

Köst, H.-P., Rüdiger, W., and Chapman, D. J. (1975). *Justus Liebigs Ann. Chem.*, 1582–1593.

Köst-Reyes, E., Köst, H.-P., and Rüdiger, W. (1975). *Justus Liebigs Ann. Chem.*, 1594–1600.

MacColl, R., and Berns, D. S. (1978). *Photochem. Photobiol.* **27**, 343–349.

MacColl, R., Habig, W., and Berns, D. S. (1973). *J. Biol. Chem.* **248**, 7080–7086.

MacColl, R., Berns, D. S., and Gibbons, O. (1976). *Arch. Biochem. Biophys.* **177**, 265–275.

Mörschel, E., and Wehrmeyer, W. (1975). *Arch. Microbiol.* **105**, 153–158.

Mörschel, E., and Wehrmeyer, W. (1977). *Arch. Microbiol.* **113**, 83–89.

Nolan, D. N., and ÓhEocha, C. (1967). *Biochem. J.* **103**, 39P–40P.

ÓCarra, P., and ÓhEocha, C. (1976). *In* "Chemistry and Biochemistry of Plant Pigments" (T. W. Goodwin, ed.), 2nd Ed., Vol. 1, pp. 328–376. Academic Press, New York.

ÓhEocha, C. (1960). *In* "Comparative Biochemistry of Photoreactive Systems" (M. B. Allen, ed.), pp. 181–203. Academic Press, New York.

ÓhEocha, C., and Raftery, M. (1959). *Nature (London)* **184**, 1049–1051.

ÓhEocha, C., ÓCarra, P., and Mitchell, D. (1964). *Proc. R. Ir. Acad., Sect. B* **63**, 191–200.

Rüdiger, W. (1970). *Angew. Chem., Int. Ed. Engl.* **9**, 473–480.

Rüdiger, W. (1975). *Ber. Dtsch. Bot. Ges.* **88**, 125–139.

Rüdiger, W., ÓCarra, P., and ÓhEocha, C. (1967). *Nature (London)* **215**, 1477–1478.

Siegelman, H. W., Chapman, D. J., and Cole, W. J. (1968). *In* "Porphyrins and Related Compounds" (T. W. Goodwin, ed.), pp. 107–120. Academic Press, New York.

Teale, F. W. J., and Dale, R. E. (1970). *Biochem. J.* **116**, 161–169.

Troxler, R. F. (1977). *In* "Chemistry and Physiology of Bile Pigments" (P. D. Beck and N. I. Berlin, eds.), Fogarty Int. Cent. Proc., No. 35, pp. 431–454. [DHEW Publ. No. (NIH) 77-1100]. U.S. Govt. Print. Off., Washington, D.C.

Troxler, R. F., Foster, J. A., Brown, A., and Franzblau, C. (1975). *Biochemistry* **14**, 268–274.

Vesk, M., and Jeffrey, S. W. (1977). *J. Phycol.* **13**, 280–288.

Wehrmeyer, W. (1970). *Arch. Microbiol.* **71**, 367–383.

Halotolerance of *Dunaliella* 6

A. D. BROWN AND LESLEY J. BOROWITZKA

I. INTRODUCTION

The invitation to write this chapter on *Dunaliella* reminded one of us of his student days and arguments about whether a certain organism should be classified as plant or animal. When the subject of this argument was, as was commonly the case, a protist, the question usually remained unresolved. With the passage of time and the acquisition of some formal qualifications in microbiology, refuge from such vexatious questions has been found in two responses, namely, (1) all schemes of classification are man-made, and (2) the organism in question is a microorganism anyway.

Nevertheless, we are intrigued by the popularity of *Dunaliella,* which is hailed as an important member of the phytoplankton by marine biologists and as a protozoon by others. Presumably its appeal to the latter lies in its

BIOCHEMISTRY AND PHYSIOLOGY OF PROTOZOA
SECOND EDITION, VOL. 1

motility and its lack of a cell wall, whereas for botanists and marine biologists its photosynthetic ability and its attendant ecological role as an aquatic primary producer are paramount. To us, of course, a microorganism is a microorganism.

There are, however, other reasons for taking an interest in *Dunaliella*. There are some whose interest is concerned primarily with its use as a feed for *Artemia* (e.g., Sick, 1976) and others who see it as a more general potential source of protein (Gibbs and Duffus, 1976). Although we are saddened by such a materialistic approach to quite an attractive alga, we are pleased to note that it performs well on both counts. The genus as a whole, however, is remarkably adaptable to change in a number of important environmental variables and, for this reason, it is unusually ubiquitous. It is common in the sea and estuaries and predominant at higher salinities. It is often the only algal genus found in salt lakes and similar environments. It thus ranges from being an important to an indispensible part of many ecosystems. Indeed, it is fair to say that, without *Dunaliella,* there would probably be no self-sustaining communities in many salt lakes, including the Great Salt Lake in Utah.

In spite of some very valuable recent publications such as that of Stewart (1974), algal physiology is not generally as well understood as that of vascular land plants. This is especially true of details of photosynthesis and associated processes. *Dunaliella* is no exception; in fact, in many respects less is known about this genus than other genera such as *Chlorella.* An outstanding and conspicuous aspect of *Dunaliella* physiology, however, is the ability, both collectively within the genus and of individual species, to survive extraordinarily large stresses engendered by sudden changes in environmental salt concentration and to thrive over a very wide range of salinities. In these two respects together, it is probably unmatched by any other organism. *Dunaliella* spp. are also quite tolerant of wide temperature ranges, very high hydrostatic pressures, and, for at least one species (*D. salina*), high light intensities.

This chapter is concerned predominantly with the physiology of halotolerance in the genus.

II. TAXONOMY

The genus *Dunaliella* belongs to the family Polyblepharidaceae, order Volvocales, and phylum Chlorophyta. Both *Dunaliella* and the closely related genus *Chlamydomonas* have salt-tolerant members, but more is known about the taxonomy and salt relations of *Dunaliella* species.

Major early studies contributing to the taxonomy of the genus *Dunaliella* include those of Téodoresco (1905, 1906) on *D. salina* and *D. viridis,* Hamburger (1905) on *D. salina,* Baas-Becking (1931) on *D. viridis,* Lerche (1938) covering many species, and Ruinen (1938), which mentions six species. The subjects of these early studies were all isolates from inland, generally saline, lakes.

A definitive review of the taxonomy is that of Butcher (1959). Two marine and ten nonmarine species are described, and a key for their identification is given. Butcher points out the difficulty of distinguishing the genus *Dunaliella,* which has no cell wall, from some species of *Chlamydomonas* with thin and elastic cell walls.

Dunaliella species grow well in many types of culture media in which they can show considerable variation in shape, size, pigmentation, and the nature of the cell contents. Of the last-mentioned, the position and appearance of the stigma and the distribution of cytoplasmic granules seem constant within a species. Since, classically, identification of species is based on morphological characters, rigid definitions of species of *Dunaliella* are of doubtful value.

Briefly, according to Butcher's system, *Dunaliella minuta* and *D. media* are separated from the rest by their shapes, respectively, fusiform and ovoid to ellipsoid. Both were isolated from saline lakes (Lerche, 1938). Most of the species referred to in this chapter are ovoid, pyriform, or cylindrical, with a rounded posterior and an acute to subacute anterior. *Dunaliella salina* (Dunal) Téodoresco is the largest and is broadly ovoid, 16–24 μm long, and 10–13 μm wide. It has been isolated from salt marshes and pools in many parts of the world. Lerche (1938) includes the *D. viridis* of Téodoresco as a variety within the species *D. salina,* and defines it as "cells 10–12 μm × 6–7 μm, ovoid and not constricted, haematochrome not developed." Hematochrome is the red colouring caused by β-carotene pigment accumulation. Butcher comments that either *D. salina* is a very variable species or there are several distinct species included under the one name. We concur with the latter conclusion and have used the name *D. viridis* Téodoresco, following Baas-Becking (1931) and Johnson *et al.* (1968), for the extremely salt-tolerant species which remains green at all salt concentrations.

Generally less than 10 μm long and ovoid to ellipsoid are *Dunaliella parva* Lerche and *D. bioculata* Butcher, both salt lake isolates, and *D. tertiolecta* Butcher, a marine isolate. *Dunaliella parva* is 7–12 μm × 4–7 μm and is usually green, but contains a small amount of β-carotene pigment prominent under conditions of limiting nutrient. In *D. tertiolecta,* β-carotene pigment development is not observed, nor, according to Butcher (1959), is sexual reproduction, although Latorella *et al.* (1974) have reported sexual reproduction in stationary phase cultures of their *D.*

tertiolecta, as in all other species of *Dunaliella. Dunaliella tertiolecta* and *D. bioculata* are both less than 10 μm long; the latter often has two stigmata.

III. MORPHOLOGY AND ULTRASTRUCTURE

Members of the genus *Dunaliella* (as defined in Butcher, 1959) have a thin and elastic cell envelope, and consequently the shape of the cells is very changeable. They are motile, with two equal flagella. There is one red eyespot and one cup-shaped chloroplast which fills about half the cell volume and contains a large pyrenoid. Starch and glycerol are the main storage products, and the chloroplast of stationary phase cells contains numerous large starch granules (Eyden, 1975). Photosynthesis and starch storage are very important, since the organism is apparently an obligate autotroph, although Eyden (1975) has suggested that some stationary phase growth may result from pinocytotic uptake of nutrients from moribund cells. Autophagic vacuoles which could affect this uptake are seen in the cytoplasm of *Dunaliella primolecta.*

Asexual reproduction is normally by longitudinal division of the motile cell. This process has been studied in detail using electron microscopy and synchronized cells of *D. bioculata;* it closely resembles the process in other members of the Volvocales and has been described in detail by Marano (1976). Sexual reproduction is much rarer and occurs in stationary-phase cultures when two motile cells fuse to form a quadriflagellate zygote. In addition, *D. salina* has been reported to produce red, haploid nonmotile aplanospores when grown in relatively low salt concentrations (Latorella *et al.,* 1974). They have thick, layered cell walls, and are probably identical with the small, nonmotile, *Chlorella*-like cells that Lerche (1938) observed in *D. salina* cultures at low salinities. To our knowledge these aplanospores have not been observed in species of *Dunaliella* other than *D. salina.*

Eyden (1975) has published detailed light and electron microscope observations of logarithmic and stationary phase cells of *D. primolecta.* In general, the fine structure of the subcellular organelles resembles that of other volvocid algae; the chloroplast and mitochondrial systems are very similar to those of *Chlamydomonas* (Eyden, 1975).

Raising or lowering salinity predictably causes the algae to shrink or swell, respectively, and to lose motility. Motility is usually recovered at about the same time as normal volume and growth. *Dunaliella viridis,* however, at a very low salt concentration (0.5 M) can recover its ability to grow and divide while remaining swollen and nonmotile (see Section VI,C,1).

The effect of salt concentration on *Dunaliella* fine structure was studied electron microscopically by Trezzi *et al.* (1965) and, more recently, by Pfeifhofer and Belton (1975). Trezzi and associates were able to preserve in fixed preparations the swelling and shrinkage caused, respectively, by decreases and increases in salinity. They concluded, however, that the "osmotic resistance" of *D. salina* is caused by ready permeability of the plasma membrane to ions as well as to water. This is unlikely to be true (see Section VI).

Pfeifhofer and Belton (1975) were primarily concerned with chloroplast structure and included freeze-fracturing among their techniques. There were no major differences in the chloroplast structure of *D. salina* grown over the range 3.5–25% NaCl, although the thylakoid membrane appeared more compressed in algae adapted to the higher salt concentration. Freeze-fracturing showed that the number of particles on a B-face end membrane was about three times as great in chloroplasts of algae grown in 25% as in 3.5% NaCl. These changes were reversible.

IV. ECOLOGY

Species of *Dunaliella* are readily isolated from the sea and from inland salt lakes with various chemical compositions and with salt concentrations ranging from below that of seawater to saturated. *Dunaliella salina, D. viridis,* and *D. parva* are commonly present in highly saline lakes and are often described as halophilic, although their halophilism is fundamentally different from that of halophilic bacteria. *Dunaliella salina* and *Halobacterium* were the first organisms shown to live in the Dead Sea (Volcani, 1936, 1944).

Brock (1975) has suggested that some species of *Dunaliella* are the most halotolerant eukaryotic microorganisms known. He showed that *Dunaliella* could be isolated from water samples containing 10% to greater than 30% NaCl from the Great Salt Lake, Utah. Enrichment cultures from the lake were set up using media with a range of salinities; at low salinities a wide range of algae grew, but only *Dunaliella* could grow at high salinities. The experiments showed the competitive advantage of *Dunaliella* in highly saline environments. Predictably, *Dunaliella* was responsible for most of the photosynthetic production in the Great Salt Lake in a spring study by Stephens and Gillespie (1976). Growth was so efficient that production was limited in late spring first by self-shading and then by nitrogen. Indeed, in the southern arm of the Great Salt Lake, populations of *D. viridis* as high as 2×10^5/ml have been encountered in late spring (Post, 1977). The *Dunaliella* population served as the major food source for the brine shrimp *Artemia salina,* as well as for the dominant component of the lake's biomass, the halophilic bacteria *Halobacterium* and *Halococcus*. *Dunaliella* is at least the major and possibly the only primary producer in the Great Salt Lake. These and other details are elaborated in an interesting account by Post (1977) of the ecology of this lake.

Dunaliella species are also an important component of marine phytoplankton populations. They are well suited to their environment; they remain viable when exposed to low temperatures and light intensity (Hellebust and Terborg, 1967) and high pressure. In fact, respiration by *D. tertiolecta* is apparently unaffected by hydrostatic pressures up to 600 atm (Pope and Berger, 1974). *Dunaliella* is more tolerant than other planktonic algae of fuel oil components; indeed, its growth was enhanced by low levels of the components which were inhibitory to other algae (Dunstan *et al.*, 1975). The reason for the enhancement is not known; unlike some bacteria, *Dunaliella* probably has no enzymes to metabolize the oil, but its presence may have a secondary effect, for instance, by changing the availability of some growth factor. This observation, however, casts some doubt on the strict autotrophy attributed to *Dunaliella* (see Section III).

Furthermore, *D. tertiolecta* can tolerate substantially higher concentrations of chlorinated hydrocarbon insecticides (Menzel *et al.*, 1970) and heavy metals (Mandelli, 1969; Davies, 1974, 1976; Bentley-Mowat and Reid, 1977) than most other marine phytoplankton. Mandelli (1969) found that *D. tertiolecta* grew in seawater with copper(II) (9.4 μg-atoms/liter), whereas diatoms and dinoflagellates were inhibited at less than half this concentration. In medium containing 5 μg-atoms/liter mercury(II), photosynthesis of *D. tertiolecta* was not inhibited (Overnell, 1975), and growth barely so, whereas 0.25 μg/atom liter was lethal for the species of *Isochrysis, Skeletonema,* and *Phaeodactylum* tested (Davies, 1974). While copper(II) causes both membrane leakage and photosynthetic inhibition, the primary effect of the group mercury(II), methylmercury, and thallium on *Dunaliella* (and *Chlamydomonas*) is photosynthetic inhibition (Overnell, 1975). The basis of *Dunaliella*'s unusual mercury tolerance is detoxication of the mercury within the cell by precipitation of an insoluble compound and, less important, the slow rate of mercury accumulation (Davies, 1976). *Dunaliella* is also reported to be relatively resistant to lead, copper, and cadmium (Bentley-Mowat and Reid, 1977).

V. GENERAL SALT RELATIONS

A. Growth and Survival

Members of the genus *Dunaliella* grow over a full spectrum of salinities, from a sodium chloride concentration less than that of seawater to saturated. *Dunaliella salina* itself can span this range; it has been grown in 2% (w/v) to 35% (w/v) NaCl (Loeblich, 1972). Table I lists salinity data for salt lake isolates of *Dunaliella* and for the marine *D. tertiolecta. Dunaliella parva* is not included because we do not know of a comprehensive study of its salinity tolerances and preferences, but Ben-Amotz and Avron (1973a) have cultured it within the range 4–12% NaCl. The growth optima listed for the salt lake isolates generally coincide with the optima for photosynthesis, and the values are remarkably similar for all the salt lake isolates except for the value 3 M NaCl. This is higher than the others, possibly because in the experiment other parameters were also optimized. Carbon dioxide-enriched air was used, and the result leads us to suggest that the other growth optima listed may be lower than the intrinsic NaCl concentration optima for the isolates, because CO_2 becomes less soluble as the NaCl concentration of the medium increases, and CO_2 limitation of photosynthesis may cause a decline in the growth rate at higher salinities.

Table I shows that neither the growth range nor the salinity of the environment from which *Dunaliella* is isolated bear much relation to the optimum for growth and photosynthesis of the isolate. It is apparently the adaptability of the organism to environments with a wide range of salinities, rather than an extremely high salt optimum, that gives *Dunaliella* a competitive advantage in saline environments. As a further

Table I General Salt Relations of *Dunaliella*[a]

| Alga | Source | NaCl concentration (M) | | Reference |
		Growth range	Growth optimum	
Dunaliella salina	Salt lake	0.3–saturated	2	Loeblich (1972)
Dunaliella viridis	Solar salt pond	1–4	0.9–2	Baas-Becking (1931); Johnson et al. (1968)
		0.3–4.8	1.0–1.5	Borowitzka et al. (1977)
Dunaliella sp.	Great Salt Lake, Utah	0.9–3.4	2.5	Brock (1975)
Dunaliella sp.	Great Salt Lake, Utah	0.06–2	ca. 3	van Auken and McNulty (1973)
Dunaliella sp.	Solar salt pond	—	1.7–2.6 M	Gibor (1956)
Dunaliella tertiolecta	Marine	0.06–>3.6	0.17–0.2	Borowitzka et al. (1977)

[a] The apparent salt relations represented in this table can be expected to be affected by temperature.

demonstration of the adaptability of *Dunaliella* species, "training" experiments extended the growth ranges of *D. tertiolecta* and *D. viridis*, although the growth optima were unchanged. By many transfers and growing cells in media containing increasing or decreasing amounts of NaCl, *D. tertiolecta* could be grown in the extended range 0.17–3.5 M, and *D. viridis* in the range of 0.3–4.8 M (Borowitzka *et al.*, 1977).

Moreover, high concentrations of sucrose were tolerated by the cells. *Dunaliella viridis*, on a single transfer from an inoculum grown in 3.4 M NaCl medium, grew in medium containing 0.05 M NaCl plus sucrose in the range of 1.0–2.0 M. Sucrose concentrations had little effect on the growth rate, but the glycerol content of the cells responded positively to sucrose concentration, as it did to NaCl concentration (Borowitzka *et al.*, 1977).

Dunaliella viridis and *D. tertiolecta* (Baas-Becking, 1931; McLachlan, 1960), and probably other *Dunaliella* species, are sensitive to high concentrations of Ca^{2+} and Mg^{2+}, dependent on their ratio, but a wide pH range, pH 6–9 (Baas-Becking, 1931) and pH 4.5–8.5 (van Auken and McNulty, 1973), is tolerated by at least two species. No vitamins are required in artificial growth media; in fact, *D. tertiolecta* releases thiamine during growth (Carlucci and Bowes, 1970).

Dunaliella species tolerate wide variations in physical growth conditions. The more salt-tolerant species *D. salina* and *D. viridis* grow optimally at higher temperatures, 30° and 37°C, respectively (Gibor, 1956; Loeblich, 1972), than the marine species *D. tertiolecta*, which grows best at 20°C (Eppley and Coatsworth, 1966). The more salt-tolerant species grow best in high-intensity light, 10,000 lx (Yurina, 1966) or higher [25,000–35,000 lx (van Auken and McNulty, 1973)], and red light is favoured for growth by *D. salina* (Drokova and Dovhoruka, 1966). There is a depression in photosynthetic activity at the blue end of the spectrum, which is more pronounced in *D. salina* cells growing in high salt concentrations than in low concentrations.

Production of β-carotene in *D. salina* varies with salinity, temperature, and light intensity; the organism characteristically becomes redder with increasing salinity; high light intensities also favour carotenoid and inhibit chlorophyll production. It has been proposed that β-carotene, which is not involved in photosynthesis, may act as a photoprotective pigment in the highly saline, brightly illuminated salt pans where the organism is normally found (Loeblich, 1972). A similar role for carotenoids has been proposed in some bacteria (Larsen, 1962; Shapiro *et al.*, 1977). It is interesting that varying the same three parameters does not lead to noticeable β-carotene production in any species but *D. salina*. Growth and pigment production of *D. salina* were the subjects of intensive research in

Russia and the Ukraine in the early 1960s, and a process for preparing vitamin A for stock feeds from β-carotene was developed (Masyuk, 1965a,b).

Growth rates of *Dunaliella* spp. range between one and two generations per day; in general, the more salt-tolerant species grow more slowly than the less salt-tolerant species. The same phenomenon occurs in yeasts; xerotolerant yeasts grow more slowly than their nontolerant counterparts (Anand and Brown, 1968). Even when *D. tertiolecta* had been "trained" to grow in the range normally the domain of *D. viridis*, it always grew faster (Borowitzka *et al.*, 1977).

Synchronized growth of *Dunaliella* can be attained by using a combination of light–dark and temperature transitions (Wegmann and Metzner, 1971); synchronized cultures have been used in both physiological and ultrastructural work (Wegmann and Metzner, 1971; Marano, 1976; Gimmler *et al.*, 1977).

B. Intracellular Composition

As a result of observations that *D. salina* cells swelled or shrank after changes in the concentration of solutes in their medium, Trezzi *et al.* (1965) proposed that the "osmotic tolerance" depended on the cell membrane's being readily permeable to water and ions, although such observations invite the opposite interpretation. These results, however, correlated with earlier work in which conductivity and freezing point determinations showed that concentrations of solutes in the cell "sap" were higher when *D. salina* was grown in 3.9 M NaCl than when it was grown in a 1.5 M NaCl medium (Marrè and Servettaz, 1959). The chloride concentration was higher in the cells grown at the higher salinity, and $^{24}Na^+$ ions appeared to flow freely into the cell when the salinity of the medium was raised from 1.5 to 3.9 M. Marrè and Servettaz concluded that *D. salina*'s tolerance of high salinity and changes in salinity was due to equilibration of internal and external solute concentrations across a highly permeable membrane. Ginzburg (1969) reported that the plasma membrane of *D. parva* was highly permeable to the large molecules inulin and sucrose, and joined the earlier workers with the suggestion that a highly permeable membrane may be of adaptive value to halophilic organisms.

A clear conclusion about intracellular salt concentration has been consistently hindered by the technical difficulties of obtaining reliable analyses. Although *Dunaliella* is very resilient osmotically, it is very fragile mechanically. For this reason, salt analyses on centrifuged algae are unreliable and are beset by uncertainty of the extent of exchange between the cells and the extracellular fluid occluded in the pellet.

There is, however, a general theoretical reason, advanced perhaps with the help of hindsight, as well as two important sets of indirect experimental evidence, which argue against the proposition that *Dunaliella* cells contain sodium chloride at high concentrations. The theoretical reason is that viable cells do not equilibrate with their environment and, if water is exchanged freely, the major extracellular solute, at least in a concentrated solution, will be largely excluded. If this were not generally true, the environmental tolerance limits of individual microbial species would be very narrow indeed.

The major evidence against salt accumulation lies, first, in the salt sensitivity of cell-free cytoplasmic enzyme preparations and some chloroplast functions (Johnson *et al.*, 1968; Ben-Amotz and Avron, 1972; Borowitzka and Brown, 1974) and, second, in the accumulation of glycerol. This compound is normally present at concentrations high enough to exclude all but minor quantities of sodium chloride (Borowitzka and Brown, 1974) and, furthermore, it is the major osmoregulator of *Dunaliella* (Craigie and McLachlan, 1964; Wegmann, 1971; Ben-Amotz

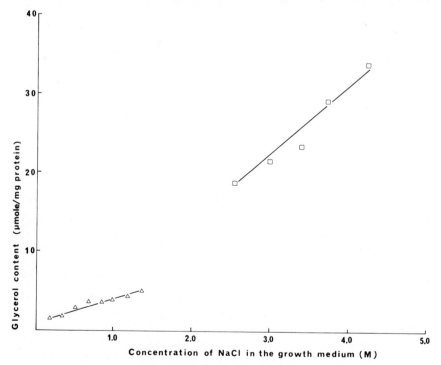

Figure 1. The effect of sodium chloride concentration on the glycerol content of growing *D. tertiolecta* (△) and *D. viridis* (□). (Results of Borowitzka and Brown, 1974.)

and Avron, 1973a); Borowitzka and Brown, 1974; Ben-Amotz, 1975; Borowitzka *et al.*, 1977). This topic is discussed further in Section VI.

The response of glycerol concentration to salinity is shown for *D. tertiolecta* and *D. viridis* in Figure 1. Values represented in this figure are equivalent to 0.3, 1.0, 2.4, and 4.4 *m* intracellular glycerol corresponding to 0.17, 1.36, 2.55, and 4.25 *M* NaCl, respectively.

The presence of glycerol therefore replaces any need for salts as osmoregulatory solutes, and, in any case, the major salt could not act as an osmoregulator. For a salt to behave as an osmoregulator in the strict sense, a salt which was a minor component of the extracellular solution must accumulate against a concentrate gradient as, classically, halophilic bacteria accumulate potassium chloride from a growth medium overwhelmingly dominated by sodium chloride. Glycerol does not inhibit cytoplasmic and membrane-bound enzymes at physiological concentrations (Brown and Simpson, 1972; Borowitzka and Brown, 1974; Heimer, 1976; Simpson, 1976; Brown, 1978). It is an interesting candidate for intracellular accumulation, since it is generally regarded as a very permeant solute. It does not leak from the cells, however, even when the sodium chloride concentration of the medium is lowered substantially. This topic is discussed more fully in Section VI.

VI. HALOPHILISM

A. Theoretical Mechanisms of Halophilism

The concept of halophilism has a meaning which is quite different for *Dunaliella* on the one hand and the extremely halophilic bacteria on the other. Although *Dunaliella* spp., halobacteria, and halococci can share a common habitat containing sodium chloride at concentrations up to saturated, the physiological mechanisms which enable this are quite different in the algae and the bacteria. There are, in fact, no eukaryotes known to be halophilic in the strict sense as it applies to bacteria; furthermore, *Dunaliella* is possibly the most halophilic of all single-celled eukaryotes [see Brock (1975)].

A highly saline environment can exert its physiological effect through direct interaction between the salts and cellular constituents, indirectly through water availability and perhaps through water structure, and through osmotically induced pressure and volume changes in a cell. These effects are not always distinguished from one another, nor are they always readily distinguishable. Moreover, their relative contributions vary during the process of adaptation to a new environment. Explanations of halophilism or halotolerance should in most cases take into account two aspects of the phenomenon, namely, the ability to survive a transition from one environment to another and the ability to thrive normally in the new environment. *Dunaliella* is outstanding in its ability to survive the first process, the transition, whereas halophilic bacteria seem to be unique in their strict salt requirements for normal growth.

Early theories of halophilism sought explanations in terms of a dilute interior, since many biologists had difficulty accepting that enzyme systems could function in solutions as concentrated as saturated brine. It is now recognized, of course, that single cells do not and cannot function in this way, and their physiology must be understood in terms of an interior which is thermodynamically adjusted to the environment. This adjustment normally includes a factor for turgor (Brown, 1964, 1976).

Brown (1976) has discussed the adjustment of a cell to a changed level of water activity or salt concentration as a process which occurs in two phases followed by a fully adapted steady state situation.

Phase 1. When a cell is transferred to a different level of water activity or salinity, there is first a rapid thermodynamic adjustment to the new situation. This adjustment, which has a time scale of minutes or even seconds, involves a water flux and a transient osmotic stress which, if sufficiently great, can be lethal.

Phase 2. If phase 1 is survived and intracellular conditions are not completely inhibitory at the new water activity, then the cell will enter a period of adaptation, the gross physical result of which will be reversion to a close approximation of its original volume and state of turgor. Phase 2 requires energy and involves changes in the balance of metabolic activities, transport processes, and levels of enzyme activity. Phase 2 is the physiological readjustment and generally can be expected to have a time scale on the order of a generation time.

Phase 3. This is the fully adapted steady state condition of growth under the new conditions, when the organism is phenotypically different from its condition in the previous environment. [Gimmler *et al.* (1977) experimentally demonstrated such a sequence with *D. parva* but subdivided into three phases the changes we have described in phases 1 and 2.]

Adaptability requires an organism to be able to negotiate the hazards of phases 1 and 2; the regulatory mechanisms discussed in Section VI,D are obviously closely integrated with the changes occurring in phase 2. The recognition of an organism as halophilic or halotolerant, however, ultimately depends on its performance in phase 3, that is, its ability to grow normally in a highly saline environment. Since in phase 3 the organism is fully adapted, osmotic stresses have little or no significance in explaining its ability to thrive under these conditions [see also Brown (1976, 1978)]. What is needed here is an explanation of the ability of enzyme systems, transport mechanisms, etc., to function under the new conditions. Brown (1976) has proposed two major types of mechanisms which, in theory, could explain phase 3 halophilism. Since we have had the pleasure of the colloquial description of these mechanisms as regurgitated by students in examination papers, there might be some merit in elevating the descriptions to the status of the printed word along with a more prosaic numerical listing.

Mechanism I (*funny proteins*). In this mechanism, the proteins of a tolerant organism are fundamentally and generally different from the corresponding proteins of a nontolerant organism and, because of the difference, are intrinsically better able to function under extreme environmental conditions. This is basically the explanation of the *requirement* of halophilic bacteria for high salt concentrations (Brown, 1976). It does not explain the physiology of *Dunaliella* and certainly does not explain the differences between species of *Dunaliella* (see Section VI,C). Moreover, there is also an element of salt *tolerance* (as distinct from a requirement) in the physiology of halobacteria; the tolerance is dependent on the second mechanism (Brown, 1976).

Mechanism II (*funny guts*). The proteins of tolerant and nontolerant organisms are essentially similar, but at low levels of water activity, enzymes can function in tolerant species because intracellular conditions are modified so that the inhibitory effects of the environment are lessened. *Dunaliella* and, apparently, all xerotolerant eukaryotic cells depend on this mechanism for their tolerance, but it does not explain the apparent requirement of some eukaryotes for diminished water activity, nor does it always explain the differences among species (Brown, 1978).

The two basic mechanisms hypothetically can be subdivided further. For example, metabolic peculiarities might confer tolerance if they are catalysed by enzymes which, in any organism, have intrinsically enhanced tolerance (mechanism IA) or because the end product of the metabolic peculiarity modifies the intracellular environment in accordance with mechanism II (mechanism IIA). We are not aware of any example of mechanism IA, but mechanism IIA is used by *Dunaliella* and all known xerotolerant eukaryotic protists. In brief, *Dunaliella* spp. can thrive at high salinities partly because they can accumulate glycerol internally to concentrations thermodynamically commensurate with these salinities. They can adapt to a remarkably wide range of salt concentrations because they can survive the initial stress (phase 1) and thereafter quite rapidly adjust their glycerol content to a new level appropriate to the changed salinity (see Section VI,C). The physiological role of glycerol under these circumstances includes the conventional, well-documented one of an osmoregulator which contributes to approximate parity between internal and external a_w while simultaneously maintaining cell volume and an appropriate level of turgor pressure. The other facet of the function of glycerol is the equally important but more exacting role of compatible solute. We assume that glycerol is distributed more or less uniformly throughout cytosol and organelles, although there is no direct evidence on this point [however, see Brown (1978)]. There are undoubtedly factors over and above glycerol accumulation involved in the halotolerance and apparent halophilism of *Dunaliella,* and especially in the different salt relations of individual species (see Section VI,C,2). There can be little doubt, however, that without glycerol *Dunaliella* could not achieve its quite remarkable degree of salinity tolerance and adaptability.

B. Compatible Solutes

The phenomenon of osmoregulation has been and is the subject of extensive investigation by plant physiologists, although it has received relatively little attention from microbiologists. It has been discussed at various levels in a number of recent reviews (e.g., Cram, 1976; Hellebust, 1976; Kauss, 1977). For the most part, however, these and other reviews have given major attention to the process in the context of a relatively dilute environment (that is, at water activities roughly equivalent to that of seawater and above). Partly for this reason, they have not been forced to look critically at the enzymological significance of osmoregulatory substances at the very high concentrations which prevail in halophilic and other xerotolerant organisms. The term "compatible solute" was introduced (Brown and Simpson, 1972; Aitken and Brown, 1972) partly to emphasize the enzymological implications of osmoregulation under these conditions. Under extreme conditions, only two compatible solutes have been identified. These are potassium chloride in extremely halophilic bacteria, and glycerol in extremely xerotolerant eukaryotes (Brown, 1976, 1978). Earlier reports (Brown and Simpson, 1972; Brown, 1974) that arabitol was the major compatible solute of xerotolerant yeasts have since been revised (Brown, 1978; Edgley and Brown, 1978) with the realization that, although arabitol is produced by these yeasts, it appears to function as a secondary or "backup" compatible solute. The primary solute, which responds as a typical osmoregulator, is glycerol.

Brown (1976, 1978) has argued that, although osmoregulation is an essential function of compatible solutes, it is not a specific function. Any solute, other than the dominant extracellular solute, whose intracellular concentration responded appropriately to changes in extracellular water activity would function as an osmoregulator. Although such nonspecific responses involving various metabolites and salts occur with small changes in water activity under dilute conditions, larger stresses always select a dominant solute. The condition of extreme halophilism or extreme xerotolerance is such that many potential osmoregulators are insufficiently soluble and most, perhaps all, are too toxic or inhibitory at the concentrations they would require at the prevailing water activity. It is in this second respect that the outstanding physiological significance of extreme solute compatibility shows itself; it is scarcely surprising that, under extreme conditions, few substances have been found to have this function.

In simple terms, a compatible solute protects enzymes against both inactivation and inhibition. The protection is conferred because, even though the compatible solute normally functions as an inhibitor (as distinct from an activator) at physiological concentrations, the inhibition it causes is slight and far less than would be caused by other solutes if it were not present at the appropriate concentration. Although some aspects of the physiology of xerotolerant eukaryotes, including *Dunaliella,* can be satisfactorily explained in terms of water activity, enzymological studies have shown that enzyme function at low levels of water activity is determined by direct interaction with the predominating solute rather than by water availability. This was shown with the salt relations of a halophilic bacterial enzyme, where the conclusion was scarcely surprising. It is worth comment, however, that even though membrane and ribosomal enzymes from *Halobacterium* have a high salt optimum (about $4 M$), which largely explains the salt *requirements* of these organisms, many soluble cytoplasmic enzymes have a salt optimum of $1 M$ or less and therefore function under conditions of inhibition at the very high salt concentrations which occur in these bacteria. Although there is little to distinguish between the interaction of a halophile isocitrate dehydrogenase with either potassium chloride (which is accumulated and is the compatible solute) or sodium chloride (which is largely excluded and is severely inhibitory) at suboptimal concentrations when both salts function as activators, at supraoptimal concentrations when they inhibit, the behaviour of the two salts is very different. Potassium chloride inhibits far less than sodium chloride, its patterns of inhibition are simpler, and it is affected far less than sodium chloride by substrate concentration (Aitken *et al.,* 1970; Aitken and Brown, 1972). The inhibitor constant K_i for isocitrate dehydrogenase–KCl is about $9.5 M$ compared with about $0.9 M$ for the corresponding sodium chloride–enzyme complex (Brown, 1976). The compatible salt has a far lower apparent affinity for the enzyme than the inhibitory but extracellularly dominant sodium chloride.

Nonelectrolytes also exert their enzymological influence in concentrated solution (that is, at low levels of water activity), by direct interaction with enzyme proteins, not indirectly through their effect on water activity (Brown, 1976). The evidence, largely derived from studies of enzyme activity, however, does not exclude indirect effects through changes in water *structure* (solvent perturbation) as distinct from water *activity*.

Direct information about the effects of a nonelectrolyte on *Dunaliella* enzymes is confined to the action of glycerol on glucose-6-phosphate dehydrogenase and glycerol dehydrogenase from *D. tertiolecta* (marine) and *D. viridis* (halophilic). Corresponding enzymes from the two species are functionally indistinguishable from one another (Borowitzka and Brown, 1974; Borowitzka *et al.*, 1977).

Although, to our knowledge, there have been no comparative studies of nonelectrolytes on any *Dunaliella* enzyme, it is now well established that salts, both sodium and potassium chloride, become very inhibitory at concentrations well below that necessary to produce the environmental water activity of even the marine species (Johnson *et al.*, 1968; Ben-Amotz and Avron, 1972; Borowitzka and Brown, 1974). Glycerol, at concentrations up to 4 M, does not inhibit glucose-6-phosphate dehydrogenase from *D. tertiolecta* or *D. viridis*, and it might even cause some degree of activation. From 4 to 10 M, there is progressive inhibition but, even in 10 M glycerol, the enzyme can retain about 15% of its maximum activity (Borowitzka and Brown, 1974). Such simple statements cannot, of course, be made about glycerol dehydrogenase, since glycerol is a substrate of this enzyme. For the present it is sufficient to state that very high glycerol concentrations do not inhibit glycerol dehydrogenase and that glycerol has a cooperative effect on the function of the enzyme. The kinetics of glycerol dehydrogenase are discussed in Section VI,C,1, dealing with the regulation of glycerol accumulation.

Comparisons of the effects of nonelectrolytes have been made with enzymes from other sources. Nevertheless there is sufficient information available both from enzymological and physicochemical studies to lend confidence to some generalizations which can be made about the relative degrees of protection conferred (or, conversely, inhibition produced) by various acyclic polyhydric alcohols and by some sugars. The interaction between nonelectrolytes and enzyme proteins is treated briefly, since it has been discussed at some length elsewhere (Brown, 1976, 1977).

An indication of the effectiveness of some acyclic polyols and sugars as potential enzyme inhibitors or, conversely, as potential compatible so-

Table II Apparent Inhibitor Constants of Yeast Isocitrate
Dehydrogenase with Polyhydric Alcohols and Selected Sugars[a]

Solute	Inhibitor constant $K_i(m)$
Methanol	90
Ethylene glycol	42
Glycerol	26
Erythritol	12
Arabitol	9
Ribitol	9
Sorbitol (glucitol)	8
Fructose	25
Glucose	9
Xylose	4
Rhamnose	3
Arabinose	2

[a] The sugars listed in the table gave linear plots of $1/V$ against sugar
concentration. All K_i values were calculated from results of Simpson
(1976) and were previously published in this form by Brown (1977).

lutes, is shown in Table II, which lists apparent inhibitor constants of the
solutes derived by Simpson (1976) for a yeast isocitrate dehydrogenase.
Not all the compounds studied by Simpson are included in this table.
Many did not give linear plots of $1/V$ versus inhibitor concentration, and the
apparent K_i could not be computed in such cases. It is noteworthy that, of
the substances that occur naturally in significant concentrations, glycerol
and fructose have essentially identical constants. Of those two compounds, only glycerol is known to occur naturally at high intracellular
concentrations, but adjustment of growth media with fructose enables
xerotolerant yeasts to attain their lowest limits of a_w tolerance. (Mono-
saccharides are not likely to accumulate intracellularly to any great
extent; they are too reactive biochemically.) True constants, obtained
with both substrates at saturating concentrations, were determined by
Simpson only for glycerol and, approximately, for sucrose. The true K_i
(glycerol) for the yeast isocitrate dehydrogenase is about 13 m and, for
sucrose, 1.0–1.5 m, values which are reminiscent of constants obtained
for the halophilic bacterial enzyme and, respectively, potassium and
sodium chloride.

The kinetics and physicochemical evidence, such as it is, suggest that
the affinity of nonelectrolytes for enzyme proteins in aqueous solutions is
determined for the acyclic polyols and some sugars by hydrophobic in-

teractions with nonpolar groups on the protein molecules [see Brown (1976, 1978) for an outline of the evidence for this statement]. For other sugars, there are apparent stereochemical factors which have not yet been evaluated. The ability of glycerol to provide long-term protection against enzyme inactivation is supported by experiments of Contaxis and Reithel (1971), Douzou (1974), and others cited by Brown (1976, 1978), as well as by the general biological observation that, since glycerol occurs in *Dunaliella,* is a natural antifreezing agent in insects and fish [see, e.g., Schmidt-Nielsen (1975)], and is commonly used to protect blood cells and spermatozoa against freezing damage during storage, it is unlikely to have any long-term inactivating action.

In view of what has been said elsewhere (Brown, 1976, 1978) about its mode of action as a compatible solute, it is sufficient to conclude this section by stating that glycerol protects enzymes because of its low affinity for them. Its affinity is predominantly for solvent water, whereas other solutes which are more inhibitory than glycerol appear to bind more tightly to proteins. At least in the case of polyols and some sugars, the binding is apparently through hydrophobic interactions. With other sugars the manner of interaction is more complex and is apparently affected in large measure by stereochemical factors. It is probably also true that the more tightly a solute seems to bind to a protein, the more effective is that solute in dehydrating the protein or in disorganizing the "bound" water.

C. The Halophilism of *Dunaliella*

1. General Observations

As we have already pointed out, the halophilism of *Dunaliella* is not the halophilism of bacteria (*Halobacterium* and *Halococcus*). Halophilic bacteria have an absolute requirement for at least 2.8 M NaCl. They cannot be trained by serial transfer to accept lower salt concentrations, nor can other solutes substitute for sodium chloride to any appreciable extent. Their requirement for salt is reflected in the salt relations of halophile enzymes, enzyme complexes, and membranes (Brown, 1964, 1976; Larsen, 1962, 1967; Lanyi, 1974). *Dunaliella* spp. can be trained to higher or lower salt concentrations (Craigie and McLachlan, 1964; Latorella and Vadas, 1973; Borowitzka *et al.,* 1977), so that the apparent limits of salinity tolerance of a marine species such as *D. tertiolecta* on the one hand, and a halophilic species such as *D. viridis* on the other (Borowitzka and Brown, 1974), reflect the history of the inoculum as much as intrinsic characteristics of the algae. Indeed, the differences between the salt relations of the two species are far from clear.

Moreover, *D. viridis* survives, in growth medium, a single-step transfer from 3.4 to 0.5 M NaCl. Following such a transfer, at least at 20°C, the algae swell, become nonmotile, and clump. They grow slowly in clumps over several weeks and often form tetrads, reminiscent in shape, but not size, of *Sarcina*. They no longer divide longitudinally as described earlier and, indeed, have no long axis. Throughout all this they lack flagella. After 3–5 weeks, some motile cells with normal morphology develop and multiply, but the bulk of the population remains clumped and usually settles on the bottom of the flask (Brown, unpublished observations). Furthermore, the apparent salt relations of actively growing *Dunaliella* can be altered substantially by incubation temperature (D. S. Kessly, unpublished observations).

Dunaliella enzymes do not require salt at concentrations commensurate with environmental salt concentrations; indeed, they are inhibited at much lower salt concentrations (Johnson *et al.*, 1968; Borowitzka and Brown, 1974), and there is no difference in this respect between corresponding enzymes obtained from marine and halophilic species (Borowitzka and Brown, 1974).

Thus, the evidence so far available suggests that the apparent halophilism of species such as *D. viridis* is really a form of tolerance. This should scarcely be surprising, given the role of glycerol in determining the salt relations of these algae. Compatible solutes characteristically confer a tolerance, not a requirement.

Moreover, the marine species *D. tertiolecta*, when trained to grow at the higher salt concentration normally associated with the halophile *D. viridis*, can grow faster than the halophile at these salt concentrations (Borowitzka *et al.*, 1977). Nevertheless *D. viridis* (trained + untrained) has a higher salt optimum (ca. 1 M) than *D. tertiolecta* (trained + untrained, ca. 0.17 M), and it is the so-called halophilic species (*D. parva, D. salina,* and *D. viridis*) which are found in highly saline environments and the marine species (*D. tertiolecta*) which is found in the sea. Furthermore, under our experimental conditions, *D. viridis* is more vigorous photosynthetically (rate of O_2 evolution) than *D. tertiolecta* (D. S. Kessly, unpublished observations; see also Section VI,C,2), a factor which correlates with its general need to produce more glycerol than the marine species. There is, however, no comprehensive explanation of the different apparent salt relations of species of *Dunaliella*, but it should not be forgotten that the growth characteristics of *D. tertiolecta* and *D. viridis* reported by Borowitzka and Brown (1974) (Figure 1) and by Borowitzka *et al.* (1977) were determined at one temperature and under one set of nutrient conditions.

2. The Physiology of Adaptation

As pointed out in Section VI,A, growth in a potentially inhibitory environment requires, on occasion, an ability to survive the stress of adaptation, as well as the ability to grow under more or less steady state conditions when the adaptation process is complete. The physiology of environmental tolerance cannot be adequately understood without some knowledge of the events of the transition from one environment to another. This is likely to be especially true of *Dunaliella,* which is outstanding in its ability to adapt to a very wide range of salt concentrations. To simplify terminology in the discussion which follows, the term "salt stress" is used to denote the stress caused by an increase in environmental salt concentration; "dilution stress" denotes the opposite situation.

The physiology of the adaptation of *Dunaliella* to a different salinity is the physiology of phase 2 (see above) adaptation. This is the period during which normal volume is restored, there are metabolic adjustments and adjustments in levels of enzyme activity until phase 3 is reached and normal growth is resumed. A salt stress in *Dunaliella* does not seem to involve the admission of substantial amounts of sodium chloride and, conversely, dilution stress does not apparently involve a substantial loss of salt from the cell. An inherent and essential part of phase 2 adaptation is adjustment of the glycerol content to a new level but, as we explain later, this in itself is not a sufficient requirement for growth to resume, at least after salt stress.

The extent to which salt penetrates *Dunaliella* has been uncertain and, to some extent, contentious for several decades [see reviews in Brown (1976, 1978)]. It is still contentious. For example, Rabinowitch *et al.* (1975) used a particle-size analyzer to measure volume changes of *D. parva* in response to concentrations of various salts and concluded, among other things, that "nonosmotic volume" accounted for 60–80% of total cell volume; this is unusually high. On the other hand, Gimmler *et al.* (1977), who also used a particle-size analyzer for a similar purpose, concluded that the nonosmotic volume of this species is 20–40% of the total volume; this is a more normal value. Nevertheless, Gimmler *et al.* (1977) contend that *D. parva* admits substantial amounts of salt. They have two major reasons for this opinion. The first is that full volume recovery after stress takes 150–180 min, whereas glycerol synthesis (citing Ben-Amotz and Avron, 1973a) takes only 90 min. Therefore, they argue, some other solute must accumulate during the remaining 60–90 min. They supplement this argument with some additional comment on the relative rates of glycerol synthesis and volume recovery.

The quantitative details of these two processes, however, show them in a somewhat different light. For example, after salt stress (a change from 0.75 to 1.2 M NaCl), there was a total volume increment of about 10 μm^3 during the recovery period. During the period, 90 min to completion (to about 180 min), the volume change was about 3 μm^3, that is, about 30% of the total change. The volume of a *Dunaliella* cell is of the order of 500 μm^3, so that the "excess" volume change (that is, the additional 3 μm^3 acquired after the completion of glycerol synthesis) was 1% or less of the total cell volume. The experimental errors inherent in determining glycerol content are such that arguments based on differences of 1% or less are totally unwarranted; there is no certainty that glycerol accumulation stops completely after 90 min, as implied by Gimmler *et al.* (see Figures 2 and 3).

Additional evidence advanced by Gimmler *et al.* (1977) in support of salt as a major contributor to the internal osmotic status of *Dunaliella* was derived from their use of the uncoupling agent carbonyl cyanide *p*-trifluoromethoxyphenylhydrazone (FCCP), which suppresses photophosphorylation and oxidative phosphorylation. It is not clear whether or not the algae were illuminated in these experiments, but FCCP (2 μM) did not appreciably affect volume recovery. Gimmler *et al.* therefore concluded that glycerol synthesis was "not the only reason for volume recovery." Extrapolation from experiments with inhibitors to a normal situation should be done only with great care. In this series of experiments, it was not demonstrated that glycerol synthesis was in fact fully inhibited in the algae subjected to volume measurement. Gimmler *et al.* (1977) cited earlier results of Gimmler which indicated complete suppression of ATP formation by FCCP at concentrations from 1 to 3 μM. Our criticism would therefore be somewhat pedantic were it not for a report by Ben-Amotz and Avron (1973a), who found only 50% suppression for adaptive glycerol synthesis by 2 μM FCCP. Nevertheless, if FCCP did in fact completely uncouple phosphorylation, the question must be asked whether plasma membrane permeability to ions was changed in the process, since uncouplers classically enhance membrane permeability to H^+ and *D. tertiolecta* has a light-dependent Na^+–H^+ exchange (Latorella and Vadas, 1973). Furthermore, it has not yet been established whether Na^+ exclusion from *Dunaliella* is achieved simply by impermeability or by an outwardly directed "pump." The latter, of course, would require energy and would cease operating if ATP generation were inhibited. Sodium ions would then diffuse into the algal cell until equilibrium was achieved (and volume restored).

Admittedly it is not certain that the glycerol content of *Dunaliella* is sufficient to explain fully the overall osmotic status of the algae, but the uncertainties are caused by insufficient analytical precision (Brown, 1978). The analyses are generally good enough, however, to allow us to conclude with confidence that glycerol together with normal metabolites can account for most of the lowering of intracellular water activity, and therefore any supplementary uptake of salts from the environment makes, at most, a minor contribution. The exact extent of this contribution is unlikely to be determined indirectly or by studies on inhibited algae. What is needed is refinement of analytical methods. In any case, there is evidence (Borowitzka and Brown, 1974), together with theoretical expectations, that any salt accumulated as part of the algal water relation is predominantly potassium chloride rather than sodium chloride.

When subjected to salt stress, *Dunaliella* synthesizes additional glycerol to a new stable level in 60–90 min (Ben-Amotz, 1975; D. S. Kessly unpublished observations). It is also of interest that, after relatively severe salt stress, *Platymonas subcordiformis* takes about the same length of time to bring about the appropriate increase in its mannitol content (Kirst, 1977). There are, however, some qualifications to be made to the general statement. Firstly, there is an apparent species difference, *D. viridis* taking about 90 min compared with 1 hr for *D. tertiolecta,* but under Kessly's experimental conditions there were big differences in the salinity change (from 0.17 to 1.6 M NaCl for *D. tertiolecta* and from 1.5 to 4.0 M NaCl for *D. viridis*). Secondly, in both species, adaptive glycerol biosynthesis takes place in either light or dark. In *D. tertiolecta* the rate of synthesis and final level of glycerol content were about the same in either light or dark (D. S. Kessly, unpublished observations, but in *D. viridis* the rate was slower in the dark and the final glycerol content was about two-thirds that obtained with illumination (Figure 2). The difference in this case was presumably caused by a starch limitation.

It has already been shown that adaptive glycerol biosynthesis can occur in the dark (Borowitzka *et al.,* 1977), but the significance of Figure 2 and Ben-Amotz's (1975) report is that they show the time interval needed to complete the process of glycerol accumulation. Moreover, Figure 2 (and Kessly's analogs results for *D. tertiolecta*) show glycerol accumulation to begin virtually immediately in response to the stress.

Thus the first essential for adaptation to salt stress, glycerol biosynthesis, can be a heterotrophic process in that it is not dependent on light. Algae vary in their need for light to provide the energy for compatible

---

Figure 2. The response of glycerol content of *D. viridis* grown in 1.5 M NaCl and subjected to a salt stress. (A) Algae transferred to 4.0 M NaCl and illuminated. (B) Algae transferred to 3.0 M NaCl and incubated in the dark. A lower salt concentration was used under these conditions, because of the limitation imposed by starch content on the amount of glycerol which can be produced. (See Borowitzka *et al.,* 1977.) Time was measured from the moment of transfer to the higher salt concentration. The different symbols denote duplicate experiments. (Previously unpublished results of D. S. Kessly.)

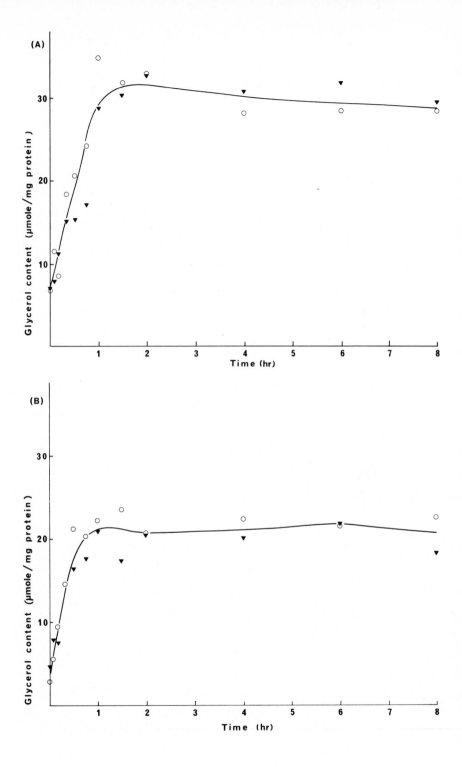

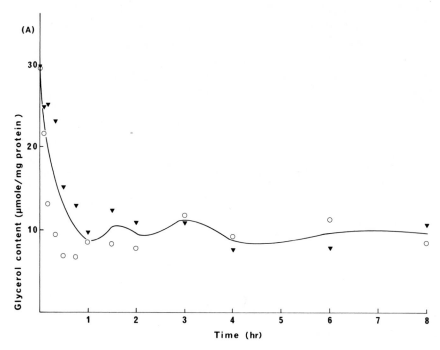

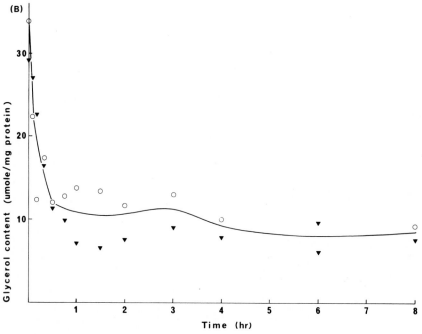

solute biosynthesis. Thus, the euryhaline flagellate, *Monochrysis lutheri,* synthesizes a cyclic polyol, cyclohexanetetrol, in response to salt stress, but can do so only in the light. The green alga *Scenedesmus obliquus* can respond to salt stress, apparently in part by accumulating sucrose, only in the light. On the other hand, the green flagellate *Platymonas subcordiformis* uses mannitol, which it can synthesize either photosynthetically or from starch (Hellebust, 1976). The disappearance of glycerol when *Dunaliella* is subjected to dilution stress also takes between 1 and 1.5 hr (Figure 3) and is also independent of light.

There is no detectable growth of the algae during the period of glycerol synthesis. This is no doubt partly because intracellular conditions are physicochemically unfavourable, but there is another important reason which is probably sufficient in itself to prevent growth. During phase 2 adaptation to salt stress, the algae divert a very large proportion of their energy supply to glycerol production. When *D. viridis* is transferred from 1.5 to 4.0 M NaCl and illuminated, glycerol is formed at a rate equivalent to about 1.8 mg glycerol/mg protein/hr. When *D. tertiolecta* is transferred from 0.17 to 1.6 M NaCl, glycerol is formed at a rate of about 0.55 mg glycerol/mg protein/hr (Figure 2; D. S. Kessly, unpublished results). The highest exponential growth rates measured under our experimental conditions were about 0.04 generation/hr for *D. viridis* and about 0.1 generation/hr for *D. tertiolecta* (Borowitzka *et al.,* 1977). Thus the rates of glycerol synthesis are orders of magnitude greater than the maximum rates of synthesis of cell substance. A similar comment might also apply to the responses after dilution stress, since it has been reported (Ben-Amotz and Avron, 1973a) that ATP formation is needed for glycerol dissimilation in *D. parva.* These observations not only emphasize the need for an organism to retain an energy supply if it is to survive water stress but also illustrate how effectively *Dunaliella* does so.

The organisms do not resume growth immediately after the glycerol content is stabilized at its new level, however, nor do they resume net photosynthetic oxygen evolution for many hours after the glycerol content has stabilized. In the light, *D. tertiolecta* requires about 36 hr after a transition from 0.17 to 1.6 M NaCl before oxygen evolution is resumed. *Dunaliella viridis* evolves oxygen about 24 hr after a transition from 1.5 to 4.0 M NaCl (D. S. Kessly, unpublished observations). During the period before net oxygen evolution resumes, the algae respire, but the amount of

---

Figure 3.   The response of the glycerol content of *D. viridis* grown in 4.0 M NaCl and subjected to dilution stress by transfer to 2.0 M NaCl. (A) Illuminated. (B) Incubated in the dark. The symbols denote duplicate experiments. Oscillations in glycerol content were common in experiments of this type. (Previously unpublished results of D. S. Kessly.)

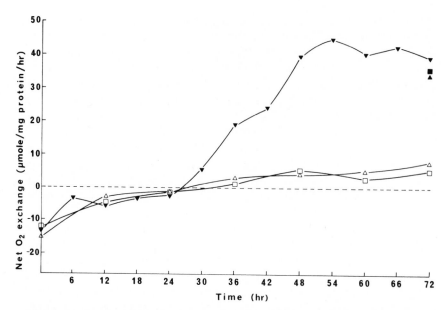

Figure 4. Recovery of oxygen evolution by *D. viridis* after salt stress. The alga was grown in 1.5 *M* NaCl and stressed by transfer to fresh medium containing 4.0 *M* NaCl; the suspension was illuminated. Positive values on the vertical axis denote oxygen evolution, and negative values denote uptake. Complete growth medium, ▼; nitrogen-free medium containing cycloheximide (2 μM), △ [see Borowitzka *et al.* (1977)]; nitrogen-free medium containing chloramphenicol (12.5 μg/ml), □; an "unstressed" suspension (in 1.5 *M* NaCl) incubated in nitrogen-free medium with chloramphenicol, ■; an unstressed suspension incubated in nitrogen-free medium with cycloheximide, ▲. (Previously unpublished results of D. S. Kessly.)

oxygen taken up diminishes progressively with time until it eventually changes to evolution (Figure 4). When the algae are incubated in the dark, under otherwise similar conditions, oxygen uptake also progressively diminishes until a minimum is reached after 30–36 hr (in both *D. viridis* and *D. tertiolecta*), after which it increases again (Figure 5). Neither the time to net oxygen evolution nor the time to minimum dark respiration is affected by inhibitors of protein synthesis (Figures 4 and 5).

Thus there is some evidence that oxygen evolution occurs earlier and at a higher rate in *D. viridis* than in *D. tertiolecta*. There is consistent evidence that light-induced oxygen evolution, expressed as a function of protein content, is more vigorous in *D. viridis* than in *D. tertiolecta,* a finding which contrasts with the consistently higher growth rates of the marine species (Borowitzka and Brown, 1974; Borowitzka *et al.,* 1977).

Although we have not measured photophosphorylation or CO_2 fixation,

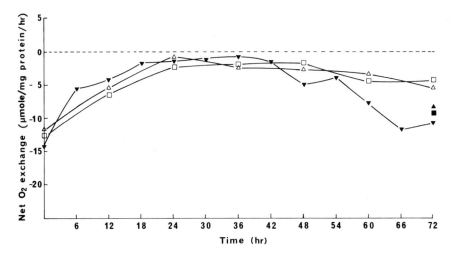

Figure 5. Changes in oxygen uptake of *D. viridis* when incubated in the dark after salt stress induced by transfer from 1.5 to 4.0 *M* NaCl. Negative values on the vertical axis denote oxygen uptake. The symbols have the same significance as in Figure 4. (Previously unpublished results of D. S. Kessly.)

and although Figures 4 and 5 provide no information at this level, Figure 2 contains evidence of photosynthetic activity during the initial 90-min period of glycerol formation. Since, under the experimental conditions, *D. viridis* adaptively produced about 30% more glycerol in the light than in the dark and necessarily did so at a faster rate, it must have been assisted by photosynthesis. The absence of such a light–dark difference in *D. tertiolecta* can be attributed to the smaller amount of glycerol it needed and the adequacy for that purpose of its starch reserves.

It is not clear from the information as it stands whether other adaptive processes, over and above glycerol production, are necessary before full photosynthetic activity is resumed or whether recovery of photosynthetic oxygen evolution is inherently much slower than glycerol biosynthesis. If there are additional adaptive processes, they do not appear to involve protein synthesis during the period up to net oxygen evolution (Figure 4). After this point there was a substantial difference in the rate of recovery of treated and untreated algae. In the presence of the inhibitors, net oxygen exchange, whether positive or negative, maintained an approximately linear rate of change with time, whereas in uninhibited algae net oxygen evolution began with a sharp increase in the slope of the oxygen exchange curve. What is fairly clear is that inhibitors of protein synthesis did not affect the time to net oxygen evolution, nor did they significantly affect the course of change of oxygen uptake in the dark, except perhaps in the final stages of the secondary increase in oxygen uptake (Figure 5).

Results of the type shown in Figure 4 do not permit a distinction to be made between mitochondrial respiration and photorespiration as contributors to oxygen uptake during the 24–36 hr after salt stress, nor do they distinguish between, on the one hand, a progressive drop in gross respiration and, on the other, a progressive diminution in net oxygen uptake caused by the gradual resumption of oxygen evolution. Figure 5 provides additional information. After salt stress induced by a change from 1.5 to 4.0 M NaCl, D. *viridis* takes 24 hr to achieve net oxygen evolution, but in the dark it needs 30–36 hr to reach minimum oxygen uptake. In the dark there is, of course, no photorespiration or oxygen evolution, and it is therefore evident that mitochondrial respiration is high immediately after stress and then progressively diminishes, only to increase again after 30–36 hr. With illumination, the sharp change in the slope of the oxygen exchange curve in the absence of inhibition of protein synthesis suggests that gross oxygen evolution did not begin much before that time. With cycloheximide and chloramphenicol, however, the much more linear rate of oxygen exchange is consistent with a process which was in operation before the transition from net uptake to net evolution. The time to net oxygen evolution for *D. tertiolecta* was about 30 hr but the general characteristics of the curve, with and without inhibition of protein synthesis, were essentially as shown for *D. viridis* in Figure 4 (D. S. Kessly, unpublished observations). In the dark *D. tertiolecta* also took 30–36 hr to reach minimum oxygen uptake, and this time was perhaps extended slightly by the inhibitors. Thus it seems that the response of respiration to stress was much the same in both species, but that *D. viridis* recovered photosynthetic oxygen evolution more rapidly.

Additional evidence that mitochondrial respiration is the major component of oxygen uptake during the recovery period is provided by the effects of bicarbonate concentration (Figure 6). In the light, a deficiency of CO_2 (or HCO_3^-) causes or enhances photorespiration; there is no evidence that this happens to any appreciable extent in either *D. tertiolecta* or *D. viridis*. It is also possible that what we have called "mitochondrial" respiration is more complex than this term suggests. Lloyd (1974) has pointed out that, in some algae, extramitochondrial respiration can occur "by way of the microsomal electron transport chain." Interestingly enough, this system can oxidize either NADH or NADPH, giving it a degree of versatility which could be very useful in the face of the importance to *Dunaliella* of NADP as a carrier of reducing equivalents (see Section VI,E).

The question of whether glycerol concentration is lowered by leakage or

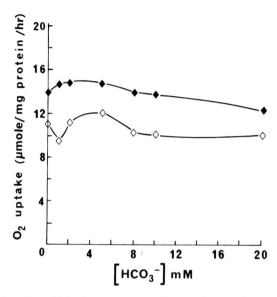

Figure 6. The effect of bicarbonate concentration on the net oxygen uptake of *D. ter-tiolecta* (◇) following salt stress induced by transfer from 0.17 to 1.60 *M* NaCl, and of *D. viridis* (◆) following a salt stress induced by transfer from 1.5 to 4.0 *M* NaCl. The suspensions were illuminated. (Previously unpublished results of D. S. Kessly.)

by metabolic dissimilation is subject to conflicting reports. Hellebust (1965) examined the excretion of metabolites by various marine phytoplankton, including the excretion of glycerol from *D. tertiolecta*. He concluded that excretion was a general property of marine algae, varied among genera, and was affected by light intensity. Since these observations were not related to salinity, their application to the question of response to dilution stress is limited. In any case it is difficult to distinguish between leakage or excretion by the whole population or lysis of part of the population. Frank and Wegmann (1974) measured glycerol (as ^{14}C) leakage from *D. tertiolecta* under various conditions and reported that a major proportion of glycerol was found in the medium. They expressed their results in several ways, but their units did not include intracellular concentrations. When their reported values are recalculated (of necessity with some assumptions) they indicate that, for example, in 14% NaCl intracellular glycerol was of the order of 90–100 mg/ml, whereas in the extracellular medium it reached about 1 mg/ml. We believe that, in fact, concentration gradients higher than these figures suggest are normal. Ben-Amotz and Avron (1973a) and Ben-Amotz (1975) reported virtually no leakage from *D. parva* grown in 1.5 *M* NaCl and transferred to any lower salt concentration greater than 0.6 M. At concentrations less than 0.6 M there was measurable leakage.

D. S. Kessly (unpublished observations; see also Figure 3) studied glycerol disappearance in *D. tertiolecta* on transfer from 1.6 to 0.53 M NaCl, and in *D. viridis* on transfer from 4.0 to 2.0 M NaCl. Light had little or no effect on the process, and there was some leakage which in both species was in the range 0.05–0.1 μmole/mg protein over the first hour. This accounted for 1.0–1.5% of the glycerol disappearance in *D. tertiolecta* and 0.3–0.5% in *D. viridis*. Thus glycerol disappearance from *Dunaliella* after even moderately severe dilution stresses is overwhelmingly the result of metabolic dissimilation rather than physical loss to the environment. Of course it is to be expected that halophilic species, such as *D. viridis* and *D. salina,* which can be exposed to very great dilution stresses, will leak significantly when this happens, but we have not attempted to delineate the circumstances of such leakage.

The fate of the metabolic conversion is unknown. Clearly low molecular weight compounds must be eliminated because, if they were retained, not only would the alga's water status remain essentially unchanged but also the organism would be worse off, since its compatible solute would have been converted to something much less benign. The possibilities open to the organism include loss of some glycerol carbon as CO_2, synthesis of starch, and leakage of low molecular weight conversion products. It seems a priori that the relative contributions of these options should be influenced by the extent to which starch reserves are used to generate ATP for the mobilization of dihydroxyacetone.

An ability to accommodate severe dilution stress without substantial leakage seems to be one of the distinguishing characteristics of *Dunaliella* and contrasts, for example, with the situation in *P. subcordiformis,* which readily and rapidly loses mannitol and other solutes under analogous conditions (Hellebust, 1976).

Dilution stresses of the magnitudes used by Kessly (from 1.60 to 0.53 M NaCl for *D. tertiolecta;* from 4.0 to 2.0 M NaCl for *D. viridis*) impaired but did not eliminate photosynthetic oxygen evolution. Recovery was achieved within 18–24 hr in both species and was not affected by cycloheximide or chloramphenicol (Figure 7). In contrast to its response to salt stress, respiration was impaired by dilution stress and, like oxygen evolution, took 18–24 hr to recover fully. Recovery was not affected by the two inhibitors of protein synthesis (Figure 8).

It is known that, although the growth rates of *Dunaliella* spp. do not respond appreciably to changes in water activity adjusted by sucrose, glycerol content does (Borowitzka *et al.,* 1977). Photosynthetic oxygen evolution in *D. tertiolecta* responds to changes in sucrose concentration in

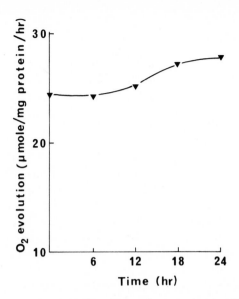

Figure 7. Changes in oxygen evolution of *D. viridis* grown in 4.0 *M* NaCl and subjected to dilution stress induced by transfer to 2.0 *M* NaCl and illuminated. The results were not affected by the presence of available nitrogen in the medium or by the addition of chloramphenicol or cycloheximide (see Figures 4 and 5). (Previously unpublished results of D. S. Kessly.)

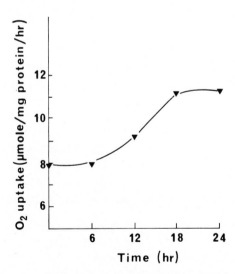

Figure 8. Changes in oxygen uptake of *D. viridis* following dilution stress (transfer from 4.0 to 2.0 *M* NaCl) and incubated in the dark. The results were not affected by the presence or absence of available nitrogen in the medium or by chloramphenicol or cycloheximide (see Figures 4 and 5). (Previously unpublished results of D. S. Kessly.)

a manner generally similar to its response to changes in salinity (D. S. Kessly, unpublished observations). Thus, again, there is evidence that the major determinant of intracellular processes (as distinct from functions in which the plasma membrane has a highly specific role) is water activity rather than salinity.

In summary then, the available evidence indicates the following physiological changes in *Dunaliella* consequent upon a salt stress. In phase 1 there is a rapid loss of water and shrinkage of the cell accompanied by or very closely followed in phase 2 by inhibition of photosynthetic oxygen evolution and stimulation of net oxygen uptake. Glycerol biosynthesis begins within minutes of the transition and proceeds to completion over 60–90 min. Glycerol biosynthesis can be supported entirely by heterotrophic metabolism, presumably involving starch breakdown, and in the very early stages of response is probably entirely dependent on this process, regardless of the state of illumination. There is evidence, however, that after a severe stress glycerol biosynthesis is assisted by photosynthesis even though the alga has not recovered its ability to evolve oxygen. Net oxygen uptake progressively declines until 24–36 hr after the transition, when net oxygen evolution resumes. The time to net oxygen evolution depends on the species and also, presumably, on the magnitude of the stress. Thus resumption of normal photosynthesis and subsequently of growth depends on the completion of adaptive processes over and above the accumulation of glycerol. These secondary adaptive processes have not yet been identified, but they do not seem to require protein synthesis to any appreciable extent.

After dilution stress changes occur in phase 1—there is a rapid uptake of water and swelling of the cell which is accompanied by or very shortly followed in phase 2 by an impairment (but not necessarily complete suppression) of net photosynthetic oxygen evolution and of mitochondrial respiration. Glycerol dissimilation begins within minutes and is complete in about an hour. Recovery of normal levels of oxygen evolution and of mitochondrial respiration takes 18–24 hr.

D. Compatible Solutes in Other Organisms

Mention has already been made in Section VI,B, of the function of glycerol as the primary compatible solute of other extremely xerotolerant eukaryotes such as yeasts and fungi, and of potassium chloride as the compatible solute of extremely halophilic bacteria. The role and physiological significance of glycerol as the primary solute and some higher polyhydric alcohols, notably arabitol, as secondary compatible solutes in xerotolerant yeasts have been discussed at length by Brown (1978). Potassium chloride has been discussed in relation to halophilic bacteria by Brown (1976).

There are, however, many microorganisms, including algae and protozoa, which from time to time experience mild osmotic or water stresses and which respond by varying the concentration of an osmoregulator which must also have the characteristics of a low-grade or intermediate compatible solute. Substances which have this function fall within a limited range of chemical types, although the range of organisms in which they occur is very wide.

Potassium ion seems to function in this way in some bacteria, but eukaryotes generally, as well as other bacteria, make use of metabolites. In nonhalophilic and nonhalotolerant organisms, the concentration of an osmoregulator rarely exceeds 500 m*M,* and at such concentrations few metabolites are likely to be significant general enzyme inhibitors. Many could act as specific inhibitors, however, and no doubt this factor limits the range of possible candidates for the role of low- and intermediate-level compatible solute. Substances identified as likely compatible solutes under mild and intermediate conditions include K^+, the amino acids glutamate, aspartate, γ-aminobutyrate, and proline, the higher polyhydric alcohols (some of which also act as antifreezing agents in insects), and relatively complex carbohydrate derivatives such as galactosylglycerol. The essential principles of their function are basically similar to that of glycerol in *Dunaliella*. Many examples of substances which act as osmoregulators under mild or moderate conditions have been listed, for example, by Craigie (1974), Cram (1976), Hellebust (1976), and Kauss (1977). The problem has been discussed from a slightly different standpoint by Brown (1976).

E. Osmoregulation in *Dunaliella*

An intriguing, physiologically important, but so far unanswered question which should be asked about *Dunaliella* concerns the means by which these algae regulate their glycerol content in response to salinity or, more specifically, to water activity. The problem can be seen in a different perspective by describing it as the response of a metabolic process to the concentration of a nonspecific, impermeant, extracellular solute. It is essentially the same phenomenon as osmoregulation in other protists and in the cells of multicellular organisms, although the cells of vascular plants and animals rarely encounter and, if they do, cannot accommodate the wide changes in water activity which *Dunaliella* can handle with such apparent ease.

There are two levels of osmoregulation which have features in common, but which might not employ identical mechanisms. There is a coarse regulation which controls major changes in compatible solute concentration during a transition, such as that discussed above, and there is also the homeostatic maintenance of compatible solute concentration in the fully adapted (phase 3) organism. Furthermore, there are two distinct and major questions concerning the mechanism of regulation. These are, (1) What is the metabolic site(s) of control, and (2) what is the signal which triggers the control mechanism? Neither question can yet be answered in detail, but there is some information, largely of a preliminary nature, primarily of relevance to the first question. The two aspects of regulation are treated separately.

1. The Metabolic Site(s) of Regulation

In order to identify the site(s) at which glycerol formation is regulated by water activity it is, of course, necessary to know in some detail the metabolic pathway by which it is formed and removed. The latter condition is important since, despite claims to the contrary, the available evidence indicates that glycerol is removed metabolically rather than by physical loss when the algae are subjected to anything less than an extreme dilution shock (see Section VI,C,2).

Unfortunately, in spite of a quite detailed understanding of the pathways of photosynthetic carbon fixation and associated processes in higher plants, the situation in unicellular algae is not nearly so well known. Some of the generalizations which have been and might be made depend in large measure on indirect evidence, assumptions, and analogy. There are some apparent contradictions in existing evidence, as we discuss later. Moreover, a comprehensive proposal for a regulatory mechanism must explain the process in both light and dark.

Glycerol has been known for some years to be an early product of photosynthesis (an early recipient of ^{14}C) and, together with starch, is a major storage product in *Dunaliella* (Craigie and McLachlan, 1964; Wegmann, 1969). Carbon dioxide fixation in *Dunaliella* occurs through the Calvin–Benson cycle, but even now the fine details of the pathway to glycerol are not known with certainty. A recent paper by Beardall *et al.* (1976) has shed some light on comparative aspects of algal photosynthesis. These workers reported short-term labelling patterns following $^{14}CO_2$ assimilation by two diatoms, a chrysophyte, a dinoflagellate, and *D. tertiolecta,* the last being used as a representative green flagellate.

Beardall *et al.* commented that the early labelling patterns in all five algae were different from those reported by Bassham and Calvin (1957) for vascular plants. In all species except *D. tertiolecta* there was rapid and substantial labelling of amino acids and amides and, in some of them, there was also substantial labelling of intermediates of the citrate cycle. In *Dunaliella* there was early labelling of sugar phosphates but, as previously reported by Wegmann (1969), glycerol was the major early recipient of the label. This and other supporting evidence was interpreted as being indicative of a primary role of phosphoenolpyruvate carboxylase in CO_2 fixation by the four algae other than *Dunaliella*. This enzyme is central to the path of carbon fixation in the so-called C_4 (vascular) plants, but algae lack the anatomical peculiarities which are also essential to the operation of the C_4 pathway. Bicarbonate, rather than CO_2, appears to be the substrate for the algal phosphoenolpyruvate carboxylase; HCO_3^- is the predominant form of carbon dioxide in the sea. In the diatoms, the bicarbonate is apparently fixed by phosphoenolpyruvate carboxylase to give oxaloacetate. There is a subsequent decarboxylation, probably of malate, and the CO_2 so released is then fixed by ribulose bisphosphate carboxylase and enters the Calvin cycle. In fact, this is a modified C_4 pathway without the anatomical complexities. In the presence of carbonic anhydrase, however, the equilibrium between HCO_3^- and CO_2 is achieved very rapidly. *Dunaliella* apparently has an active carbonic anhydrase [see also Latorella and Vadas (1973)] and a relatively low level of phosphoenolpyruvate carboxylase activity. In this case, carbon dioxide is evidently fixed "conventionally" as CO_2 by ribulose bisphosphate carboxylase and the Calvin cycle. It is not clear whether the algae used in these experiments were in a physiological steady state.

The early steps of carbon fixation are unlikely to be involved in the regulation by water stress of glycerol production, however, especially when it is remembered that the regulatory mechanism (at least during phase-2 adaptation) can function in the dark. The site(s) of regulation should be much closer to glycerol itself and presumably on a pathway common to photosynthesis and the heterotrophic degradation of starch.

Two metabolic pathways immediately preceding and leading to glycerol production have been proposed for *Dunaliella*. One, which is espoused by Wegmann (1969, 1971), Frank and Wegmann (1974), and Mutschler and Wegmann (1974), is the same as glycerol fermentation in yeast, namely,

$$\text{NADH} \quad \text{NAD}^+ \qquad\qquad P_i$$
$$\text{Dihydroxyacetone phosphate} \rightleftharpoons \text{glycerol 3-phosphate} \longrightarrow \text{glycerol}$$

Wegmann and his associates, however, have presented no convincing evidence that this is a major pathway in *Dunaliella*, although it might make a supplementary contribution to glycerol metabolism.

On the other hand, *Dunaliella* spp. possess a very active NADP-specific glycerol dehydrogenase (Ben-Amotz and Avron, 1973b; Borowitzka and

Brown, 1974; Borowitzka *et al.*, 1977). This enzyme can, incidentally, supply reducing equivalents via NADP to a nitrate reductase (Heimer, 1976). The glycerol dehydrogenase has apparent Michaelis constants for all four reactants similar to their likely or known intracellular concentrations, and the reaction is freely reversible (Borowitzka and Brown, 1974; Borowitzka *et al.*, 1977). Furthermore, the use of NADPH rather than NADH for the reduction step is to be expected in an organism which is actively photosynthesizing. Although the reduction of NAD^+ by NADPH via a transhydrogenase is possible, the extra step is less easily reconciled with the very early labelling of glycerol reported by Wegmann (1969) and by Beardall *et al.* (1976). *Dunaliella parva* has also been shown to contain a dihydroxyacetone kinase (Lerner and Avron, 1977) which presumably functions in glycerol dissimilation. Thus there is little doubt that, at least in the light, the final steps to glycerol production are

$$\text{Dihydroxyacetone phosphate} \xrightarrow{\quad P_i \quad} \text{dihydroxyacetone} \underset{\text{NADPH}}{\overset{\text{NADP}^+}{\rightleftharpoons}} \text{glycerol}$$

The formation of glycerol in the dark raises additional questions, since there could be a priori reasons for assuming that the degradation of starch was glycolytic, in which case NAD rather than NADP would be the likely currency for the transfer of reducing equivalents. However, if glycolytic degradation of starch proceeded only to the point of triose phosphate, there would be no generation of reducing equivalents at all. To obtain the capacity to reduce C_3 compounds from the level of triose phosphate to that of glycerol it would be necessary for the glycolytic pathway to continue further to pyruvate. It probably does this (see, e.g., Lloyd, 1974) but not at a sufficiently fast rate to generate enough reducing capacity to explain the large amounts of glycerol the algae synthesize when stressed. Kwon and Grant (1971) have claimed that metabolism of uniformly labelled [^{14}C]glucose by cell-free preparations of *D. tertiolecta* produces a distribution of label consistent with the operation of both glycolysis and the hexosemonophosphate shunt (the pentose phosphate cycle). They also state that, even in the presence of added NADP, "more than 82% of the glucose" was metabolized by glycolysis. Their results do not support this claim, however, since they used uniformly labelled glucose and most of the label accumulated at the level of triose phosphate. Their results are equally consistent with the pentose phosphate cycle. Furthermore, by using [1-^{14}C] glucose they demonstrated a large accumulation of label in CO_2, stimulated about 50% by NAD and about sixfold by NADP. [6-^{14}C]-Glucose gave negligible labelling of CO_2. These results are entirely consistent with a mechanism of glucose degradation which occurs overwhelmingly by the pentose phosphate cycle. They also indicate, as Kwon and Grant pointed out, a very small contribution to decarboxylation by the citrate cycle. The pentose phosphate cycle produces 12 molecules of NADPH for every molecule of glucose oxidized to CO_2 and water, and it maintains a catalytic pool of intermediates at the level of triose phosphate. It is therefore well placed to furnish the reducing capacity as NADPH for the further reduction of triose to glycerol.

Little is known about regulation of the pentose phosphate cycle in algae. In yeast, however, the first point of regulation is glucose-6-phosphate dehydrogenase (Osmond and Rees, 1969). In rat liver this enzyme is inhibited by NADPH at physiological concentrations, and the inhibition is relieved by oxidized glutathione (Eggleston and Krebs, 1974). In general, aeration or oxidizing conditions tend to favour the pentose phosphate cycle, and there are specific reports, for example, by Woodhead and Walker (1975), who noted that vigorous aeration increased the quantitative importance of the pentose phosphate cycle in *Penicillium expansum*. If these general properties are also found in *Dunaliella*, and they probably are, then it is easy to reconcile the function of the pentose phosphate cycle with the probable status of $NADP^+$ or NADPH under various physiological conditions. Thus, in the light, concentrations of NADPH or, more precisely, $NADPH/NADP^+$ ratios should be high enough to inhibit glucose-6-phosphate dehydrogenase and shut off the pentose phosphate cycle. In the dark the cycle is the major source of NADPH. As we understand them, the physiological characteristics of fully adapted phase 3 algae are consistent with a moderate turnover of the pentose phosphate cycle (starch reserves permitting) but are controlled by the $NADPH/NADP^+$ ratio. During phase 2 adaptation to salinity stress, however, glycerol synthesis is very rapid (see above) and can probably use NADPH about as fast as it is formed. Under these conditions the cycle probably runs close to its maximum speed.

Other evidence relevant to the competing claims of glycolysis and the pentose phosphate cycle as the primary source of glycerol in the dark are the demonstration of glucose-6-phosphate dehydrogenase (Johnson et al., 1968; Borowitzka and Brown, 1974), 6-phosphogluconate dehydrogenase (Borowitzka, unpublished observations), pentose phosphate isomerase and hexose phosphate isomerase (Johnson et al., 1968), triose phosphate dehydrogenase (R. M. Lilley and A. Duong, unpublished observations), and lactate dehydrogenase (Ben-Amotz and Avron, 1972). Furthermore, *D. tertiolecta* has an active NADP-specific isocitrate dehydrogenase but very feeble if any activity of NAD-specific isocitrate dehydrogenase (R. M. Lilley and A. Duong, unpublished observations). In a eukaryotic context, this is consistent with a lack of or deficiency in the citrate cycle, and as such is consistent with the labelling experiments of Kwon and Grant (1971) but is an apparent contradiction of the mitochondrial respiration described in Section VI,E,1, unless this respiration is in fact extramitochondrial. Moreover, it continues to emphasize NADP as a major carrier of reducing equivalents in these algae during heterotrophic metabolism.

Experiments with inhibitors have not been very informative. Glycerol metabolism, both synthesis and dissimilation, is inhibited in *D. parva* by the uncoupling agent, FCCP (Ben-Amotz and Avron, 1973a), but it is not clear whether or not the algae were illuminated during these experiments. The inhibitor of photosynthesis, dichlorophenyldimethylurea (DCMU), is reported to inhibit glycerol formation in illuminated *D. tertiolecta,* whereas the respiratory inhibitor, antimycin A, does not (Frank and Wegmann, 1974). Frank and Wegmann used antimycin in the light, however, under which conditions the results were irrelevant and, in any case, its effect on respiration was not determined under their experimental conditions. Branched respiratory chains with pathways insensitive to antimycin A have been reported in some algae (Lloyd, 1974).

There are two other observations that are somewhat incidental but illustrate the importance of NADP as a carrier of reducing equivalents and also demonstrate what appears to be an ability of *Dunaliella* to accumulate excess reducing capacity under some circumstances. *Dunaliella parva* can use either NADH or NADPH to reduce nitrate via a nitrate reductase (Heimer, 1976). Nitrate reduction normally uses only NADH and, as Heimer has commented, this type of versatility is very unusual (according to Heimer, "unique") even among algae. The second observation is one we have made repeatedly when *D. tertiolecta* cultures are grown at 20°C under laboratory illumination; the cultures can produce H_2S. Sulfide production occurring together with oxygen evolution seems unusual, but we have not yet examined the mechanism for it. A useful working hypothesis, however, is that reducing capacity is accumulated either at the level of reduced ferredoxin or NADPH faster than it can be used for carbon dioxide fixation, and some excess is channelled off to be used for sulfate reduction.

It is scarcely surprising that an alga with a highly developed mechanism for generating a reducing potential should use this potential for accumulating a reduced metabolite, glycerol. The oxidative pentose phosphate cycle produces NADPH and triose phosphate, a characteristic which it shares with the photosynthetic reductive pentose phosphate cycle. From triose phosphate on, however, the formation of glycerol is evidently the same in either case, and it is in this segment that regulation of glycerol production is likely to occur. An important factor in regulation seems to be glycerol dehydrogenase itself.

Before discussing some of the regulatory properties of glycerol dehydrogenase, two other matters of a more general nature should be considered. The first is whether control of glycerol production occurs at the level of regulation of existing enzymes or by induction–repression of enzyme synthesis. The second is concerned with the possible role of equilibrium displacement as a regulatory mechanism.

In *Dunaliella,* both the time of response of glycerol formation to water stress and the failure of inhibitors of protein synthesis to impair the regulatory process show that short-term response to water stress is controlled at the level of preformed enzymes (Borowitzka *et al.,* 1977). This is also true of *Ochromonas* (Kauss, 1977). Of course this does not exclude regulation of enzyme formation as a factor in long-term response and, to some extent, in homeostatic control. Indeed it would be astonishing if levels of enzyme activity did not reflect the salinity to which the algae were adapted.

The second question, concerning the possible role of equilibrium displacement in the control process invites two answers, one concerned with phase 2 adaptation and the other with the phase 3 fully adapted state. Phase 2 is characterized in its early stages by rapid formation or removal of glycerol, and, no doubt, major compensating adjustments of the concentrations of other metabolites. Such conditions are very far from equilibrium, and regulatory mechanisms associated with them must be assumed to be entirely kinetic. This is scarcely surprising; living systems are classically far from equilibrium. Nevertheless, in fully adapted phase 3 algae, the question should be looked at a little more closely.

In fully adapted cells, glycerol concentration is constant (more or less, ignoring minor fluctuations). Strictly speaking, it is, of course, a steady state situation in which glycerol is continually being formed to compensate for the volume increase which accompanies cell growth. Nevertheless, if ever a set of circumstances approached equilibrium in a healthy, growing organism, this is likely to be such a case. Firstly, the glycerol concentration is very high relative to that of all other metabolites (at least three orders of magnitude higher than that of most of them). Secondly, it is at a metabolic dead end and is removed via the reversible reaction by which it is formed. Thirdly, growth rate, hence the rate of glycerol synthesis needed for volume compensation, is slow compared with the rate at which glycerol concentration can be changed in response to stress. In other words, the rate of the steady state process—the formation of glycerol to compensate for growth—is slow compared with the reaction velocities potentially available to the reversible reaction by which it is synthesized.

Over and above this, Krebs has argued that, even within the context of a total steady state situation, individual reactions or reaction sequences can approximate equilibrium conditions, if necessary by coupling with other reactions which brings the complex into a state quite close to equilibrium (Krebs, 1973, 1975; Eggleston and Krebs, 1974). Given this possibility, equilibrium displacement (mass action effects) should be recognized as a possible, even probable, means of regulating glycerol content in the homeostatic condition of phase 3 cells.

Let us now return to phase 2 adaptation and assess the evidence for regulation of glycerol dehydrogenase which, as we have said, should be wholly kinetic under these conditions. The solute present at the highest concentration in *Dunaliella* under most conditions, and subject to the greatest absolute changes in concentration during adaptation, is glycerol. It should therefore not be too astonishing to find that the interaction between glycerol and its dehydrogenase is more complex than a simple enzyme–substrate interaction. Moreover, there is evidence from other sources (reviewed in Brown, 1978) that glycerol can modify a variety of biochemical processes.

The kinetics of glycerol dehydrogenation are indeed complex and in several respects are indicative of an enzyme with regulatory capabilities. The enzyme does not have simple Michaelis–Menten kinetics under some important conditions. In what we have called the forward reaction, that is, the dehydrogenation of glycerol, double reciprocal plots with glycerol as variable substrate were linear only when $NADP^+$ was at high, virtually saturating concentrations. At lower concentrations the plots were concave up and apparently parabolic. The nonlinearity and effect of the $NADP^+$ concentration were greater at pH 7.5, which is optimal for dehydrogenation, then at pH 9.0, which is optimal for the reverse reaction. The nonlinearity in this case implies a positive cooperative role of glycerol. In other words, glycerol adds more than once (twice if the curves are in fact parabolic) in the reaction sequence, once as a substrate and once as an effector which is not changed during the reaction. The other substrate, $NADP^+$, does not behave in this way but, as we have said, its concentration affects the shape of the plots, and it can be inferred therefore that it also affects the binding of effector glycerol (Borowitzka and Brown, 1974).

In the reverse direction, double reciprocal plots with dihydroxyacetone as variable substrate were also nonlinear. In this case they were concave down, however, but again apparently parabolic. The second substrate, NADPH, restored linearity when it was present at sufficiently high con-

$$G \cdot E \rightleftharpoons G \cdot E \cdot N \rightleftharpoons G \cdot E \cdot N \cdot G \rightleftharpoons G \cdot E \cdot N' \cdot D \rightleftharpoons G \cdot E \cdot N' \rightleftharpoons G \cdot E \quad (1)$$

$$E \rightleftharpoons E \cdot N \rightleftharpoons E \cdot N \cdot G \rightleftharpoons E \cdot N' \cdot D \rightleftharpoons E \cdot N' \rightleftharpoons E \quad (2)$$

$$D \cdot E \rightleftharpoons D \cdot E \cdot N \rightleftharpoons D \cdot E \cdot N \cdot G \rightleftharpoons D \cdot E \cdot N' \cdot D \rightleftharpoons D \cdot E \cdot N' \rightleftharpoons D \cdot E \quad (3)$$

Scheme 1. D, Dihydroxyacetone; E, enzyme; G, glycerol; N, $NADP^+$; N', NADPH. Effector molecules are placed before the enzyme and reactants after the enzyme. (From Borowitzka *et al.*, 1977, reproduced by permission.)

centrations (Borowitzka *et al.*, 1977). Double reciprocal plots which are concave down can imply that more than one enzyme catalyses the same reaction, but in this case the evidence favoured the alternative interpretation of a negatively cooperative participation of dihydroxyacetone in the reaction sequence (Borowitzka *et al.*, 1977). In other words, in addition to its role as a substrate, dihydroxyacetone also participated in the reaction as a negative effector. A mechanism proposed for the reaction sequence is shown in Scheme 1.

This mechanism shows three possible channels, two of which are distinguished by the involvement of an effector. During initial velocity measurements and presumably also *in vivo* during periods of rapid synthesis or removal of glycerol, as in phase 2, two channels should dominate. They are reactions (1) and (2) in the forward direction and reactions (2) and (3) in the reverse direction. Under these conditions, of course, regulation is achieved kinetically. The significance of this is that, at $NADP^+$ concentrations substantially less than saturating, raising the glycerol concentration accelerates the rate of glycerol disappearance at a rate which is less than predictable simply from its role as a substrate. Conversely, in the reverse direction, at NADPH concentrations substantially less than saturating, increasing the concentration of dihydroxyacetone accelerates the rate of glycerol formation to an extent greater than predictable simply from its role as a substrate. Thus, the evidence to this point suggests that the characteristics of glycerol dehydrogenase are weighted in favour of glycerol formation rather than its dissimilation.

Of course the real situation is not so simple. First, as we have said, the nonlinearity in the kinetics can be eliminated by high concentrations of $NADP^+$ or NADPH and, furthermore, the reaction velocities in each direction are quite sensitive to hydrogen ion concentration (Figure 9). Nevertheless, if it is assumed that the enzyme normally has an environment in the vicinity of pH 8, then a relatively small decrease in pH will favour glycerol formation, and vice versa.

If, as we suppose, the homeostatic circumstances of fully adapted algae involve little more than minor disturbance of an equilibrium, then kinetic

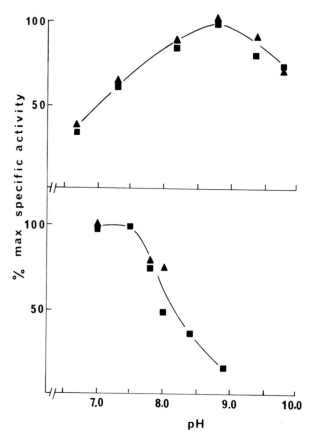

Figure 9.   The effect of pH on the glycerol dehydrogenase of *D. tertiolecta* (■) and *D. viridis* (▲). The upper panel represents glycerol dehydrogenation, and the lower panel dihydroxyacetone reduction. (Results of Borowitzka *et al.*, 1977.)

effects such as those just described are no longer so important. Nevertheless, H$^+$ remains an important factor in determining the composition of the equilibrium mixture. The reaction we are discussing is

$$\text{Glycerol} + \text{NADP}^+ \rightleftharpoons \text{dihydroxyacetone} + \text{NADPH} + \text{H}^+$$

An increase in H$^+$ (a decrease in pH) shifts the equilibrium to the left, in favour of glycerol formation. In other words, a change in pH shifts both equilibrium composition and relative reaction rate in the same direction.

Although NADP$^+$/NADPH does not act kinetically as an effector for glycerol dehydrogenase, the absolute concentration of the coenzyme is important because of its influence on the effector role of glycerol or dihydroxyacetone. Furthermore, the $NADP^+/NADPH$ ratio is also important, firstly because of the kinetic effect of reactant concentration on reaction velocity and secondly because of the mass action effect in a near-equilibrium situation. In metabolically active animal tissues and yeast, cytoplasmic $NAD^+/NADH$ ratios are of the order of 10^3 (Krebs, 1973) and are thus very sensitive to small changes in the concentration of NADH. The situation in plants is more obscure. Little is known at this level about algae, and virtually nothing is known about *Dunaliella*. In several higher plants the $NADP^+/NADPH$ ratios in chloroplasts have been reported to be about 3 in the light and about 0.7 in the dark. Cytoplasmic ratios were in the range 0–0.4 and essentially independent of light (Heber and Santarius, 1965). In extracts of whole *Chlorella*, $NADP^+/NADPH$ ratios ranged from less than 1 to about 2, depending on conditions of illumination (Green and Israelstam, 1970). Thus, in these plants the relevant nucleotide coenzyme ratios are not nearly as susceptible as those in yeast to changes in the concentration of the reduced form of the coenzyme but are more sensitive to changes in the concentration of the oxidized form. Although we have no specific information about *Dunaliella,* the inference drawn from the early comments about H_2S production is that high $NADP^+/NADPH$ ratios are unlikely, at least during periods of illumination in fully adapted phase 3 algae.

If the ratios of oxidized to reduced pyridine nucleotides constitute a significant factor in the regulation of glycerol production, then there are a priori reasons for suspecting that ATP/ADP ratios will also be influential. Once again, the analogous situation in heterotrophs is understood better. For example, Krebs (1973) has noted that the ATP/ADP ratio is directly proportional to the $NAD^+/NADH$ ratio in the cytosol of animal cells but inversely proportional to it in mitochondria. Presumably there is a correlation in *Dunaliella* between the ATP/ADP and $NADP^+/NADPH$ ratios, although we are not aware of specific information at this level. Nevertheless, there are reasons, as there are in yeast (Brown, 1978), for looking to ATP/ADP ratios as a factor potentially involved in the regulation of glycerol production in *Dunaliella*.

This far we have discussed a number of levels at which glycerol production seems potentially able to be regulated. Certainly glycerol dehydrogenase has regulatory properties, and there are reasons for supposing that the $NADP^+/NADPH$ and ATP/ADP ratios, as well as H^+, are all involved both kinetically and in equilibrium displacement. We are, however, unable to propose a precise mechanism for the regulation. This should not be (and no doubt is not) too astonishing. Firstly, notwithstanding some moderately detailed kinetics for glycerol dehydrogenase, we do not have enough information to permit the recognition of even a simple feedback loop. Secondly, on the evidence we do have, control seems to involve more than a simple feedback process. Indeed, simple feedback loops, such as occur in a number of fairly well-defined biosynthetic pathways, are probably a special case of biochemical regulation. In steady state open systems, there are many more factors involved, and these factors are quite difficult to identify. Perhaps the techniques of systems analysis will have some application in such cases when sufficient detailed information becomes available. Furthermore, in the present instance there are reasons, such as we have discussed, for believing that two types of regulation are involved, the choice between them depending on the physiological state of the organism. Broadly speaking, those mechanisms are kinetic control and equilibrium displacement.

2. The Regulatory Signal

The ensuing discussion of the regulatory signal is even less complete and more speculative than the preceding section on metabolic aspects of regulation. First, however, there is the question of the location of glycerol dehydrogenase within the algae. If we accept that this enzyme is an important site for regulating glycerol metabolism, then understanding the process requires that the cellular location of the enzyme be known. There are several fairly obvious reasons for this. Among them is the need to know whether the NADP resides in the chloroplast or the cytosol, hence the immediate source of its reducing equivalents, whether glycerol migrates as such into the cytosol, or whether dihydroxyacetone phosphate is transported from the chloroplast and subsequently reduced in the cytosol. Perhaps most important is the need to know whether the signal must be transmitted solely across the plasma membrane or plasma membrane plus chloroplast envelope (or mitochondrial envelope, although this is less likely). Whether there are one or two boundaries reflects the probable nature of the signal.

A priori, the involvement of NADP (rather than NAD), together with the ability of Dunaliella to use starch for glycerol production, suggest the chloroplast as the site of the enzyme, but there is no direct experimental evidence to support this assumption. The problem is currently under investigation, but a major difficulty lies in the virtual impossibility of isolating, by any means currently at our disposal, intact functional chloroplasts from Dunaliella (R. M. Lilley and A. Duong, unpublished observations).

Regulation of glycerol content in Dunaliella has characteristics in common with osmoregulation in other organisms. Since there is no direct and very little indirect evidence about the nature of the regulatory signal(s) in Dunaliella, there are reasons for looking to other organisms for useful comparisons and analogies. Care is needed, however, since the essential facts of osmoregulation are not universal and the regulatory mechanisms should also be different. The following major types of osmoregulation have been recognized.

1. Accumulation of a salt (usually potassium chloride) from the extracellular fluid. This often occurs by a "pump and leak" mechanism, in which there is an active uptake and a passive efflux of the relevant solute. Eventually a steady state is reached when the efflux balances the active uptake. This mechanism is common; it has been identified, for example, in erythrocytes and is probably the means by which potassium chloride accumulates in the extreme case of *Halobacterium* and *Halococcus*. Cram (1976) has commented that in this type of system "there is nothing that would be called 'regulation'."

2. The analogous situation involving a metabolite is encountered in the yeast *Saccharomyces cerevisiae*. This organism, which is not xerotolerant, responds to water stress by varying its glycerol content. It achieves this by controlling glycerol synthesis, while glycerol efflux is apparently not controlled (Edgley and Brown, 1978; see also Brown, 1978).

3. A variant of mechanism 2 involves responding to solute stress by synthesizing and retaining more compatible solute, whereas dilution stress elicits the simple response of leakage of the solute from the cell. This is what happens, for example, in *P. subcordiformis* (Hellebust, 1976), and is the mechanism attributed by Frank and Wegmann (1974) to *Dunaliella* but which we dispute (see also Ben-Amotz, 1975). The essence of our contention is that leakage is an important factor only after a major dilution stress and not in relatively minor or moderate readjustments.

4. The xerotolerant yeast, *Saccharomyces rouxii,* regulates glycerol content in response to water activity by a mechanism which is predominantly biophysical rather than biochemical. At least down to moderate levels of water activity it synthesizes an approximately constant amount of glycerol but controls the amount it excretes into the growth medium (Edgley and Brown, 1978; see also Brown, 1978). Under more extreme conditions of low water availability it probably makes more use of metabolic regulation.

5. Complete metabolic regulation in which the cellular content of the compatible solute is controlled entirely by metabolic interconversions; physical loss from the cell is either insignificant or constant. This is apparently the method used, for example, by *Ochromonas* (Kauss, 1977) and by *Dunaliella*.

It is unlikely that the same type of signal functions in each of these circumstances, and the extent to which analogy might be useful is limited accordingly. There are at least six types of osmoregulatory signals which either have been proposed or warrant consideration. Readers interested in a more comprehensive theoretical treatment of osmoregulation within the framework of control theory are referred to a recent review by Cram (1976). Theoretically, the concentration of an extracellular solute should be able to regulate an intracellular metabolic process in one or more of the following ways: (a) The solute could react with a specific receptor site on the cell surface and thereby transmit a signal to an intracellular messenger, such as cyclic AMP, which in turn is responsible for the metabolic regulation. This is the type of process encountered, for example, in hormone action and, at another level, with cholera toxin. Reissig (1974) has reviewed this kind of event in microorganisms. (b) Changes in salinity (or water activity) might induce an abnormal uptake of a solute, probably an ion, which affects the relevant biochemical process. Changes in turgor can influence the uptake of a specific solute. For example, when the turgor of the giant marine alga *Valonia* is eliminated by the insertion of capillaries into the cell, the potassium ion influx increases fourfold; it can be returned to normal by the artificial application of internal hydrostatic pressure (Cram, 1976). (c) A change in water activity can modify the solute fluxes which normally accompany energy metabolism, and the altered flux constitutes the signal. A potentially important solute in this sense is H^+ which is subject to major fluxes and which, as we have already discussed, can influence both the kinetics of glycerol dehydrogenase and the equilibrium of the reaction this enzyme catalyses. Kauss (1977) has suggested changes in cellular pH as a signal in the control of galactosylglycerol accumulation in *Ochromonas*. (d) Cell volume is homeostatically controlled, and volume changes provide the signals for regulating glycerol metabolism. Regulation of cell volume, mainly in animals, has been recently reviewed by MacKnight and Leaf (1977) and discussed largely from a theoretical standpoint by Cram (1976). (e) Turgor pressure is homeostatically controlled and provides the signal for regulating glycerol metabolism. This has also been discussed by Cram (1976). (f) The thickness, and hence the electrical resistance and conductivity, as well as the general permeability characteristics of the plasma membrane are changed as a result of membrane stretching, electrical potential, or variation in hydrostatic pressure. The signals for regulating glycerol metabolism are expressed by the changed membrane properties (see, e.g., Coster *et al.*, 1977).

There is insufficient experimental evidence to allow the signal in *Dunaliella* to be identified, but some preliminary sorting out is possibly justified. Thus mechanism a, the interaction between a specific substance and a specific receptor site on the cell surface is unlikely in view of the significance of water activity rather than salinity in determining glycerol content. On the other hand, there is a degree of specifity in the effect of sodium chloride concentration on growth rate (Borowitzka *et al.*, 1977), but this is more likely explained by membrane integrity and conformation rather than by a specific receptor site for Na^+, Cl^-, or $NaCl$. It is, in any case, a different problem.

We are not aware of any evidence for or against possibility b—that an "abnormal" uptake of a solute induced directly by a change in salinity or water activity regulates glycerol metabolism in *Dunaliella*. The possibility that modification of energy-induced proton fluxes might provide the primary signal has been subjected to a preliminary investigation; the results so far suggest that they do not. Kessly (unpublished observations) has observed that there is a sharp increase in the pH of the medium when suspensions of *Dunaliella* are illuminated; this is normal. The pH changes (and the corresponding titre of H^+) diminish when the algae are exposed either to salt stress or dilution stress. Furthermore, the changes are predominantly the result of HCO_3^- uptake and are virtually eliminated when CO_2 and HCO_3^- are excluded. Thus, although changes in intracellular H^+ concentration can undoubtedly influence glycerol metabolism (see above), modification of a proton flux does not seem on present evidence to be the primary controlling signal.

Mechanisms d, e, and f are closely interrelated and are best considered together. There now seems little reason to doubt that in different organisms either cell volume or turgor pressure can be regulated. Cram (1976) states it more bluntly when he says, "The primary signal for these adaptive responses must be a function of volume in wall-less cells while it is most probably a function of turgor pressure in walled cells." Cram cites direct evidence that turgor is not the osmoregulatory signal and that it is not homeostatically controlled in some algae, for example, *Nitella*.

Dunaliella does not have a rigid wall, and we can assume therefore that volume is likely to be of more importance than turgor pressure as a signal in osmoregulation. It is possibly relevant that the response of glycerol content to stress is essentially instantaneous, whereas time lags have been reported for some organisms in which the signal is turgor pressure (Cram, 1976; Kauss, 1977). If volume is indeed of primary and major importance in controlling glycerol metabolism, then there are two levels at which it might function. The first is its effect on the concentration of intracellular metabolites, some of which might be regulatory. This mechanism is not as simple as it might at first seem. The complications are perhaps best illustrated by restating the sequence of events, as they affect intracellular solutes, following salt stress. The immediate result of the stress is a water efflux accompanied by an increase in the concentration of intracellular solutes and cell shrinkage. This process continues essentially to equilibrium with the environment, glycerol synthesis ensues, and water returns to the cell at a rate which maintains a_w parity with the outside. In other words, during the phase 2 adaptation internal water activity remains constant and about equal to the outside. During this period volume is progressively restored, so that the absolute concentration of solutes (other than glycerol) again drops but, because a_w remains about constant, the thermodynamic activity of the internal solutes does not parallel the concentration. An important question is therefore suggested: If an internal solute provides the signal, is the relevant parameter its concentration or its thermodynamic activity? This question has little practical significance under dilute conditions but becomes important under the extreme conditions tolerated by halophilic species of *Dunaliella*. We should add the additional comment that the process of volume recovery continues until the algae have resumed approximately their original volume and are in *a state of turgor*. In other words the recovery process continues past the point of simple restoration of volume.

The second way in which volume changes might provide a signal is through changed membrane properties [primarily thickness, see Coster *et al.* 1977], but here again it is difficult to distinguish clearly between the effects of volume and turgor on the membrane, especially when membrane stretching is involved.

We have some difficulty in accepting that, if volume is regulated, turgor pressure has no regulatory function at all, even in a wall-less alga such as *Dunaliella*. If volume is the primary reference, there seems to be some reason for accepting turgor as a supplementary one. Indeed, if we give free reign to our intuition, we find merit in the thought that volume might provide the coarse control during phase 2 transitions, whereas turgor pressure provides the signal for the finer homeostatic control in the fully adapted algae.

Finally, it should not be forgotten that the primary site of regulation might be within an organelle, in which case the preceding comments need some minor modifications to be fully applicable.

The whole problem of osmoregulation, especially with regard to identification of the signal, is clearly difficult and complex and is unlikely to yield to any simple explanation which purports to identify a single biochemical reaction and a single signal as the whole mechanism. For example, in *Ochromonas,* there are probably at least four steps in the metabolic pathways associated with galactosylglycerol metabolism, which are regulated by water activity (Kauss, 1977). In the specific case of *Dunaliella* the difficulties are compounded by an extreme paucity of facts although, as we all know, this can be a great asset when speculating. The physiological ramifications of osmoregulation and salinity tolerance, however, are such as to warrant a continuing careful study of the problem.

VII. RECAPITULATION

As we said at the beginning of this chapter, adaptability to a very wide range of salinities is one of the outstanding characteristics of *Dunaliella*. The genus is not halophilic in the sense that halobacteria are, and although they cohabit with halophilic bacteria in environments saturated with salt, the relationship is one of tolerance on the part of *Dunaliella* not a requirement for the salt. On the other hand some species are equally happy at marine salinities and less, at which concentrations halobacteria cannot remain intact, let alone grow.

The salt tolerance of *Dunaliella* spp. depends primarily on the accumulation of the compatible solute, glycerol, to concentrations thermodynamically commensurate with environmental salinity. The adaptability depends on several factors. First, when subjected to dilution stress, the algae swell extensively but rarely lyse. Glycerol leaks to some extent on severe stress, but there are few quantitative details available about this. On moderate stress, however, glycerol content is diminished by metabolic dissimilation; the energy for this process is apparently obtained largely from respiration. Survival of dilution stress thus depends on membrane elasticity and the retention of a metabolic capability.

Salt stress does not challenge the cell with the threat of lysis; here the problem lies in the physiological and biochemical consequences of dehydration. Salt stress impairs photosynthetic oxygen evolution for about 24 hr but the organisms can nevertheless respond immediately to the stress by synthesizing glycerol. The energy for this is derived predominantly, perhaps wholly in some cases, from respiration. Notwithstanding the impairment of net oxygen evolution, however, there is evidence that photosynthesis can assist glycerol synthesis after severe stress. Glycerol synthesis continues for about 90 min, after which normal volume is regained. Adaptability to salt stress thus depends largely on the retention of some metabolic functions under conditions of dehydration with inadequate concentrations of a compatible solute.

There is conflicting evidence about the admission of salt during normal growth and following salt stress but, on balance, it seems that salt is excluded in all but minor amounts and that osmoregulation is achieved overwhelmingly with glycerol. There are substantial differences among the overall salt relations of the various species, but the physiological reasons for these differences remain obscure.

A fundamental characteristic of *Dunaliella* physiology is the ability to regulate glycerol production in response to salinity. One probable regulatory site is the NADP-specific glycerol dehydrogenase. This enzyme, which catalyses a freely reversible reaction, has characteristics which suggest that it could participate in the kinetic control of glycerol metabolism during the period of recovery from stress. There are also reasons for believing that homeostatic control of glycerol content during normal growth might include some degree of equilibrium displacement. There is, as yet, no direct information about the regulatory signal governing glycerol metabolism, but there are reasons for suspecting cell volume as a primary signal with a supplementary contribution to homeostasis from turgor pressure.

ACKNOWLEDGMENT

We are greatly indebted to Mr. D. S. Kessly for permission to use his results before publication.

REFERENCES

Aitken, D. M., and Brown, A. D. (1972). *Biochem. J.* **130**, 645–666.
Aitken, D. M., Wicken, A. J., and Brown, A. D. (1970). *Biochem. J.* **116**, 125–134.
Anand, J. C., and Brown, A. D. (1968). *J. Gen. Microbiol.* **52**, 205–212.
Baas-Becking, L. G. M. (1931). *J. Gen. Physiol.* **14**, 765–779.
Bassham, J. A., and Calvin, M. (1957). "The Path of Carbon in Photosynthesis." Prentice-Hall, Englewood Cliffs, New Jersey.
Beardall, J., Mukerji, D., Glover, H. E., and Morris, I. (1976). *J. Phycol.* **12**, 409–417.
Ben-Amotz, A. (1975). *J. Phycol.* **11**, 50–54.
Ben-Amotz, A., and Avron, M. (1972). *Plant Physiol.* **49**, 240–243.
Ben-Amotz, A., and Avron, M. (1973a). *Plant Physiol.* **51**, 875–878.
Ben-Amotz, A., and Avron, M. (1973b). *FEBS Lett.* **29**, 153–155.
Bentley-Mowat, J. A., and Reid, S. M. (1977). *J. Exp. Mar. Biol. Ecol.* **26**, 249–264.
Borowitzka, L. J., and Brown, A. D. (1974). *Arch. Microbiol.* **96**, 37–52.
Borowitzka, L. J., Kessly, D. S., and Brown, A. D. (1977). *Arch. Microbiol.* **113**, 131–138.
Brock, T. D. (1975). *J. Gen. Microbiol.* **89**, 285–292.
Brown, A. D. (1964). *Bacteriol. Rev.* **28**, 296–329.
Brown, A. D. (1974). *J. Bacteriol.* **118**, 769–777.
Brown, A. D. (1976). *Bacteriol. Rev.* **40**, 803–846.
Brown, A. D. (1978). *Adv. Microb. Physiol.* **17**, 181–242.
Brown, A. D., and Simpson, J. R. (1972). *J. Gen. Microbiol.* **72**, 589–591.
Butcher, R. W. (1959). *Minist. Agric. Fish Food (G.B.) Ser. IV* Part 1, pp. 1–74.
Carlucci, A. F., and Bowes, P. M. (1970). *J. Phycol.* **6**, 393–400.
Contaxis, C. C., and Reithel, F. J. (1971). *J. Biol. Chem.* **246**, 677–685.
Coster, H. G. L., Steudle, E., and Zimmermann, U. (1977). *Plant Physiol.* **58**, 636–643.
Craigie, J. S. (1974). *In* "Algal Physiology and Biochemistry" (W. D. P. Stewart, ed.), pp. 206–235. Blackwell, Oxford.
Craigie, J. S., and McLachlan, J. (1964). *Can. J. Bot.* **42**, 777–778.
Cram, W. J. (1976). *In* "Encyclopedia of Plant Physiology" (U. Lüttge and M. G. Pitman, eds.), New Series, Vol. II, Part A, pp. 284–316. Springer-Verlag, Berlin and New York.

Davies, A. G. (1974). *J. Mar. Biol. Assoc. U.K.* **54**, 157–169.
Davies, A. G. (1976). *J. Mar. Biol. Assoc. U.K.* **56**, 39–57.
Douzou, P. (1974). *Methods Biochem. Anal.* **22**, 401–512.
Drokova, I. H., and Dovhoruka, S. I. (1966). *Ukr. Bot. Zh.* **23**, 59–62.
Dunstan, W. M., Atkinson, L. P., and Natoli, J. (1975). *Mar. Biol.* **31**, 305–310.
Edgley, M., and Brown, A. D. (1978). *J. Gen. Microbiol.* **104**, 343–345.
Eggleston, L. V., and Krebs, H. A. (1974). *Biochem. J.* **138**, 425–435.
Eppley, R. W., and Coatsworth, J. L. (1966). *Arch. Mikrobiol.* **55**, 66–80.
Eyden, B. P. (1975). *J. Protozool.* **22**, 336–344.
Frank, G., and Wegmann, K. (1974). *Biol. Zentralbl.* **93**, 707–723.
Gibbs, N., and Duffus, C. M. (1976). *Appl. Environ. Microbiol.* **31**, 602–604.
Gibor, A. (1956). *Biol. Bull. (Woods Hole, Mass.)* **111**, 223–229.
Gimmler, H., Schirling, R., and Tobler, U. (1977). *Z. Pflanzenphysiol.* **83**, 145–158.
Ginzburg, M. (1969). *Biochim. Biophys. Acta* **173**, 370–376.
Green, W. G. E., and Israelstam, G. F. (1970). *Physiol. Plant.* **23**, 217–231.
Hamburger, C. (1905). *Arch. Protistenkd.* **6**, 111–130.
Heber, U. W., and Santarius, K. A. (1965). *Biochim. Biophys. Acta* **109**, 390–408.
Heimer, Y. M. (1976). *Plant Physiol.* **58**, 57–59.
Hellebust, J. A. (1965). *Limnol. Oceanogr.* **10**, 192–206.
Hellebust, J. A. (1976). *Annu. Rev. Plant Physiol.* **27**, 485–505.
Hellebust, J. A., and Terborgh, J. (1967). *Limnol. Oceanogr.* **12**, 559–567.
Johnson, M. K., Johnson, E. J., MacElroy, R. D., Speer, H. L., and Bruff, B. S. (1968). *J. Bacteriol.* **95**, 1461–1468.
Kauss, H. (1977). *Int. Rev. Biochem., Plant Biochem. II* **13**, 119–140.
Kirst, G. O. (1977). *Planta* **135**, 69–75.
Krebs, H. A. (1973). *In* "Rate Control of Biological Processes" (D. D. Davies, ed.), Symposia of the Society for Experimental Biology, Vol. 27, pp. 299–318. Cambridge Univ. Press, London and New York.
Krebs, H. A. (1975). *Adv. Enzyme Regul.* **13**, 449–472.
Kwon, Y. M., and Grant, B. R. (1971). *Plant Cell Physiol.* **12**, 29–39.
Lanyi, J. K. (1974). *Bacteriol. Rev.* **38**, 272–290.
Larsen, H. (1962). *In* "The Bacteria" (I. C. Gunsalus and R. Y. Stanier, eds.) Vol. 4, pp. 297–342. Academic Press, New York.
Larsen, H. (1967). *Adv. Microb. Physiol.* **1**, 97–132.
Latorella, A. H., and Vadas, R. L. (1973). *J. Phycol.* **9**, 273–277.
Latorella, A. H., Regan, P. R., Eustice, D., and Ritter, E. (1974). *Proc. Rochester Acad. Sci.* **12**, 411.
Lerche, W. (1938). *Arch. Protistenkd.* **88**, 236–268.
Lerner, H. R., and Avron, M. (1977). *Plant Physiol.* **59**, 15–17.
Lloyd, D. (1974). *In* "Algal Physiology and Biochemistry" (W. D. P. Stewart, ed.), pp. 505–529. Blackwell, Oxford.
Loeblich, L. A. (1972). Ph.D. Thesis, Univ. of California, San Diego.
MacKnight, A. D. C., and Leaf, A. (1977). *Physiol. Rev.* **57**, 510–573.
McLachlan, J. (1960). *Can. J. Microbiol.* **6**, 369–379.
Mandelli, E. F. (1969). *Tex. Univ., Contrib. Mar. Sci.* **14**, 47–57.
Marano, F. (1976). *J. Microsc. Biol. Cell.* **25**, 279–282.
Marrè, E., and Servettaz, O. (1959). *Atti Accad. Naz. Lincei, Cl. Sci. Fis., Mat. Nat., Rend.* **26**, 272–278.
Masyuk, N. P. (1965a). *Ukr. Bot. Zh.* **22**, 3–11.
Masyuk, N. P. (1965b). *Ukr. Bot. Zh.* **22**, 18–22.

Menzel, D. W., Anderson, J., and Randtke, A. (1970). *Science* **167**, 1724–1726.
Mutschler, D., and Wegmann, K. (1974). *Biol. Zentralbl.* **93**, 725–729.
Osmond, C. B., and Rees, T. A. (1969). *Biochim. Biophys. Acta* **184**, 35–42.
Overnell, J. (1975). *Mar. Biol.* **29**, 99–103.
Pfeifhofer, A. O., and Belton, J. C. (1975). *J. Cell Sci.* **18**, 287–299.
Pope, D. H., and Berger, L. R. (1974). *Can. J. Bot.* **52**, 2375–2379.
Post, F. J. (1977). *Microb. Ecol.* **3**, 143–165.
Rabinowitch, S., Grover, N. B., and Ginzburg, B. Z. (1975). *J. Membr. Biol.* **22**, 211–230.
Reissig, J. L. (1974). *Curr. Top. Microbiol. Immunol.* **67**, 43–96.
Ruinen, J. (1938). *Arch. Protistenkd.* **90**, 210–258.
Schmidt-Nielson, K. (1975). "Animal Physiology." Cambridge Univ. Press, London and New York.
Shapiro, A., DiLello, D., Loudis, M. C., Keller, D. E., and Hutner, S. H. (1977). *Appl. Environ. Microbiol.* **33**, 1129–1133.
Sick, L. V. (1976). *Mar. Biol.* **35**, 69–78.
Simpson, J. R. (1976). Ph.D. Thesis, Univ. of New South Wales, Australia.
Stephens, D. W., and Gillespie, D. M. (1976). *Limnol. Oceanogr.* **21**, 74–87.
Stewart, W. D. P., ed. (1974). "Algal Physiology and Biochemistry," Botanical Monographs, Vol. 10. Blackwell, Oxford.
Téodoresco, E. C. (1905). *Beih. Bot. Zentralbl.* **18**, 215–232.
Téodoresco, E. C. (1906). *Rev. Gen. Bot.* **18**, 353–371.
Trezzi, F., Galli, M. G., and Bellini, E. (1965). *G. Bot. Ital.* **72**, 255–263.
van Auken, O. W., and McNulty, I. B. (1973). *Biol. Bull. (Woods Hole, Mass.)* **145**, 210–222.
Volcani, B. E. (1936). *Nature (London)* **138**, 467.
Volcani, B. E. (1944). *In* "Papers Collected to Commemorate the 70th Anniversary of Dr. Chaim Weizmann," pp. 71–85. Daniel Sieff Res. Inst., Rehovoth.
Wegmann, K. (1969). *Prog. Photosynth. Res.* **3**, 1559–1564.
Wegmann, K. (1971). *Biochim. Biophys. Acta* **234**, 317–323.
Wegmann, K., and Metzner, H. (1971). *Arch. Mikrobiol.* **78**, 360–367.
Woodhead, S., and Walker, T. A. (1975). *J. Gen. Microbiol.* **89**, 327–336.
Yurina, E. V. (1966). *Vestn. Mosk. Univ.* **21**, 76–83.

Physiology of Coccolithophorids

7

DAG KLAVENESS AND EYSTEIN PAASCHE

I. INTRODUCTION

The treatment of coccolithophorids as a separate group of algae (or of protozoa, if a zoological classification is preferred) is motivated by their ability to produce surface scales encrusted with calcium carbonate, so-called coccoliths (Figures 2, 8, and 10). Although we treat coccolithophorids as one group in this chapter, a number of recent investigations have indicated that some of them, notably *Emiliania huxleyi* (Lohm.) Hay & Mohler, are more closely related to certain genera of noncalcified haptophycean flagellates than to the more typical coccolithophorids (references in Parke and Green, 1976). Moreover, cell types that produce no calcium carbonate are the predominant ones in the life cycle of certain species (e.g., Figure 9).

The study of coccolithophorids was left mainly in the hands of biological oceanographers and micropaleontologists until they were adopted, about 15 years ago, as interesting test subjects by workers primarily interested in calcification in animal systems. During the last few years it has become evident that calcification in these organisms cannot be studied in a profitable way except in the context of the general physiology of the coccolithophorid cell. In this chapter we review evidence showing that

BIOCHEMISTRY AND PHYSIOLOGY OF PROTOZOA
SECOND EDITION, VOL. 1

formation of the mineralized parts of coccoliths is merely one of the last steps in a series of events involving most of the organelles in the cell.

The coccolithophorid species used in experimental studies are necessarily of the "laboratory weed" category, and in many respects not representative of the group as a whole. The interested reader should consult one of the modern taxonomic papers, e.g., the one by Okada and McIntyre (1977), if he wishes to obtain an impression of form variation in the group. Although it is still true that most living species inhabit tropical or subtropical offshore waters, the painstaking studies by Manton and others (e.g., Manton *et al.*, 1977) revealed a flora of cold-water species that had been overlooked by earlier investigators. Despite steady progress in the art of growing plankton algae, efforts to obtain cultures of representative species of offshore coccolithophorids have met with failure thus far, so that many interesting questions concerning their physiology must remain unresolved for a long time to come.

II. INTERNAL STRUCTURE OF THE CELL

Recently there has been an emphasis on structural as well as functional aspects of calcification in coccolithophorids, and it is necessary to present a brief outline of the internal organization of the coccolithophorid cell as a general background. The intracellular fine structure of the group of organisms to which coccolithophorids belong (class Haptophyceae = Prymnesiophyceae in the botanical system) is well understood. The essentials are conveniently presented by Dodge (1973), and in a more summary fashion by Hibberd (1976). We omit aspects that are not relevant to the following discussion.

Figure 1. *Emiliania huxleyi.* Section of a cell with an extracellular coccolith cover and an intracellular coccolith. The large chloroplast (Chl) with a pyrenoid (Py) partly encloses the nucleus (N). The coccoliths are made one by one in a vesicle adjacent to the nucleus. M, Mitochondrion. The cell was treated with lead salts during fixation (Klaveness, 1976), and the section is unstained. Magnification: ×20,000.

Figure 2. *Emiliania huxleyi.* Scanning electron micrograph showing the extracellular coccolith cover. Magnification: ×20,000.

Figure 3. *Emiliania huxleyi.* Intracellular coccolith of a cell treated to preserve the mineral (From Klaveness, 1976). Magnification: ×37,000.

Figure 4. *Emiliania huxleyi.* Intracellular coccolith stained selectively to visualize the matrix material (From Klaveness, 1976). The mineral inside the matrix was lost during processing. Another coccolith is under preparation beneath the mature coccolith, seen here as the matrices of two rhombohedral calcite crystals only. Magnification: ×27,000.

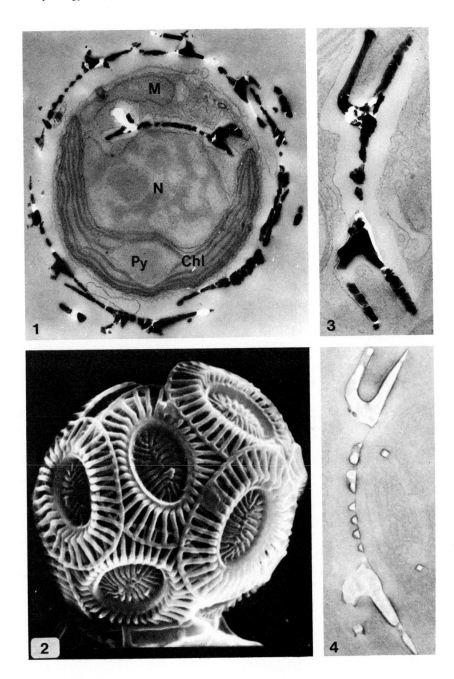

Figure 1 and 5 show sections of coccolithophorids, *E. huxleyi* and *Calyptrosphaera sphaeroidea* Schiller, respectively. In the latter, the coccoliths have been lost during processing for electron microscopy. The nucleus is conspicuous and located centrally, and there is continuity between the nuclear envelope, the outer plastidial envelope, and the endoplasmic reticulum. The endomembrane system (Morré and Mollenhauer, 1974) also comprises the Golgi apparatus, usually derived from endoplasmic reticulum or an amplexus through vesicle flow. The peculiar biosynthetic activity of the endomembrane system in coccolithophorids is a major issue of this chapter, as coccoliths are made inside Golgi vesicles or in the endoplasmic reticulum.

The chloroplasts are usually one or two in number, are parietal, and have a pyrenoid. The latter may vary from an inconspicuous, interlamellar pyrenoid, as in *E. huxleyi* (Figure 1), to a conspicuous, bulging one visible with the light microscope, as in *Ochrosphaera* (Schwarz, 1932; Klaveness, unpublished observations) and some *Hymenomonas* species (Mills, 1975, Pienaar, 1976b). The pyrenoid is frequently traversed by one (Figure 5) or a few chloroplast lamellae with a reduced number of thylakoids and may show a crystal lattice substructure (Klaveness, 1971). The chloroplast lamellae consist of usually three thylakoids. Girdle lamellae are lacking within the Haptophyceae (Massalski and Leedale, 1969). The mitochondria are several in number in each cell, are elongated with tubular cristae, and sometimes contain fibrillar material (Pienaar, 1969b; Klaveness, 1971).

In motile stages, coccolithophorids have two flagella and sometimes a more or less well-developed third one, the haptonema. The true flagella are smooth, are of equal to slightly unequal length, and have the general 9 + 2 pattern of axoneme microtubular organization, while the haptonema has a strongly deviating structure. Development of the haptonema in the flagellated stages of coccolithophorids varies within wide limits (e.g., Klaveness, 1973; Gayral and Fresnel, 1976). No haptonema has been found in *E. huxleyi* (Klaveness, 1972a,b).

In close association with the kinetosomes (flagellar bases) there may be bundles of microtubules. Indiscriminate use of the term "flagellar roots" in describing microtubular bundles in haptophycean cells merely serves to confuse the many differentiated and perhaps highly specialized functions of these structures. Authors have indicated participation of these bundles in plasma rotation and Golgi vesicle excretion during coccolith and scale formation and distribution (Brown, 1974a,b; Brown and Romanovicz, 1976), in cytoplasmic cleavage (Manton and Peterfi, 1969), in influencing flagellar movement (Gayral and Fresnel, 1976), and in several other functions (Brown and Franke, 1971) in the haptophycean cell. The ontogeny of microtubular bundles in coccolithophorid cells is obscure, but their function is certainly not related only to the flagellar apparatus.

Figure 5 also shows the peripheral sacs, containing a highly stainable granular material (Klaveness, 1973). The function of these sacs may be related to excess scale digestion and/or plasma rotation (see Section V).

III. LIFE CYCLES

Coccolithophorids show a diversity in regard to their modes of sexual and vegetative reproduction that is truly remarkable, especially considering that not more than a handful of the 200-odd species have been studied. The subject was reviewed about 10 years ago (Paasche, 1968b), and only newer findings that may be of interest to experimentalists are treated here in some detail.

In *Pleurochrysis,* the predominating stage is nonmotile and consists of sessile colonies of cells (cf. Figure 9). These vegetative cells produce organic scales (cf. Figure 8) in great profusion (Brown, 1969), but they do not form coccoliths. In liquid media they may give rise to one or the other of two morphologically different flagellate cell types. One of these (the zoospores) does not have coccoliths, although it produces organic surface scales, whereas the other one carries coccoliths (Figure 10) as well as scales (Parke, 1961; Leadbeater, 1971). The exact relationship between these cell types is not clear, although it would be reasonable to assume that the coccolithophorid stage occupies the same position in the life cycle of *Pleurochrysis* as in that of the closely related genus *Cricosphaera* (Leadbeater, 1971). Transition to the coccolithophorid stage may be brought about by experimental manipulation of the nitrogen supply, as shown by Brown and Romanovicz (1976). In cultures maintained on agar containing inorganic salts, L-serine, L-alanine, and L-asparagine, the *Cricosphaera*-like, i.e., coccolithophorid, phase could be induced by transferring zoospores to agarized media containing serine as the sole organic nitrogen source followed by flooding of the plates with serine-containing liquid media.

The life history of *Cricosphaera carterae* (Braarud & Fagerland) Braarud was interpreted by von Stosch (1967) and Leadbeater (1970). In *Cricosphaera* the sessile (benthic) phase is slightly more differentiated and branched than in *Pleurochrysis* (cf. Figure 9). The potential for phase transitions in *C. carterae* and similar species seems to be dependent on the

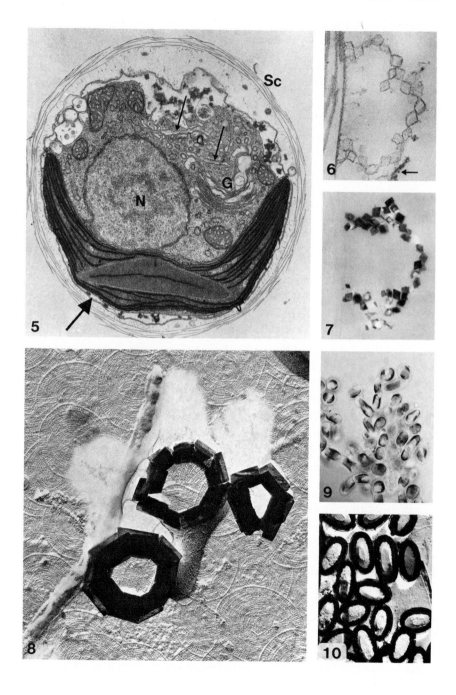

strain or isolate. Some of the earlier cytological investigations (Rayns, 1962; von Stosch, 1967) suggested that the free-living coccolithophorid phase was diploid, while the multicellular, sessile phase was haploid. Although subsequent investigations (Leadbeater, 1970; Gayral *et al.*, 1972) confirmed the existence of a morphological alternation between motile and sessile stages, there are diverging views with regard to the interpretation of events taking place. To make the situation even more complicated, Lefort (1971, 1975) suggested that a morphologically distinct coccolithophorid previously described as *Ochrosphaera verrucosa* Schussnig (Figure 8) may be another stage in the life history of *Cricosphaera* and *Hymenomonas* species. Certain strains of *Cricosphaera*-like coccolithophorids in the Plymouth Culture Collection, believed not to be true *C. carterae*, also produce an *Ochrosphaera* stage (Parke and Green, 1976).

So far, only two representatives of truly planktonic species are known from laboratory studies to have a heteromorphic life history: *Coccolithus pelagicus* (Wallich) Schiller (Parke and Adams, 1960) and *E. huxleyi* (Klaveness, 1972b). The life cycle involves two stages, a nonmotile coccolithophorid stage and a motile flagellate stage, both reproducing their own kind by binary fission when kept separately in pure culture. The flagellate stage of *C. pelagicus* was previously known as *Crystallolithus hyalinus* Gaarder & Markali and carries a simple kind of coccolith on the cell surface. The motile stage of *E. huxleyi* has uncalcified organic scales only (Klaveness and Paasche, 1971; Klaveness, 1972b). Wilbur and

Figure 5. *Calyptrosphaera sphaeroidea.* Section of cell showing the nucleus (N), Golgi apparatus (G), and other cell organelles such as the conspicuous chloroplast and several small mitochondria (From Klaveness, 1973). The smaller arrows point to Golgi vesicles containing organic scales under construction, while the large arrow indicates the peripheral sac (see text). The extracellular organic scale covering (Sc) is seen, but the coccoliths (external to the scales) have been lost during processing for electron microscopy. Magnification: ×15,000.

Figure 6. *Calyptrosphaera sphaeroidea.* Organic matrix of a coccolith constructed upon the organic baseplate (From Klaveness, 1973). Arrow points to part of the skin (see text). Ruthenium red-stabilized, unstained section. Magnification: ×41,000.

Figure 7. *Calyptrosphaera sphaeroidea.* Section of lead acetate-stabilized coccolith (From Klaveness, 1973; see also Carasso and Favard, 1966) showing the single rhombohedral calcite crystals. The organic baseplate scale can barely be seen because of lack of staining. Magnification: ×29,000.

Figure 8. *Ochrosphaera verrucosa.* Shadow-cast preparation of coccoliths and organic scales showing the radial and concentric organization of fibrils in the latter. Magnification: ×35,000.

Figure 9. Sessile (benthic) phase of a haptophycean flagellate related to *Cricosphaera*. The cell wall is entirely constructed of organic scales, and coccoliths are absent. Light micrograph. Magnification: approximately ×300.

Figure 10. Coccoliths of *Cricosphaera* sp. The coccoliths of *Pleurochrysis* are very similar. Magnification: ×9000.

Watabe (1963) induced a phase transition from a scaly (flagellated; see Klaveness and Paasche, 1971) cell to a coccolithophorid cell by growing cultures in nitrogen-deficient media.

IV. MORPHOLOGY OF COCCOLITHS AND ASSOCIATED STRUCTURES

The mineral parts of coccoliths are constructed of units, each representing a crystal of calcium carbonate (Paasche, 1968b; Blackwelder et al., 1976). Each unit is enclosed within an organic envelope, a matrix (Figures 4 and 6; see also Braarud et al., 1953; Outka and Williams, 1971).

The coccoliths of C. sphaeroidea are so-called holococcoliths; they are built of rhombohedral units (Gaarder, 1962; Klaveness, 1973) which exhibit the native crystal form of calcite (Figures 6 and 7). According to Gaarder (1962), the previous assumption of hexagonal prisms as the crystal form in certain holococcoliths (Halldal and Markali, 1954, 1955; Black, 1963) was due to a misinterpretation of electron micrographs. Photographs of Calyptrosphaera coccolith matrices (Figure 6) corroborate Gaarder's opinion.

In Calyptrosphaera the cell is covered by a ''skin'' (Figure 6) located externally to the coccolith layer. A similar skin has been found in the Crystallolithus phase of C. pelagicus; in this case also the coccoliths are of the holococcolith type (Manton and Leedale, 1963). The skin is highly ruthenium red-reactive, indicating a pectic nature.

The formation and functional significance of the skin have not been elucidated. In Crystallolithus, several superimposed systems of skin, coccoliths, and scales may be present (Manton and Leedale, 1963; Klaveness, 1973).

The coccoliths of E. huxleyi (which are of the placolith type, i.e., radial flat elements at two levels connected by a central tube; Figures 1–4) are composed of 20 to 40 structural units consisting of an upper and a lower element together with a sector of the central connecting tube (Braarud et al., 1953). Each unit yields an electron diffraction pattern characteristic of a single crystal (Watabe, 1967). In this case, it is apparent that the cell exerts very precise control over the process of crystal growth.

In C. carterae the oval coccoliths are constructed of alternating elements of two different shapes, as was beautifully illustrated by Outka and Williams (1971). The coccoliths of other Cricosphaera species (Figure 10) are organized in a similar way (Gayral and Fresnel, 1976).

The calcium carbonate in coccoliths occurs mainly as calcite (Paasche, 1968b; Blackwelder et al., 1976). However, early experimental work by

Wilbur and Watabe (1963) indicated an influence of nitrogen nutrition on the mineral constitution of the coccoliths of *E. huxleyi*. Similarly, Blackwelder *et al.* (1976) found a pronounced influence of the magnesium content in the medium upon the aragonite/calcite ratio in the coccoliths of *C. carterae*. As noted previously by Lewin and Chow (1961), strontium had no effect on the mineral polymorph deposited, and was strongly discriminated against with regard to incorporation into coccoliths (Blackwelder *et al.*, 1976).

Micropaleontologists and biological oceanographers distinguish between warm- and cold-water forms of *E. huxleyi* (e.g., Okada and Honjo, 1973). A certain amount of temperature-induced variability was observed in experiments on this species by Watabe and Wilbur (1966), but the effect was not so great that it alone could explain the natural variability. Unless morphologically different strains occur in nature, environmental factors other than temperature may be more important than is realized (Wilbur and Watabe, 1967). The amount of carbonate deposited per cell did not vary much with temperature in experiments conducted with *E. huxleyi* (Paasche, 1968a).

In arctic waters the fraction of uncalcified coccolithophorid specimens was astonishingly high, and the absence of matrix calcification seems to be "environmentally induced" (Manton *et al.*, 1976). In fossil *Discoaster* species an evolutionary reduction of the amount of calcite deposited may be interpretable as indicating a cooling of the oceanic waters (Bukry, 1971).

"Dithecatism," in which coccolithophorids have two layers of different coccoliths, is normally encountered within, for example, the genus *Syracosphaera* (Gaarder and Heimdal, 1977; Okada and McIntyre, 1977). A more abnormal and hitherto unexplained phenomenon is the occasional presence of coccoliths of two different species (and genera) on a single cell (references in Clocchiatti, 1971).

A firm interlocking of the coccoliths in species with placoliths gives mechanical self-support and rigidity to the coccolith cover. Dixon (1900) viewed the curved profile and oval outline of placoliths as adaptations allowing interlocking and rearrangement among the external coccoliths upon the extrusion of a new one. In many species the coccoliths do not interlock, and it seems likely that they are stuck to the cell surface by an adhesive material. This material is difficult to stain and is rarely encountered in electron micrographs, where coccoliths may appear to be floating freely. Polysaccharide stains may sometimes reveal a sparse, fibrillar material between or underneath coccoliths; this may be a denatured form of the adhesive material which is then no longer in its native state of a "viscous gel."

In *Calyptrosphaera* and *Cricosphaera,* as well as in *C. pelagicus* (including the *Crystallolithus* phase), the calcified parts of the coccolith rest on a flat scale or baseplate (Figure 6) composed of organic matter. A baseplate is probably an integral part of the coccolith in most coccolithophorids, the only known exception being *E. huxleyi* (Figures 3 and 4; see also Klaveness, 1972a). The morphology and chemical composition of the baseplate are treated in Section V. In addition, most species carry one or several layers of uncalcified scales (Figures 5, 6, and 8) underneath the coccolith layer (Manton and Leedale, 1969; Klaveness, 1973; Leadbeater and Morton, 1973). These scales are usually smaller than, though otherwise very similar to, the baseplates.

V. COCCOLITH FORMATION AND CALCIFICATION

The intracellular origin of coccoliths was suggested by Dixon (1900), who sectioned and stained cells of *C. pelagicus* by classical methods. Lohmann (1902) described intracellular coccoliths in other species, and subsequent authors were also aware of coccoliths inside the cell. A further indication of the intracellular origin of coccoliths was provided by Parke and Adams (1960), Paasche (1962), and Lavine *et al.* (1962), subsequently to be corroborated by a number of fine-structure (Figure 1) and physiological investigations.

The concept of a cellular endomembrane system was introduced by Morré and Mollenhauer (1974) to visualize the integrated efforts in cellular protein–polysaccharide biosynthesis. This concept plays an important part in attempts to understand the process of coccolith formation. The coccolith is constructed of several components of different chemical composition: (1) The organic baseplate (Manton and Leedale, 1969; Pienaar, 1969a), usually found beneath the calcified structure; this organic baseplate may be lacking, as in the coccoliths of *E. huxleyi* (Klaveness, 1972a); (2) the organic matrix (Braarud *et al.,* 1953; Wilbur and Watabe, 1963), presently considered a prerequisite for controlled crystal growth in biological systems (cf. Eastoe, 1968; Towe, 1972); and (3) the mineral, calcium carbonate (Braarud *et al.,* 1953; Isenberg *et al.,* 1963; Wilbur and Watabe, 1963). The biosynthesis of coccoliths involves synthesis of the components, and the subsequent assembly and transport of the final product to the cell surface.

Since 1955 a number of marine haptophycean flagellates have been described, carrying characteristic organic surface scales (see Boney, 1970, for review). The organic scales are of species-specific morphology, consisting of a radial and/or concentric pattern of fibrils adorned with rims, spines, etc. These organic scales with rims and/or spines were considered by Manton and Leedale (1969) to be homologous to coccoliths, the organic scale corresponding to the subtending baseplate, and the rim to the calcified organic matrix of the coccolith. "On this interpretation, the fact of calcification would be regarded as a biochemical complication which is not essential to the basic scale morphology" (Manton and Leedale, 1969). This hypothesis gained strength with the subsequent discovery of cases of a close morphological parallelism between coccoliths and noncalcified haptophycean scales (Leadbeater and Morton, 1973; Thomsen, 1977), as well as of a number of weakly calcified coccolithophorid species (Tangen, 1972; Manton *et al.*, 1977).

The biosynthetic pathways and biochemical composition of organic scales in coccolithophorids have recently been investigated in detail. In accordance with the homology between coccoliths and organic scales with rims, we draw substantially upon these investigations in trying to understand the biochemistry behind the process of coccolith formation.

The organic scales of the benthic noncalcifying phase of *Pleurochrysis scherffelii* Pringsheim were investigated chemically by Brown and co-workers (Brown, 1969; Brown *et al.*, 1969, 1970; Herth *et al.*, 1972; Romanovicz and Brown, 1976). Thorough treatment with acid and alkali left behind a scale fraction consisting of radial and concentric fibrils. Further alkali treatment gave rise to short microfibrils which, following total hydrolysis, yielded glucose as the only monomer. Further confirmatory tests showed that the microfibrillar network was cellulosic. As this discovery was the first direct evidence of cellulose biosynthesis in the Golgi region of any plant species, further evidence was provided (Herth *et al.*, 1972; Romanovicz and Brown 1976).

Simultaneously, Franke and Brown (1971) investigated the scales and coccoliths of the coccolithophorid *C. carterae*. Radial and concentric fibrils were present in both the organic scales and the coccolith baseplates. The substructure of the fibrils, the occurrence of kinking sites, and the fibrillar resistance to alkaline and weak acid treatment strongly resembled that of the scales of *P. scherffelii*. Therefore, Franke and Brown (1971) hypothesized that the fibrillar network of the scales and the coccolith baseplates of both species was structurally and chemically identical, and that the fibrillar material consisted of a celluloselike polysaccharide.

The fibrillar network of organic scales and coccolith baseplates is covered by or intermingled with amorphous or nonfibrillar material (Manton and Leedale, 1969; Franke and Brown, 1971; Brown *et al.*, 1969; Herth *et al.*, 1972; Romanovicz and Brown, 1976). This material was removed by an extensive, strong-alkali treatment of scales of *Pleurochrysis,* but was included in the analysis if untreated or mildly alkali-treated scales were hydrolyzed (Brown *et al.*, 1969, 1970; Herth *et al.*, 1972). The latter procedure gave rise to a number of sugar residues, such as galactose, ribose, glucose, and arabinose, in addition to nitrogen-containing material. Herth *et al.* (1972) found cellobiose in an alkali-treated scale fraction and a number of amino acids of which serine was dominant. Up to 9% of the dry weight of the scale fraction was protein, presumably covalently linked to sugar residues as oligopeptides (Herth *et al.*, 1972). In an authoritative paper by Romanovicz and Brown (1976), *Pleurochrysis* scales were analyzed by advanced physical and chemical means. By carefully controlled serial extraction procedures it was possible to determine the composition of the various scale subcomponents: the radial microfibrils, spiral microfibrils, and amorphous polysaccharide coating. A protocol of the extraction procedure, the fraction solubilized, and the analysis of each fraction are given in Table I.

Several research groups have tried to analyze isolated coccoliths or coccolith fractions. Early work by Isenberg *et al.* (1965) demonstrated the presence of acid polysaccharides in coccoliths (including organic baseplates and uncalcified scales) of *Cricosphaera* sp. In addition, they found a number of amino acids, of which hydroxyproline was especially interesting. Previously, hydroxyproline was rarely found outside collagen or collagenlike proteins (which are important components of calcifying vertebrate matrices), but recently hydroxyproline has been shown to be an important constituent of the plant cell wall (Lamport, 1970; Loewus, 1973) and of the cell wall of algae from different classes (e.g., Gotelli and Cleland, 1968). *Cricosphaera* is reported to have a high content of hydroxyproline (Siegel and Siegel, 1973), and in *Pleurochrysis* scales the coating material of the cellulosic spiral microfibrils contains hydroxyproline as the most abundant amino acid residue (Lamport, cited in Romanovicz and Brown, 1976).

The coccoliths of *E. huxleyi* are particularly interesting subjects for analysis, since they carry no organic baseplate (Klaveness, 1972a). Westbroek *et al.* (1973) isolated soluble organic material from coccoliths of this species and found a protein-free polysaccharide containing about 26% uronic acid, some methylpentose, and no amino sugars. Further work along these lines with an EDTA-extractable polysaccharide fraction showed the existence, in addition to the uronic acids, of unidentified, more strongly acidic groups (De Jong *et al.*, 1976). The polysaccharide bound calcium ions at two different types of sites, which were tentatively identified with the uronic acid groups and the unknown acid groups, respectively. This polysaccharide, isolated from the calcified structural components of coccoliths, may be related to the matrix material operating during the process of calcification. In the first report (Westbroek *et al.*, 1973) it

Table I Serial Extraction Procedure to Determine the Chemical Composition of *Pleurochrysis* Organic Scales[a]

Fraction	Content	Method of analysis	Result	Treatment of centrifugate[b]
Scale fraction A	Untreated scales	Staining with Alcian blue at pH 2.5 and 1.0, methylation and saponification, ultrastructural localization with cationized ferritin	Pectinlike acid polysaccharides, sulfated and carboxylated, present; cationized ferritin only bound along radial microfibrils	Extract with water, pH 5, 37°C, 18 hr
Supernatant 1	Amorphous polysaccharide	Thin-layer chromatography of hydrolyzate, ninhydrin test for free amino acids, phenol–sulfuric acid test for sugars	Six percent protein, 94% carbohydrate; principal monosaccharides were galactose, glucose, and fucose	Extract with 0.5 N TFA, 37°C, 3 hr
Supernatant 2	Radial microfibrils	As above	Seven percent protein, 93% carbohydrate; principal monosaccharides were galactose, arabinose, and fucose	Extract with 2 N TFA, 121°C, 3 hr
Supernatant 3	Coating of spiral microfibrils	As above	Nine percent protein, 91% carbohydrate, of which principal monosaccharides were galactose, glucose, and mannose	Hydrolyze with 6 N HCl, 100°C, 24 hr
Supernatant 4	Spiral microfibrils	As above	Thirty-seven percent protein, 63% carbohydrate, where principal monosaccharide was glucose	

[a] From Romanovicz and Brown (1976).
[b] TFA, Trifluoroacetic acid.

was stated to be located *inside* the calcium carbonate crystals, but according to the more recent work, this assumption is unfounded (De Jong *et al.*, 1976).

The chemical evidence suggests that the various components of the coccoliths are assembled in a definite sequence, and this is supported by morphological observations and cytochemical tests. The origin of the organic scale precursors has been traced back to the endoplasmic reticulum fraction of the endomembrane system by silver methenamine staining, which with certain precautions reacts specifically with certain scale precursors (Brown *et al.*, 1973; Brown, 1974a). "Since it is well known that the RER is a site of protein synthesis, the story becomes all the more interesting in view of the fact that the RER in *Pleurochrysis* is an extension of the outer nuclear membrane and that specific topographical regions of RER could be interpreted as a manifestation of the specific genomic control of cell wall constituents" (Brown, 1974a).

By standard electron microscope techniques the scale formation in *Pleurochrysis* can be localized to Golgi vesicles immediately distal to "peculiar Golgi" vesicles (Parke *et al.*, 1959; Manton, 1967), that is, expanded polymerization centers within the Golgi (Brown *et al.*, 1970). The scale radial microfibrils are stainable and appear in a folded configuration, subsequently to become unfolded through a "Z" stage in the Golgi cisterna and then to expand into the quadriradial orientation characteristic of the mature scale (Brown *et al.*, 1973; Brown, 1974a). Pienaar's (1969b, 1971) interpretation of events in *Cricosphaera* is strongly at variance with this, as he considered that the orientation of the microfibrils depended on the development of special tubules within the Golgi vesicles. During the latter part of the formation of the radial microfibrillar part of the *Pleurochrysis* scale, a pronounced aryl sulfatase activity is present, marking the deposition of sulfated acid polysaccharides (Brown and Romanovicz, 1976).

More distally in the Golgi vesicle stack, a spiral band of five to eight parallel cellulosic microfibrils is deposited on the radial skeleton of the scale (Brown, 1974a). During this process the vesicle is connected distally to a tubule whose opposite end expands into a dilated cisternal region. This "central feeding tubule" was tentatively advanced as the site of cellulosic spiral microfibril synthesis. The spiral microfibrils are associated with a peptide rich in hydroxyproline (Brown and Romanovicz, 1976).

As seen in the electron microscope, assembly of the coccolith organic baseplate in certain coccolithophorids and of the organic scales in *Pleurochrysis* seems to take place in a very similar manner. The crucial evidence with regard to matrix formation and calcification, presented in the following discussion, is mainly morphological.

In *Cricosphaera* construction of the organic matrix, the eventual site of calcification, apparently takes place by a peculiar accretion of minute, highly stainable bodies called "coccolithosomes" (Outka and Williams, 1971). Coccolithosomes have also been reported by Manton and Leedale (1969), Pienaar (1971), and Gayral and Fresnel (1976) in species of *Cricosphaera,* and by Leadbeater (1971) in the coccolithophorid stage of *Pleurochrysis.* Coccolithosomes have a clearly resolvable granular substructure, as does the fully assembled coccolith matrix (Outka and Williams, 1971; cf. also Manton and Peterfi, 1969; Gayral and Fresnel, 1976). Coccolithosomes are found in ears or pockets of the coccolith vesicle, from which they apparently move into the vesicle and fuse into a morphologically determined shape, the matrix, along the organic baseplate rim. It is uncertain if and how the vesicle membrane influences the shape and size of the growing matrix.

The origin of coccolithosomes is not quite clear, but they are "undoubtedly a specific Golgi elaboration" (Outka and Williams, 1971). It has not been possible to decide whether they are synthesized at the ends of the same cisterna that contains the baseplate or in separate vesicles. If the latter is true, then the vesicles must eventually fuse with the base vesicles (Outka and Williams, 1971). In *Pleurochrysis,* amorphous polysaccharide originating from budding vesicles of the rough endoplasmic reticulum migrates to and fuses with the periphery of the distal Golgi vesicles. By this process the scale surface is covered by amorphous, acidic polysaccharide (Brown and Romanovicz, 1976).

In contrast with the previous species, *C. pelagicus* synthesizes each coccolith in a Golgi cisterna adjacent to the vesicle stack; the coccolith vesicle is T-shaped and the stem of the T retains its position in the stack. Only one coccolith is produced at a time. The organic matrix and the subtending baseplate of the coccolith are very thin and inconspicuous, but the radial microfibrils in the baseplate can be seen in favorable grazing sections. Calcification is initiated as a marginal deposit within the matrix material along the baseplate rim (Manton and Leedale, 1969). Coccolith formation has been more thoroughly investigated in *E. huxleyi,* with which Wilbur and Watabe (1963) pioneered modern morphological investigations of coccolithophorids. By sectioning they found the coccolith to be formed in a separate "organic matrix region". As the growth of the mineral advanced, membranous organic material appeared to enclose the centers of calcification. A reinvestigation of coccolith formation in *E. huxleyi* has extended Wilbur and Watabe's observations (Klaveness, 1972a, 1976). In this species coccolith formation takes place in a vesicle separate from the Golgi vesicle stack but possibly connected to the endoplasmic reticulum (the "reticular body" of Wilbur and Watabe, 1963; see also Klaveness and

Paasche, 1971; Klaveness, 1972a, 1976). No scales are produced by the coccolith-forming stage of this species, and no baseplate is present in the coccolith. The organic matrix material is entirely amorphous and without visible substructure. Cytochemical reactivity indicates that the matrix material is a glucoprotein and that it originates within the endoplasmic reticulum in connection with the coccolith vesicle (Klaveness, 1976; cf. also Wilbur and Watabe, 1967).

A similar reactivity is shown by the skin and coccolith matrix material of *Calyptrosphaera,* which also show a marked increase in structural stability when exposed to ruthenium red (Klaveness, 1973). Ruthenium red is known to bind specifically to adjacent negative charges 4.2 Å apart, provided steric hindrance is absent; these provisions are met by pectic acids (Sterling, 1970).

In the absence of precautions to prevent dissolution of mineral, calcified structures appear as holes (white areas) in sections of embedded material. The relationship between matrix and mineral is difficult to interpret, as the progressive dissolution of mineral is initiated immediately upon fixation of the cell. The matrix–mineral relationship is of general interest, as calcification theory prescribes the matrix to "be in existence before whatever is bred" (Eastoe, 1968). Klaveness (1973, 1976) employed reagents to prevent as much as possible the dissolution of mineral during processing. He came to the conclusion, by comparing the deposition of matrix material and mineral, that matrix formation and mineral deposition are simultaneous events in *E. huxleyi.* The initiation of crystal growth is not topographically related to a baseplate (which is lacking in this species). Each individual unit, while surrounded all the time by a complementary matrix, grows from a small rhombohedral calcite crystal into its final complicated shape.

This is different from *Cricosphaera,* where the subtending baseplate apparently predetermines size relationships before matrix and mineral deposition commence. No micrograph published to date shows a size increase of the single coccolith elements in this species, proceeding from a rhombohedral shape to the complicated alternating elements of the coccolith. Rather, coccolithosomes appear to fuse directly into the correct shape and final size of the mature element matrix along the rim of the baseplate (Outka and Williams, 1971). The question of simultaneousness with regard to matrix formation and mineralization in *Cricosphaera* is less easily resolved.

However, we believe that, in the coccolithophorids investigated so far, the matrix formation (excluding the baseplate) and mineralization are simultaneous events. Shape is determined by the coccolith vesicle membrane, functioning as a "local mold" during morphogenesis (cf. Fauré-

Fremiet *et al.*, 1968; Klaveness, 1976; see also Outka and Williams, 1971). The function of the matrix material is obscure (cf. Towe, 1972; Klaveness, 1976).

Enzymatic activities pinpointed by cytochemical techniques support the morphological and biochemical evidence. Alkaline phosphatase activity was found close to the forming face in the Golgi of *Pleurochrysis*, in connection with folded microfibril synthesis. This was interpreted as evidence for the hydrolysis of sugar phosphate monomers during polysaccharide synthesis (Brown and Romanovicz 1976). Acid phosphatase activity was also localized in the region of radial microfibril synthesis with one out of three substrates tested, but with the other substrates activity was located in the distal cisternae of the Golgi and in the subsurface cisternae.

Phosphatases are highly interesting enzymes in the field of calcification in general, as they may hydrolyze organic phosphorus compounds known to inhibit crystal growth (e.g., Simkiss, 1975).

The final product, the coccolith, is transported toward the cell surface and deposited in an ordered manner. With the exception of the technically amazing time-lapse exposures taken at the Institut für den Wissenschaftlichen Film, Göttingen, West Germany (Brown, 1974b), documentation of intracellular transport and extrusion (exocytosis) of coccoliths and scales is rare. Plasma rotation has been demonstrated in *Pleurochrysis* and *Cricosphaera* (Brown, 1974b), and appears to be the means by which coccoliths and scales are deposited evenly over the cell surface in these forms. The interpretation of the peripheral sac as the "ball-bearing" or flexible surface upon which the protoplast rotates (Figure 5; see also Brown, 1974b; Brown *et al.*, 1973) can easily be extended to a number of other haptophycean species. However, the interpretation is not always clear-cut, as the single peripheral sac of *Pleurochrysis* may in other species be represented by more than one cavity (Klaveness, 1973) and appears to be lacking in yet other species.

Vesicles or organelles in *Cricosphaera* species termed "intracellular coccolith precursors" and believed to be involved in coccolith formation (Isenberg *et al.*, 1966; Pienaar, 1969b) or, alternatively, "lamellar bodies" of unknown function (Olson *et al.*, 1967), were later interpreted as autophagic vesicles (Pienaar, 1971; Outka and Williams, 1971). This has recently been confirmed in *Pleurochrysis* by Brown and Romanovicz (1976). When scale exocytosis was prevented by treatment with colchicine, the scales were transported to autophagic vesicles via the lumen of the subsurface cisternal sac. In *Cricosphaera*, scale exocytosis was blocked by cobalt ion, resulting in an accumulation of scales in vesicles (Blankenship and Wilbur, 1975).

Cell wall formation is intimately connected with cytokinesis in many algae, and there is some indication in *Pleurochrysis* that the production of organic scales is greater just after cytoplasmic cleavage than at other times (Brown, 1969). In view of the marked dependence of calcification on light (see below), it seems unlikely that the same is true of calcified coccoliths. There is some indirect evidence from *E. huxleyi* that the cells are able to cover up newly formed cell surfaces by the rearrangement of coccoliths already on the outside; despite the very low rate of dark calcification (Paasche, 1966), cell division in the dark results in cells that are completely covered by coccoliths (Paasche, 1967). Pienaar (1976a) has published some observations on *Cricosphaera* that may be relevant.

Under culture conditions coccolithophorids tend to lose their calcifying capacity. Uncalcified coccoliths are produced consisting of the organic matrix envelope only, or entirely "naked" cells emerge (Mjaaland, 1956; Paasche, 1963; Manton and Leedale, 1963; Manton and Peterfi, 1969; Klaveness, 1973; Blankley, 1976). In *E. huxleyi,* the noncalcifying (N) cell type was shown to produce a coccolith-forming vesicle (minus the coccolith), as well as associated parts of the endoplasmic reticulum (Klaveness and Paasche, 1971); this suggests that blockage of calcification occurs fairly close to the step in which calcium carbonate and the matrix are formed. One interesting aspect of the loss of calcifying ability is that it seems to be a gradual process, so that it is possible to grow cultures in which each cell carries only one or a few coccoliths. The loss of coccolith calcification may be a response to the unnaturally high nutrient concentrations in standard growth media. Low-nutrient media are preferable for maintaining a continuous, high rate of calcification (R. Guillard, personal communication). In at least some naked strains of *E. huxleyi,* calcification is resumed in phosphorus-deficient media, although the resulting coccoliths may be of abnormal shape (O. K. Andersen, personal communication). It seems likely at present that nitrogen nutrition affects coccolith formation in cultures mainly in an indirect way via transitions between the different stages in coccolithophorid life cycles (see Section III).

A very striking feature of calcification in coccolithophorids is its strong dependence on light. The literature on this subject was reviewed by Wilbur and Watabe (1967), Paasche (1968b), and Darley (1973). More recently, the effect of light was again confirmed by Dorigan and Wilbur (1973). Since both photosynthesis and calcium carbonate deposition require an influx of inorganic carbon from the surrounding water into the cell, a close interdependence of these two processes has been visualized by several workers. Bicarbonate utilization in photosynthesis in aquatic plants frequently leads to carbonate precipitation (Raven, 1970). In many cases, including calcifying marine algae, this calcification is probably incidental to the release of hydroxyl ions at the cell surface. In an analogous way, the deposition of carbonate in coccoliths could be viewed as an intracellular mechanism for neutralizing hydroxyl ions arising from a bicarbonate-based photosynthetic carbon assimilation. The evidence for and against this interpretation has already been discussed in the above-mentioned reviews and can be very briefly summarized as follows.

Calcification in coccolithophorids is catalyzed by light absorbed by chloroplast pigments and is maintained at a high rate only as long as the cells are illuminated. In *E. huxleyi*, there is some evidence that bicarbonate ions are a source of carbon in photosynthesis as well as in calcification, and the stoichiometry of the two processes sometimes agrees with the sum of the following two equations.

$$HCO_3^- \rightarrow CO_2 \text{ (used in photosynthesis)} + OH^- \tag{1}$$
$$HCO_3^- + OH^- \rightarrow CO_3^{2-} \text{ (used in calcification)} + H_2O \tag{2}$$

However, this simple picture is complicated by a number of further observations:

1. Photosynthesis continues at a normal rate even if calcification is blocked by a lack of calcium in the medium (Paasche, 1964).

2. There is no evidence that the supply of inorganic carbon to the chloroplast follows a different route in naked strains in which the ability to calcify has been permanently lost. In the N cell type of *E. huxleyi*, the kinetics of photosynthesis was shown to be essentially the same as in the calcium carbonate-producing C cell type, although the loss of calcification was apparently accompanied by a slight reduction in the size of the chloroplast (Paasche and Klaveness, 1970).

3. The light saturation kinetics of carbonate precipitation is different from that of photosynthesis (Paasche, 1964, 1969; Blankley, 1971).

4. In extreme cases, several times more carbon is deposited in the coccoliths than is assimilated in photosynthesis (Paasche, 1969).

5. Calcified coccoliths are formed in the dark at least in some strains and/or types of medium (Paasche, 1966; Blankley, 1971; Ariovich and Pienaar, personal communication).

6. Light-dependent coccolith formation continues in the presence of chlorophenyldimethylurea (CMU) which effectively blocks photosynthetic carbon assimilation (Paasche, 1964, 1965).

Not all these observations have been confirmed in other laboratories. Crenshaw (1964), for example, found that dichlorophenyldimethylurea (DCMU) inhibited calcification and photosynthesis to an equal extent, and Dorigan and Wilbur (1973) were unable to demonstrate calcification in the dark.

It seems likely that the intracellular deposition of calcium carbonate is stimulated by simultaneous photosynthesis, which tends to raise the pH in the cell regardless of whether carbon dioxide or bicarbonate is the carbon source. On the other hand, it is clear that not all the carbonate precipitated in coccoliths is produced in response to a need for neutralizing hydroxyl ions. An alternative explanation for the effect of light is that there is an energy requirement in calcification that can be satisfied by ATP from photosynthetic phosphorylation. This hypothesis, which was proposed 15 years ago (Paasche, 1964, 1965) but still has not been put to a critical test, is not out of step with present-day concepts of the role of chloroplasts as cellular ATP factories (Raven, 1976). Possibly this need for ATP arises in connection with calcium ion transport to the calcifying sites; Klaveness (1976) presented tentative evidence for ATPase activity at the boundary membrane of the coccolith vesicle in $E.$ $huxleyi.$ Calcification in *Cricosphaera* is inhibited by oligomycin, an ATPase inhibitor (Dorigan and Wilbur, 1973). It may be significant that high activities of a calcium-activated ATPase have been reported in a calcareous red alga (Okazaki, 1977). Light-dependent transport systems for bicarbonate and hydroxyl ions have been identified in green algae, but they appear to be linked to photosynthesis in ways other than via photosynthetic phosphorylation (Raven, 1970; MacRobbie, 1973).

Although it is very clear that coccolith calcification is intimately connected with chloroplast reactions, the question of how this interaction takes place is essentially unresolved. The lack of experimental work on light-dependent coccolith calcification in recent years is deplorable, especially in view of the great advances in our knowledge of the biosynthesis and assembly of the organic components of coccoliths.

ACKNOWLEDGMENTS

The permission of Mrs. Astrid Dick and Mrs. Berit Heimdal to use Figure 2 is gratefully acknowledged. Figures 3, 4, and 5–7 have been reproduced with permission from *Protistologica* and *Norw. J. Bot.*, respectively.

REFERENCES

Black, J. (1963). *Proc. Linn. Soc. London* **174**, 41.
Blackwelder, P. L., Weiss, R. E., and Wilbur, K. M. (1976). *Mar. Biol.* **34**, 11.
Blankenship, J. L., and Wilbur, K. M. (1975). *J. Phycol.* **11**, 211.
Blankley, W. F. (1971). Ph.D. Thesis, Univ. of California, San Diego.
Blankley, W. F. (1976). Am. Soc. Limnol. Oceanogr. Congr., Ga. Abstract.

Boney, A. D. (1970). *Oceanogr. Mar. Biol.* **8**, 251.
Braarud, T., Ringdal Gaarder, K., Markali, J., and Nordli, E. (1953). *Nytt Mag. Bot.* **1**, 129.
Brown, R. M. (1969). *J. Cell Biol.* **41**, 109.
Brown, R. M. (1974a). *Port. Acta Biol., Ser. A* **14**(1/2), 369–384.
Brown, R. M. (1974b). Wiss. Film C 1071. Inst. Wiss. Film, Göttingen.
Brown, R. M., and Franke, W. W. (1971). *Planta* **96**, 354.
Brown, R. M., and Romanovicz, D. K. (1976). *Appl. Polym. Symp.* **28**, 537.
Brown, R. M., Franke, W. W., Kleinig, H., Falk, H., and Sitte, P. (1969). *Science* **166**, 894.
Brown, R. M., Franke, W. W., Kleinig, H., Falk, H., and Sitte, P. (1970). *J. Cell Biol.* **45**, 246.
Brown, R. M., Herth, W., Franke, W. W., and Romanovicz, D. K. (1973). *In* "Biogenesis of Plant Cell Wall Polysaccharides" (F. Loewus, ed.), p. 207. Academic Press, New York.
Bukry, D. (1971). *Micropaleontology* **17**, 43.
Carasso, N., and Favard, P. (1966). *J. Microsc. (Paris)* **5**, 759.
Clocchiatti, M. (1971). *C. R. Acad. Sci., Ser. D.* **273**, 318.
Crenshaw, M. A. (1964). Ph.D. Thesis, Duke Univ., Durham, North Carolina.
Darley, W. J. (1973). *In* "Algal Physiology and Biochemistry" (W. D. P. Stewart, ed.), p. 655. Blackwell, Oxford.
De Jong, E. W., Lendert, B., and Westbroek, P. (1976). *Eur. J. Biochem.* **70**, 611.
Dixon, H. H. (1900). *Proc. R. Soc. London* **66**, 305.
Dodge, J. D. (1973). "The Fine Structure of Algal Cells." Academic Press, New York.
Dorigan, J. L., and Wilbur, K. M. (1973). *J. Phycol.* **9**, 450.
Eastoe, J. E. (1968). *Calcif. Tissue Res.* **2**, 1.
Faure-Fremiet, E., André, J., and Ganier, M. C. (1968). *J. Microsc. (Paris)* **7**, 693.
Franke, W. W., and Brown, R. M. (1971). *Arch. Mikrobiol.* **77**, 12.
Gaarder, K. R. (1962). *Nytt Mag. Bot.* **10**, 35.
Gaarder, K. R., and Heimdal, B. R. (1977). *"Meteor" Forschungsergeb., Reihe D* **24**, 54.
Gayral, P., and Fresnel, J. (1976). *Phycologia* **15**, 339.
Gayral, P., Haas, C., and Lepailleur, H. (1972). *Soc. Bot. Fr., Mem. 1972,* p. 215.
Gotelli, I. B., and Cleland, R. (1968). *Am. J. Bot.* **55**, 907.
Halldal, P., and Markali, J. (1954). *Nytt Mag. Bot.* **2**, 117.
Halldal, P., and Markali, J. (1955). *Avh. Nor. Vidensk-Akad. Oslo, Mat.-Naturvidensk Kl.* No. 1.
Herth, W., Franke, W. W., Stadtler, J., Bittiger, H., Keilich, G., and Brown, R. M. (1972). *Planta* **105**, 79.
Hibberd, D. J. (1976). *Bot. J. Linn. Soc.* **72**, 55.
Isenberg, H. D., Lavine, L. S., Moss, M. L., Kupferstein, D., and Lear, P. E. (1963). *Ann. N.Y. Acad Sci.* **109**, 49.
Isenberg, H. D., Lavine, L. S., Mandell, C., and Weissfellner, H. (1965). *Nature (London)* **206**, 1153.
Isenberg, H. D., Douglas, S. D., Lavine, L. S., Spicer, S. S., and Weissfellner, H. (1966). *Ann. N.Y. Acad. Sci.* **136**, 155.
Klaveness, D. (1971). Cand. Real. Thesis, Univ. of Oslo, Oslo.
Klaveness, D. (1972a). *Protistologica* **8**, 335.
Klaveness, D. (1972b). *Br. Phycol. J.* **7**, 309.
Klaveness, D. (1973). *Norw. J. Bot.* **20**, 151.
Klaveness, D. (1976). *Protistologica* **12**, 217.
Klaveness, D., and Paasche, E. (1971). *Arch. Mikrobiol.* **75**, 382.
Lamport, D. T. A. (1970). *Annu. Rev. Plant Physiol.* **21**, 235.
Lavine, L. S., Isenberg, H. D., and Moss, M. L. (1962). *Nature (London)* **196**, 78.

Leadbeater, B. S. C. (1970). *Br. Phycol. J.* **5**, 57.
Leadbeater, B. S. C. (1971). *Ann. Bot. (London)* **35**, 429.
Leadbeater, B. S. C., and Morton, C. (1973). *Nova Hedwigia Z. Kryptogamenkd.* **24**, 207.
Lefort, F. (1971). *C. R. Acad. Sci., Ser. D* **272**, 2540.
Lefort, F. (1975). *Cah. Biol. Mar.* **16**, 213.
Lewin, R. A., and Chow, T. J. (1961). *Plant Cell Physiol.* **2**, 203.
Loewus, F., ed. (1973). "Biogenesis of Plant Cell Wall Polysaccharides." Academic Press, New York.
Lohmann, H. (1902). *Arch. Protistenkd.* **1**, 89.
MacRobbie, E. A. C. (1973). *In* "Algal Physiology and Biochemistry" (W. D. P. Stewart, ed.), p. 676. Blackwell, Oxford.
Manton, F. (1967). *J. Cell Sci.* **2**, 265.
Manton, I., and Leedale, G. F. (1963). *Arch. Mikrobiol.* **47**, 115.
Manton, I., and Leedale, G. F. (1969). *J. Mar. Biol. Assoc. U.K.* **49**, 1.
Manton, I., and Peterfi, L. S. (1969). *Proc. R. Soc. London, Ser. B* **172**, 1.
Manton, I., Sutherland, J., and Oates, K. (1976). *Proc. R. Soc. London, Ser. B* **194**, 179.
Manton, I., Sutherland, J., and Oates, K. (1977). *Proc. R. Soc. London, Ser. B* **197**, 145.
Massalski, A., and Leedale, G. F. (1969). *Br. Phycol. J.* **4**, 159.
Mills, J. T. (1975). *J. Phycol.* **11**, 149.
Mjaaland, G. (1956). *Oikos* **7**, 251.
Morré, D. J., and Mollenhauer, H. H. (1974). *In* "Dynamic Aspects of Plant Ultrastructure" (A. W. Robard, ed.), p. 84. McGraw-Hill, New York.
Okada, Ч., and Honjo, S. (1973). *Deep-Sea Res.* **20**, 355.
Okada, H., and McIntyre, A. (1977). *Micropaleontology* **23**, 1.
Okazaki, M. (1977). *Bot. Mar.* **20**, 347.
Olson, R. A., Jennings, W. H., and Allen, M. B. (1967). *J. Cell Physiol.* **70**, 133.
Outka, D. E., and Williams, D. C. (1971). *J. Protozool.* **18**, 285.
Paasche, E. (1962). *Nature (London)* **193**, 1094.
Paasche, E. (1963). *Physiol. Plant.* **16**, 186.
Paasche, E. (1964). *Physiol. Plant., Suppl.* **3**.
Paasche, E. (1965). *Physiol. Plant.* **18**, 138.
Paasche, E. (1966). *Physiol. Plant.* **19**, 271.
Paasche, E. (1967). *Physiol. Plant.* **20**, 946.
Paasche, E. (1968a). *Limnol. Oceanogr.* **13**, 178.
Paasche, E. (1968b). *Annu. Rev. Microbiol.* **22**, 71.
Paasche, E. (1969). *Arch. Mikrobiol.* **67**, 199.
Paasche, E., and Klaveness, D. (1970). *Arch. Mikrobiol.* **73**, 143.
Parke, M. (1961). *Br. Phycol. Bull.* **2**, 47.
Parke, M., and Adams, I. (1960). *J. Mar. Biol. Assoc. U.K.* **39**, 263.
Parke, M., and Green, J. (1976). *J. Mar. Biol. Assoc. U.K.* **56**, 551.
Parke, M., Manton, I., and Clarke, B. (1959). *J. Mar. Biol. Assoc. U.K.* **38**, 169.
Pienaar, R. N. (1969a). *J. Cell Sci.* **4**, 561.
Pienaar, R. N. (1969b). *J. Phycol.* **5**, 321.
Pienaar, R. N. (1971). *Protoplasma* **73**, 217.
Pienaar, R. N. (1976a). *In* "The Mechanisms of Mineralization in the Invertebrates and Plants" (N. Watabe and K. J. Wilbur, eds.), p. 203. Univ. of South Carolina Press, Columbia.
Pienaar, R. N. (1976b). *J. Mar. Biol. Assoc. U.K.* **56**, 1.
Raven, J. A. (1970). *Biol. Rev. Cambridge Philos. Soc.* **45**, 167.

Raven, J. A. (1976). *In* "Topics in Photosynthesis. Vol. 1: The Intact Chloroplast" (J. Barber, ed.), p. 403. Elsevier, Amsterdam.

Rayns, D. G. (1962). *J. Mar. Biol. Assoc. U.K.* **42,** 481.

Romanovicz, D. K., and Brown, R. M. (1976). *Appl. Polym. Symp.* **28,** 587.

Schwarz, E. (1932). *Arch. Protistenkd.* **77,** 434.

Siegel, B. Z., and Siegel, S. M. (1973). *Crit. Rev. Microbiol.* **3,** 1.

Simkiss, K. (1975). "Bone and Biomineralization," Studies in Biology, No. 53. Arnold, London.

Sterling, C. (1970). *Am. J. Bot.* **57,** 172.

Tangen, K. (1972). *Norw. J. Bot.* **19,** 171.

Thomsen, H. A. (1977). *Bot. Not.* **130,** 147.

Towe, K. M. (1972). *Biominer. Res. Rep.* **4,** 1.

von Stosch, H. A. (1967). *In* "Handbuch des Pflanzenphysiologie" (W. Ruhland, ed.), Vol. 18, p. 646. Springer-Verlag, Berlin and New York.

Watabe, N. (1967). *Calcif. Tissue Res.* **1,** 114.

Watabe, N., and Wilbur, K. M. (1966). *Limnol. Oceanogr.* **11,** 567.

Westbroek, P., de Jong, E. W., Dam, W., and Bosch, L. (1973). *Calcif. Tissue Res.* **12,** 227.

Wilbur, K. M., and Watabe, N. (1963). *Ann. N.Y. Acad. Sci.* **109,** 82.

Wilbur, K. M., and Watabe, N. (1967). *Stud. Trop. Oceanogr.* **5,** 133.

Small Amebas and Ameboflagellates

8

FREDERICK L. SCHUSTER

BIOCHEMISTRY AND PHYSIOLOGY OF PROTOZOA
SECOND EDITION, VOL. 1

I am pleased to dedicate this chapter to Dr. William Balamuth, whose interests in pathogenic and free-living amebas stimulated many of his students to explore this area of protozoology.

I. INTRODUCTION

The size of an organism is a relative property and a poor way to delimit a group of protozoa. Characterizing the forms treated in this chapter as small amebas leads to an inexact definition of the group. These organisms are often referred to as soil amebas, another descriptive term that tends to mislead. While found in soil, they also occur in aqueous habitats, as tissue culture contaminants, and as pathogens or potential pathogens in animal organisms. The group is diverse not only in regard to kinds of amebas, but also in regard to the ecological niches occupied by its individual members. We deal here with organisms that are (1) readily recovered from soil environments, (2) pygmies compared to ameboid giants such as *Amoeba* and *Pelomyxa,* (3) capable of morphogenetic transformation into a dormant cyst stage in all cases and, in some cases, a flagellate stage, (4) free-living but able, in several species, to develop as lethal pathogens when the opportunity presents itself and, finally, (5) difficult to define taxonomically because of a paucity of morphological features.

Problems involving their identification and classification have kept these forms in relative obscurity, except for studies on soil ecology and differentiation. Attention shifted dramatically to this group over the past decade when some of these organisms were recognized as human pathogens. The accompanying notoriety, along with their utility in biochemical and developmental studies, have earned the small amebas a niche of their very own, in contrast to their having shared a chapter with *Entamoeba* in the first edition of this series (Balamuth and Thompson, 1955). Excluded from consideration in this chapter are the cellular and true slime molds. The former are dealt with elsewhere in this series, and the latter are separated from the small amebas by a complex life cycle and other distinguishing features (Gray and Alexopoulos, 1968; Olive, 1975). The parasitic amebas, *Entamoeba* and its allies, are not treated in this chapter, nor are the various small amebas from marine habitats. Structural and functional aspects of small ameba membranes are treated in detail in a later volume.

A. Scope of the Group

Small amebas as a group have come to include the *Acanthamoeba–Hartmannella* complex as representative ameboid forms. These organisms are widely found in soil and in both freshwater and marine habitats, either as trophic amebas or as cysts. *Naegleria* is another genus included in this group. It too has a cosmopolitan distribution in soil and freshwater habitats. In addition to the ameboid and cyst stages, *Naegleria* amebas have the capacity to differentiate into flagellates upon appropriate stimulation (Fulton, 1977a), a feature which sets them apart from the *Acanthamoeba–Hartmannella* grouping.

Other forms—some amebas and some ameboflagellates—are encountered which fit into the group, including *Tetramitus, Vahlkampfia,* and *Adelphamoeba;* these are all members of the family Vahlkampfiidae as defined by Page (1967a). This listing does not include some additional genera recognized by Page (1974a) and by Singh (1952, 1975) in their respective taxonomic treatments of the small amebas. Thus, the selection of organisms treated here is somewhat arbitrary and restricted to those which are readily obtainable from soil, water, or established culture collections and, more to the point, those which are actively used in research.

B. Life Cycles

Already alluded to, life cycles are a means of distinguishing among the soil amebas. For the *Acanthamoeba–Hartmannella* complex, the cycle is

$$\text{ameba} \rightleftharpoons \text{cyst}$$

a simple alternation between a trophic or feeding ameba and a dormant cyst. An added dimension is found in the ameboflagellate life cycle:

$$\text{flagellate} \rightleftharpoons \text{ameba} \rightleftharpoons \text{cyst}$$

with the flagellate stage capable of feeding and dividing as in *Tetramitus* (Rafalko, 1951), or only dividing as in *Tetramastigamoeba* (Singh and Hanumaiah, 1977a), or neither feeding or dividing as in *Naegleria* (Page, 1967a; Rafalko, 1947). Also, in *Tetramitus,* reversion from the flagellate to the ameboid stage may not occur (Outka, 1965), and in some strains and/or species of ameboflagellates, flagellation may be difficult to induce (Darbyshire *et al.,* 1976). The presence of these life cycles with their alternative stages permits critical analysis of events involved in cellular differentiation, an experimental approach that has been exploited with success in both *Acanthamoeba–Hartmannella* in the study of encystment and excystment of the ameba (Weisman, 1976) and in *Naegleria* and *Tet-*

ramitus with their flagellation and reversion to the ameboid stage (Fulton, 1977a).

C. Taxonomic Criteria

Because of a scarcity of morphological detail, the taxonomy of the small amebas has had its share of difficulties leading to, for biologists working with these amebas, more than its share of confusion. The problems of taxonomy of these amebas are rooted, however, in contradictory, imprecise, or poor descriptions firmly established in the literature. Initial descriptions were often made on organisms grown in richly organic media with undefined or indigenous bacteria as a food source, a circumstance that can lead to pleomorphy of the organisms. One suspects, too, that descriptions were not based on clonal isolates from nature, another source of difficulty in sorting out species. Chiefly because of the efforts of Singh (1952) and Page (1967a,b), who have meticulously defined conditions of growth for the organisms they have described or redescribed, a better understanding of the taxonomy of the group is now at hand, as is an understanding of the phylogenetic relationships among these small amebas. The bases for most taxonomic treatments of these organisms are manner of movement of the trophozite, shape of the pseudopods, pattern of nuclear division, morphology of the cyst as well as manner of excystment of the amebas and, in genera where flagellate formation occurs, morphology of the flagellate stage. These criteria have been particularly useful in distinguishing among genera, though less useful at the species level.

Meticulous descriptions notwithstanding, the patience of the novice and even the expert is often greatly tried when one is called upon to identify small amebas. Nuclear division, a major factor in study of the group, is difficult to observe, and hundreds of amebas must be examined before the events of mitosis can be reconstructed. In addition, some of the staining techniques used for metazoan cells and other protozoan organisms give only mediocre results with small amebas. Feulgen staining is, of course, ideal for the study of mitotic figures but may require modification to accentuate the appearance of chromatin in many of these species (Rafalko, 1946). Other stains, however, have been successfully employed (Fulton and Guerrini, 1969). Amebas have also been subjected to a program of heat shocks to increase the proportion of dividers in the population (Fulton and Guerrini, 1969; Napolitano and Smith, 1975). Even the terminology of the dividing nucleus is confusing, since the components emphasized by some researchers in division figures (e.g., the so-called interzonal body) are thought by other workers to be of limited value in describing genera (Page, 1967a; Singh and Das, 1970). Some of the more important genera are compared in Table I on the basis of morphologies of the trophic ameba and cyst, pattern of mitosis, and several other features of use in distinguishing genera from one another. These descriptions are not definitive; the continued examination of extant stocks and the isolation of new strains and species has led and will doubtlessly continue to lead to additional changes in the systematics of the group.

Generic nomenclature used in this chapter generally follows that proposed by Page (1967a,b) and is reflected in the designations used in Table I. Name substitutions have been made where such changes are in accordance with current usage. Thus, *Hartmannella culbertsoni* has been changed to *Acanthamoeba culbertsoni*. In keeping with Page's outline, *Schizopyrenus russelli* has been changed to *Vahlkampfia russelli*.

A number of previously distinct species (*Acanthamoeba* sp., *Hartmannella* sp., *H. rhysodes,* and *Mayorella palestinensis*) referred to in this chapter are now regarded as strains of *Acanthamoeba castellanii* (Page, 1967b), having previously passed through a period as *Hartmannella castellanii* (Adam, 1964b). Strain differences do exist, however, and where necessary for identification or clarity, such differences are noted in the text. The organism once known as *Hartmannella rhysodes* is now regarded as the HR strain of *A. castellanii* (Band and Mohrlok, 1973b). *Mayorella palestinensis* is considered by some to be the MP strain of *A. castellanii* (Band and Mohrlok, 1973b) but, based on what appear to be unique morphological features, others regard it as a distinct species, *Acanthamoeba palestinensis* (Lasman, 1977). *Hartmannella glebae* is referred to in the text as *Acanthamoeba glebae,* though Adam (1964b) considers it a strain of *A. castellanii.* Recent attempts have been made to distinguish among representative species of *Acanthamoeba* by use of serological analyses (Stevens *et al.,* 1977a) and physiological features (Griffiths *et al.,* 1978).

What features should be looked for in the isolation of amebas from soil or water samples? There is no general agreement, even among experts, about what features should have prime importance. Some place emphasis on patterns of movement and pseudopodial types, others rely upon the mitotic apparatus, and still others use a complex of features. For the novice faced with an unknown ameba from nature, the following guidelines are useful for distinguishing amebas: (1) pattern of motion and type of pseudopods formed (separating *Acanthamoeba* from *Hartmannella* and the ameboflagellates), (2) presence of a flagellate stage (separating *Naegleria* and other ameboflagellates from all other genera), and (3) appearance of the cyst (separating amebas at both the generic and specific levels). These traits provide a relatively simple means for identification and preliminary screening without resorting to a tedious search for mitotic figures in living or stained material, a necessary task but of secondary impor-

Table 1 Comparative Features of Common Genera of Soil Amebas

Feature	Family Acanthamoebidae	Family Hartmannellidae	Family Vahlkampfiidae		
	Acanthamoeba	*Hartmannella*	*Naegleria*	*Tetramitus*	*Vahlkampfia*
Pattern of locomotion	Noneruptive; ectoplasmic pseudopod	Noneruptive; monopodial; constant	Eruptive; monopodial; sinuous	Eruptive; monopodial	Eruptive; monopodial; straight line or sinuous
Pseudopodial pattern	Filamentous acanthopodia; broad hyaline lobopodium	Broad anterior hyaline pseudopod	Anterior hyaline pseudopod	Anterior hyaline pseudopod	Anterior hyaline pseudopod
Nuclear type	Vesicular; central Feulgen-negative nucleolus	Vesicular	Vesicular	Vesicular	Vesicular
Mitotic pattern	Nucleolus and nuclear envelope disintegrate; centriolelike structures reported (metamitosis)	Nucleolus disintegrates; nuclear envelope persists to metaphase–anaphase; no centrioles (mesomitosis)	Nucleolus and nuclear envelope persist; no centrioles (promitosis)	Promitosis; mitosis in ameba and flagellate stages	Promitosis
Cyst structure	Double-walled; with pores and plugs; wall remains after excystment	Smooth; no pores; cyst stage may not form in some cultured species	Smooth, rough, and angular types; double-walled; with pores and plugs; wall remains after excystment	Smooth; thin wall; no pores; wall digested during excystment	Smooth; thin wall; no pores; wall partially digested during excystment
Flagellate stage	None	None	Generally with two to four flagella; temporary	With four flagella; cytostome; divides	None
Sources of description and comments	Based on Page (1967b), Pussard (1973), and Sawyer and Griffin (1975); genus not recognized by Singh (1952, 1975) or Singh and Das (1970)	Based on Page (1967a); imprecisely defined in literature and includes forms generally regarded as *Acanthamoeba* spp. (cf. Singh and Das, 1970)	Based on Page (1967a, 1975); includes *Didascalus* of Singh (1952)	Based on Rafalko (1951)	Based on Page (1967a) equivalent to *Schizopyrenus* of Singh (1952)

tance. Page (1976) has compiled an illustrated guide to the amebas that can be of invaluable assistance in distinguishing different amebas to the species level. Finally, a cautionary note about identifications. Many of these features exhibited by amebas tend to vary, particularly with culture conditions. Pattern of movement, type of pseudopod, or size and shape of the cyst, for example, will be influenced to some extent by whether the amebas are growing axenically or in the presence of bacteria, and whether they are growing in a fluid medium or upon an agar surface. The ease of flagellate induction is another variable; some strains of ameboflagellates produce flagellates readily, others readily but erratically, and still others only with great difficulty. Again, culture conditions and, in this case, age of the cultures, may be responsible for variability or recalcitrance.

Considerable success has already been achieved in species and strain characterization through the use of biochemical and immunochemical techniques. These applications are still novel in this area of protozoology but hold promise for ameba identifications based on their proteins. As methods improve for mass axenic cultivation of amebas, DNA composition may yet prove to be a valuable tool establishing the systematics of the small amebas.

D. Phylogeny

Evidence is still too fragmented to allow a reconstructed phylogeny of the small amebas. It is safe to say that the assemblage is polyphyletic, with representative types having originated from different ancestral lineages. The presence of a flagellate stage—transitory in the case of *Naegleria* and permanent or semipermanent in *Tetramitus*—is supportive of a flagellate ancestry for these amebas. Phagotrophy, well-developed in flagellates, does not necessarily require the presence of a cytostome. A number of chrysomonads (e.g., *Ochromonas*) are able to ingest particulate food by vacuole formation, as do amebas, in spite of the lack of a structurally distinct oral apparatus. Furthermore, in their flagellate stage some amebas (e.g., *Tetramitus*) possess a cytostome. The presence of photosynthetic pigments in flagellate stocks and their absence in amebas is no obstacle, since the loss of pigmentation has occurred in numerous flagellate lineages. Presumably, ameboflagellates represent forms which have retained the ability to revert to this ancestral stage, giving rise to plastic flagellates (as in *Naegleria*) or flagellates with a highly organized cytoskeletal framework and a cytostome (as in *Tetramitus* and *Paratetramitus*). This ability to flagellate is also found in representatives of the slime molds (Olive, 1975). There is even less to say concerning the ancestry of *Acanthamoeba* and *Hartmannella*. Their mitotic patterns suggest an origin from a flagellate stock different from the one that gave rise to the ameboflagellates.

II. THE TROPHIC AMEBA

For most of the genera covered in this chapter, the ameboid stage is the sole reproductive and feeding phase of the life cycle. Examination of these amebas with either a light or electron microscope reveals a disappointing sameness among all the genera, in regard to the various types of organelles present and the general organization of cyto- and nucleoplasm.

A. Nuclear and Cytoplasmic Morphology

In all these forms, the nucleus is the most readily visible structure present in the ameba. Described as a vesicular nucleus, the name reflects its bubblelike appearance (Plate I, Figures 1 and 2). Located centrally is a large Feulgen-negative nucleolus or endosome; the latter term has been used to distinguish a nucleolus that does not disintegrate but remains more or less intact and visible throughout karyokinesis (Page, 1967a; Pussard, 1973). A typical nuclear envelope is present around the structure. At the light microscope level, the outer face of the envelope is decorated with refractile spheres which are infrequently observed in the electron microscope. These spheres are apparently of lipoid material, as demonstrated by Pittam (1963), and are probably dissolved during the preparation of cells for electron microscopy. Chromatin, present in the nucleoplasm between the nucleolus and nuclear envelope, cannot be reliably identified at either the light or electron microscope level, but can be distinguished by Feulgen staining (Page, 1967a; Rafalko, 1947).

Plate I: Figure 1. *Naegleria gruberi* trophozoites from an axenic culture. A characteristic ectoplasmic pseudopod is evident in the ameba on the left; both amebas exhibit filaments at the posterior (uroid) end. The arrow points to a vesicular nucleus. Interference-contrast. Magnification: ×2000.

Figure 2. Cysts of *N. gruberi* from an axenic culture. The arrow points to a pore in a cyst wall. Interference-contrast. Magnification: ×3200.

Figure 3. Trophozoites from a bacterized culture of *Naegleria* as seen *in situ* on an agar surface. The amebas form a characteristic sheet of cells. The arrow indicates one of several contractile vacuoles seen in the amebas. Phase-contrast. Magnification: ×1000.

Figure 4. *Acanthamoeba* trophozoites from a bacterized culture photographed *in situ* on an agar surface. A contractile vacuole (arrow) is evident as a clear bubble seen in some of the cells. Nuclei are also evident. Interference-contrast. Magnification: ×1000.

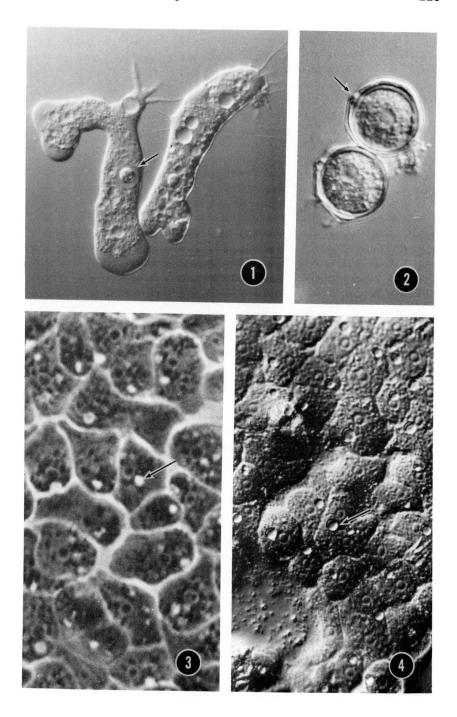

Patterns of nuclear division in the small amebas have been described as promitotic, mesomitotic, or metamitotic by Pussard (1973) in an updating of Chatton's (1953) groupings. Promitosis is characterized by the persistence of the nucleolus and nuclear envelope throughout karyokinesis, and is found in *Naegleria* and *Tetramitus*. The nucleolus forms a dumbbell-shaped structure which ultimately pulls apart to give two daughter nucleoli. In mesomitosis (*Hartmannella*), the nuclear envelope persists until the beginning of anaphase, but the nucleolus disintegrates. Metamitosis, which is found in *Acanthamoeba*, is characterized by early breakdown of the nuclear envelope (by prophase or metaphase), as well as disintegration of the nucleolus. This latter mitotic pattern is similar to what is seen in dividing cells of higher animals. Centrioles are not present in *Naegleria*, either as kinetosome precursors (Fulton and Dingle, 1971) or during mitosis (Schuster, 1975). *Acanthamoeba*, however, possesses a structure which has a centriolelike appearance at the electron microscope level; it is electron-dense, and microtubules are seen radiating outward from the structure (Bowers and Korn, 1968; Willaert *et al.*, 1978b). Mitotic figures have not been reported ultrastructurally, so the functional relationship of this organelle to the division figure is unknown; Pussard (1972, 1973), however, has reported centrioles at the poles of mitotic figures in *Acanthamoeba*. An unusual amitotic division can be induced in *A. castellanii*, strain HR, though the division products are nonviable (Band and Mohrlok, 1973a,b).

Cytoplasm tends to have a sparse appearance in these amebas, with minimal amounts of endoplasmic reticulum. These vesicles show no tendency to concentrate in any particular region of the cell. A Golgi complex is present in some of the amebas (*Acanthamoeba*), but is lacking in other genera. In amebas that lack an organized Golgi apparatus, a functional equivalent may be the vesicles of smooth endoplasmic reticulum scattered about the cytoplasm. Mitochondria represent another major type of or-

---

Plate II: Comparisons of mitochondria from trophic and cyst stages of soil amebas. Magnification: ×35,000.

Figure 5. Mitochondria from a trophic *Acanthamoeba*, showing tubular cristae.

Figure 6. Mitochondrion from an encysted *Acanthamoeba* with a characteristic intracristal structure.

Figure 7. Mitochondria from a trophic *Naegleria*. Cristae appear as irregular electron-lucent areas within the matrix.

Figure 8. Mitochondria from an encysting *Naegleria*. Organelles appear smaller in size and more regular in shape than those of trophic ameba.

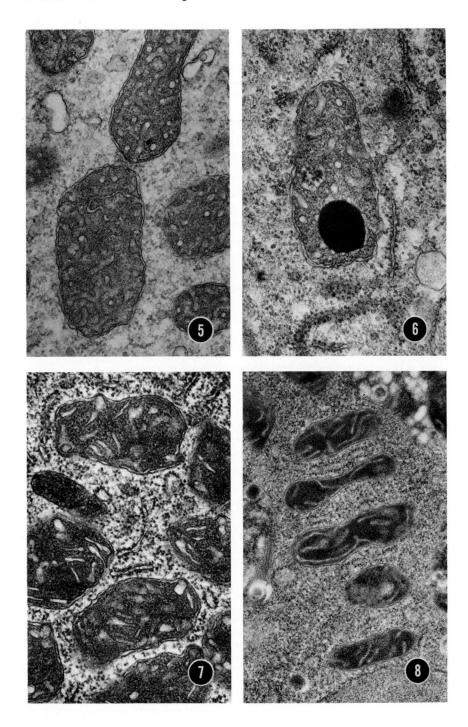

ganelle. These show subtle differences in the various genera of amebas. In some organisms, the cristae are decidedly fingerlike and tortuous (as in *Acanthamoeba;* Plate II, Figure 5), as seen in profile in the mitochondrial matrix; in other amebas, the cristae are more saccate and less well defined (as in *Naegleria;* Plate II, Figure 7). In at least one instance, mitochondrial differences have been cited as a feature distinguishing *Naegleria gruberi* from *N. fowleri* (Carter, 1970).

Located at the uroid (posterior or tail region) is the contractile vacuole complex, usually characterized by several small vacuoles which, with continued fluid uptake, fuse to form a single large vacuole immediately prior to vacuolar emptying or systole. Fluid contents are emptied without benefit of a formed pore at the surface.

Among the small amebas pseudopods are either of the lobose ectoplasmic variety or of the tapering hyaline type. Lobose pseudopods are typical of the limacine amebas (*Hartmannella, Naegleria,* and *Tetramitus*), whereas tapering pseudopods are found in *Acanthamoeba* and are generally referred to as acanthopodia (Page, 1967b). Ectoplasmic pseudopods are recognizable in the electron microscope as distinct areas not containing any of the organelles seen in the endoplasm (Plate III, Figures 9 and 10). There may or may not be microfilaments and microtubules (Holberton, 1969) in the cytoplasm of these amebas, though actin and actinlike filaments have been reported in *Naegleria* (Lastovica, 1976) and *Acanthamoeba* (Bowers and Korn, 1968).

B. Phagocytosis and Pinocytosis

Food vacuoles are present in all amebas, whether grown on agar with bacteria as a food source or axenically in a nonparticulate fluid medium (Ryter and Bowers, 1976). This is presumably a major pathway for uptake of nutrients into the organism. Ingestion of bacteria into food vacuoles results in the production of myelinlike membranous whorls (Plate III, Figure 9) as digestion of the bacteria proceeds (Fulton, 1970a; Schuster, 1963a). Whorls are ultimately expelled from the organism and appear to

---

Plate III: Comparison of eruptive and noneruptive ectoplasmic pseudopods of small amebas.

Figure 9. *Tetramitus rostratus* ameba with an eruptive pseudopod. A sharp distinction is seen between the dense endoplasm and the less dense ectoplasm. Myelinlike figures in vacuoles (F) are bacteria undergoing digestion. Magnification: ×14,000.

Figure 10. Noneruptive ectoplasmic pseudopod of a soil ameba (PR 9). Some membranous vesicles are present in the ectoplasmic region, but otherwise no organelles are seen in the pseudopod. Magnification: ×19,000.

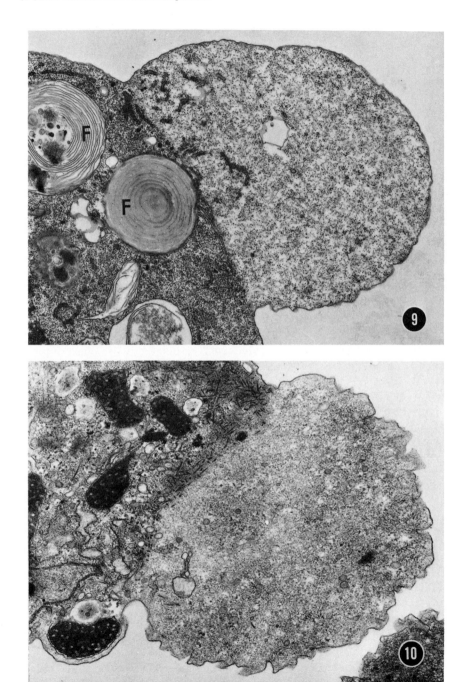

fade away once released into the extracellular medium. Of particular interest regarding phagocytosis by amebas is the observation by Weisman and Korn (1967) that *Acanthamoeba* exhibits remarkable selectivity in particle ingestion. Latex beads suspended in a medium containing radioactive glucose or albumin are ingested by the amebas, to the virtual exclusion of tracer molecules.

Although there are no specialized organelles for food uptake, limacine amebas use their posterior or tail region for this purpose (Plate I, Figure 1). As the ameba advances, fine strands of protoplasm at the posterior remain attached to the substrate. These are apparently sticky, as indicated by the tendency for bacteria and other particles of small size to adhere to them. With continued forward progression, these strands break free from the substrate and are resorbed into the ameba, along with the accompanying bacteria. Thus, food intake in limacine amebas appears restricted to this uroid region. Blood cells and products of cellular breakdown are ingested by *N. fowleri* when it invades the central nervous system of experimental animals and presumably humans. In brain tissue, amebas exhibit small digitiform pseudopods which may facilitate penetration between the cells and cell processes making up the neuropil (Visvesvara and Callaway, 1974).

Biochemical and cytochemical events associated with phagocytosis have been studied by Ryter and Bowers (1976) and by Wetzel and Korn (1969). Though lysosomes have not been found in *A. castellanii* (Ryter and Bowers, 1976) or *A. culbertsoni* (Childs, 1973), Müller (1969) and Lasman (1967, 1975) have detected hydrolase activity in *Acanthamoeba* spp. Pinocytotic activity and associated membrane turnover have been followed in *Acanthamoeba* using various tracer molecules (Batzri and Korn, 1975; Bowers and Olszewski, 1972; Chlapowski and Band, 1971b). Comparisons have also been made of the relative degrees of pinocytotic and phagocytic activity (Bowers, 1977; Chambers and Thompson, 1976).

C. Contractile Proteins

Considerable interest has developed in nonmuscle contractile systems (Korn, 1978), particularly in ameboid organisms where such components would have an obvious role in locomotion, as well as in food, pinocytotic, and contractile vacuole formation. Weihing and Korn (1971) isolated and characterized *Acanthamoeba* actin, representing ca. 0.2% of cell protein. They found that it polymerized from 3 S globular monomers to form 30 S fibrous polymer, that it combined with muscle myosin to form actomyosin, and that it activated Mg^{2+}-ATPase activity of muscle heavy meromyosin. It differed from other actins in having ϵ-N-methyllysine, but otherwise resembled muscle actin in its amino acid composition, including the presence of a rare amino acid, 3-methylhistidine.

In subsequent studies, a myosin ATPase was isolated from *Acanthamoeba,* accounting for ca. 0.3% of cell protein (Pollard and Korn, 1973a). It had a molecular weight of 180,000 and was composed of three polypeptide chains, one heavy and two light. In what appears to be the first example of two different myosins present in the same cell, Maruta and Korn (1977b) isolated a second myosin (myosin II to distinguish it from myosin I) from *Acanthamoeba.* Myosin II has a molecular weight of ca. 400,000 and consists of four chains, two heavy and two light pairs. Myosin I is more active as a K^+-ATPase than myosin II, which is more active as a Ca^{2+}-ATPase. Myosin II is a double-headed myosin, and myosin I is a single-headed myosin. Another characteristic distinguishing the two myosins is the requirement for cofactor protein to activate the Mg^{2+}-ATPase of *Acanthamoeba* myosin I (Pollard and Korn, 1973b), but not II. Cofactor has been purified (Maruta and Korn, 1977c) and found to be a myosin kinase which catalyzes phosphorylation of the single heavy chain of myosin I.

Contractile proteins were extracted from crude *Acanthamoeba* homogenates using cold sucrose (Pollard, 1976); when warmed to 25°C, these gelled and, upon shrinkage, soluble proteins were squeezed out. Gel formation was temperature dependent, regulated by ATP and Mg^{2+}, and accelerated by low Ca^{2+} concentrations. Gelation appears, however, to be a property of four proteins, each of which alone causes F actin to gel (Maruta and Korn, 1977a).

Myosin thick filaments were not observed in *Acanthamoeba* (Bowers and Korn, 1968; Pollard and Korn, 1973a), though Pollard *et al.* (1970) studied cytoplasmic microfilaments *in situ* and purified F actin from *Acanthamoeba.* Purified actin appeared as beaded filaments measuring 60 Å in width and up to 3 μm in length. These filaments reacted with heavy meromyosin to give characteristic "arrowheads" with 370 Å spacing. Thus, the contractile system of *Acanthamoeba* consists of two myosins, myosin kinase cofactor protein for actin activation of myosin I, actin, and four gelation factors. What is not clear is how these are arranged within the cell to provide for the various types of cellular changes (ameboid movement) and morphogenetic changes (such as encystment). Pollard and Korn (1973c) observed F-actin filaments attached to the inner face of the *Acanthamoeba* plasma membrane and suggested that the membrane was analogous to the Z line of striated muscle. Their model has myosin cross-linking neighboring F-actin membrane filaments which, in the presence of cofactor and ATP, produce a sliding motion not unlike that found in muscle.

Superprecipitation of an actomyosin protein complex from *N. gruberi* amebas was used as an assay for its presence by Lastovica and Dingle (1971). Complex formation was dependent upon ATP concentration, stimulated by Mg^{2+}, and inhibited by Ca^{2+} and by Salyrgan (which binds to sulfhydryl groups). In ameboid motion of *N. fowleri* Lastovica (1976) has postulated a role of two types of microfilaments: thin (5–7 nm) and actinlike, and thick (17–19 nm) and myosinlike. Unpublished evidence exists (cited in Fulton, 1977a,b) that abundant actin is present in *Naegleria* amebas, where it serves as the basis for movement in the ameba stage but not in the flagellate stage.

D. Plasma Membrane and Cell Clumping

Extensive studies of the plasma membrane of *Acanthamoeba* have been carried out following the development of techniques for large-scale isolation by Ulsamer *et al.* (1971) and by Schultz and Thompson (1969). Korn and co-workers (Bowers and Korn, 1973, 1974; Dearborn *et al.*, 1976; Korn and Olivecrona, 1971; Korn and Wright, 1973) have conducted biochemical and cytochemical analyses of the plasma membrane. Dykstra and Aldrich (1978) have demonstrated cell coats (using a combination of Alcian blue and ruthenium red in conjunction with electron microscopy) in *Acanthamoeba* and *Naegleria*, which differ from the mucoid glycocalyx typical of some of the large amebas. Areas of calcium ion concentration have been demonstrated in the plasma membrane of *Acanthamoeba* by Sobota *et al.* (1977). These formed even in the presence of inhibitors of glycolysis or oxidative metabolism.

Band and Mohrlock (1969a) studied adhesion patterns in *A. palestinensis* (a clumper) and *A. castellanii*, strain HR (a nonclumper), and concluded that clumping was dependent upon an undetermined metabolic factor as well as temperature. Hoover (1974) performed electrophoresis on isolated surface membrane of these two amebas and detected differences in their proteins. In the presence of concanavalin A (Con A), the nonclumpers adhered, but not the clumping strain. Spies *et al.* (1972, 1975b) employed lysolecithin and Con A to study, with the electron microscope, cell clumping of *A. castellanii*. Cell contacts were characterized by tight or gap junctionlike regions separated by a space of 60–200 Å, though these were not observed in the strain examined by Band and Mohrlok (1969a). It has been suggested that such contact areas might function in exchange of macromolecular materials between cells in culture (Pigon and Morita, 1973). Spies *et al.* (1975b) obtained maximal clumping from log phase cells, with virtually no clumping occurring in post-log phase cells. Exposing cells to ferritin-labeled Con A revealed a decrease in binding sites upon the surface of post-log phase cells. Bowers (1977) reported that the size of *Acanthamoeba* clumps varied with the Con A concentration used. No difference in Con A-induced agglutination was observed by Stevens and Stein (1977) between the avirulent *A. castellanii* and the virulent *A. culbertsoni*, which is contrary to what had been reported earlier (Stevens and Kaufman, 1974).

Con A-induced agglutination (100 μg/ml) of *Naegleria* spp. can be used to distinguish between the innocuous *N. gruberi*, which clumps (>50%) in the presence of Con A, and the pathogenic *N. fowleri*, which does not (Josephson *et al.*, 1977). A conjugate of fluorescein isothiocyanate–Con A has been used to label cell surface receptor sites in *Naegleria* (Preston and O'Dell, 1974; Preston *et al.*, 1975). At low (15 μg/ml) Con A levels, motility of the amebas resumed after an initial rounding of cells, and the label was taken up into pinosomes. At high (200 μg/ml) levels, the fluorescence became localized in the uroid of the ameba and was eventually shed, followed by agglutination, rounding, and vacuolation of cells. Reversing or blocking of the Con A effect was possible through the use of α-methyl-D-mannopyranoside, indicating specificity of binding. This is taken as evidence for the presence of mannopyranose or glucopyranose at the membrane surface. Phytohemagglutinin, however, produced no effect, indicating the absence of exposed galactosamine groups. *Naegleria jadini* and *Tetramitus rostratus* amebas do not respond to Con A treatment (Preston *et al.*, 1975). As the lectin conjugate is endocytosed by the amebas, no additional binding sites become available, supporting the concept of membrane conservation (O'Dell *et al.*, 1976). Binding of Con A is not associated with the loose cell coat present on the plasma membrane, but rather with the membrane itself; cells placed on a Con A-filmed glass surface become immobilized, unless their Con A-binding sites have been previously shed (Preston *et al.*, 1975). Fluorescein-labeled cationic ferritin combines with anionic sites on the *Naegleria* membrane in a manner similar to that of Con A. The label passes to the uroid region, where it is ultimately cast off as a residue. Anionic binding sites, however, are different from those that bind Con A, though the identity of the negatively charged locus is speculative (King and Preston, 1977).

Acanthamoeba spp. have been separated from one another on the basis of their electrokinetic mobility (Band and Irvine, 1965). Precyst stages showed an increased negative surface charge. The advancing pseudopod of *Naegleria* was found to have a higher negative surface charge density than the rest of the cell surface (Forrester *et al.*, 1967). After aligning themselves in the electrophoretic field, amebas migrated toward the anode and changed direction by 180° upon reversal of the field. A novel approach using reflection–interference and scanning microscopy was employed by

Preston and King (1978a,b) to examine areas of contact between *Naegleria* trophozoites and a cover glass substrate. In deionized water, a diffuse area of contact was maintained with a space of ca. 100 nm, from which filopods extended to attach to the surface as focal contacts. With forward progression, the filopods passed to the uroid where resorption took place, leaving behind the point of focal attachment. Perfusion of amebas with a medium of increased ionic strength (10 mM KCl) shifted the ameba–substrate association to a focal type with a gap of ca. 20 nm.

Most, if not all, soil amebas produce a wall-to-wall sheet of cells when growing upon an agar surface (Plate I, Figures 3 and 4), which Willmer (1970) has described as an ''epithelial'' growth pattern.

E. Osmotic and Ionic Balance

The contractile vacuole, present in virtually all ameba isolates from soil and fresh water, is generally assumed to have an osmoregulatory, nonexcretory function (Pal, 1972). Like most freshwater organisms, the small amebas are faced with two problems: (1) uptake of water by osmosis from a dilute environment, and (2) loss of internal salts by diffusion. Pal (1972) has examined the performance of the *A. castellanii* contractile vacuole as a function of age of the culture and various osmotic and ionic changes in the medium. Internal osmotic pressure and ionic concentration have been studied by Drainville and Gagnon (1973) and by Larochelle and Gagnon (1976, 1978).

Isolated *Acanthamoeba* mitochondria are tolerant of low tonicity for prolonged periods of time (Klein and Neff, 1960). Volume changes in the mitochondria are nonosmotic but are due to colloidal swelling or shrinkage. Mitochondrial swelling leads to increased respiratory activity, making energy available for water or ion pumping. Calcium ion causes shrinkage of the mitochondrion and a reduction in the rate of succinate oxidation by ca. 90%. Vickerman (1962) has suggested that the intracristal bodies seen in the mitochondria of encysting *Acanthamoeba* (Plate II, Figure 6) may be a means of reducing the volume of the mitochondrial lumen and causing a decrease in respiratory rate.

According to Klein (1961), *Acanthamoeba* maintains intracellular K^+ and Na^+ levels by cation binding. More Na^+ is bound than K^+, and the binding of Na^+ may compensate for the lack of a sodium pumping mechanism to maintain a low intracellular Na^+ level. The factor responsible for binding appears to be in the supernatant cell fraction, rather than associated with organelles such as mitochondria. Shifting of K^+ between exchangeable and nonexchangeable fractions was caused by the application of low temperature (5°C), hypertonicity, CO_2, and dinitrophenol (Klein, 1964).

Osmotic and ionic balance have an important role in formation of the flagellate stage of ameboflagellates; discussion of this topic is reserved for Section IV,B.

F. The Amebic Genome

The DNA composition of the small amebas has been of some value, in conjunction with morphological and physiological characteristics, in understanding phylogenetic relationships between genera and species. Adam *et al.* (1969) studied the DNA base composition of *Acanthamoeba* spp. For *A. castellanii*, the guanine + cytosine (GC) content was 61%. This figure varied for other representatives of the genus from 57% for *A. polyphaga* to 62% for *A. palestinensis*. Two species of DNA are recognized on the basis of buoyant density: a major component (1.720 gm/cm³) and a minor component (1.693 gm/cm³), the latter being of mitochondrial origin. Adam and Blewett (1974) conclude that the difference in content between *A. castellanii* and *A. palestinensis* is not significant, but that differences between other species are. These results were supported by examination of base sequence homologies. Marzzoco and Colli (1974) isolated nuclei from *A. castellanii* and characterized DNA (an average of 10.6 pg/nucleus) as 58% GC. In studies on whole-cell DNA, major (1.717) and minor components (1.692) were recognized (Marzzoco and Colli, 1975). The minor peak included both mitochondrial and extramitochondrial DNA. The examination of mitochondrial DNA from *Acanthamoeba* revealed open circular molecules (Bohnert, 1973).

The presence of extramitochondrial DNA in *Acanthamoeba* spp. was indicated also by electron microscope autoradiography of amebas labeled with [³H]thymidine (Ito *et al.*, 1969; McIntosh and Chang, 1971). Ito and co-workers (1969) suggested that the cytoplasmic DNA was associated with endosymbionts, present (since they could not be visualized in the electron microscope) as defective viruses or episomelike elements.

The DNA of *N. gruberi*, strain NEG, has a main component (1.693) and two minor components (1.683 and 1.702), according to Fulton (1977a). The GC content of the main peak, which is nuclear, is 34%; of the minor components, one is nuclear and the other mitochondrial. Fulton (1970a) reported that strain NEG is haploid and that another strain, NB-1, is diploid. For the NEG and NB-1 strains, the DNA content is 0.17 and 0.32 pg/ameba, respectively. Ploidy characterizations were based on cell and nuclear volume, DNA content, patterns of ploidal inheritance, and ease in producing mutants. For the LEE strain of *Naegleria fowleri,* the DNA content is 0.2 pg/ameba (Weik and John, 1978). DNA content doubles at the onset of the stationary phase, when the number of nuclei increases.

Sexual stages have been looked for in *Naegleria,* but none has been found. Droop (1962) has, however, described a mating system for the ameboflagellate *Heteramoeba clara,* based upon indirect evidence of segregation occurring in the isolate. Evidence for sexual cycles in the nonameboflagellate soil amebas discussed here is generally lacking.

III. THE CYST

The environment for an organism living in the soil, as might be expected, is one of wildly fluctuating growth conditions: cycling from dry to wet, and back to dry, and an abundant bacterial food supply followed by food scarcity. Survival of soil amebas requires some means by which they can pass from one period of favorable growth conditions to the next. The dilemma is resolved quite easily for these organisms with the formation of a thick-walled cyst or dormant stage which protects them against desiccation and insufficiency of food. Cysts also form in fluid media, so drying alone is not a trigger for encystment, and in amebas growing on agar surfaces cyst formation occurs with depletion of the bacterial food supply. Other factors that may influence encystment are pH changes, waste accumulation, oxygen deficiency, and crowding of cells in culture. Several comprehensive reviews on the *Acanthamoeba* cyst are available (Griffiths, 1970; Neff *et al.,* 1964b; Neff and Neff, 1969; Weisman, 1976).

Cysts exist in many shapes and sizes, with single or double walls, and with or without pores. None of the cysts produced by amebas covered in this chapter are reproductive structures, as is the case with some types of ciliate cysts (Grell, 1973) and cellular slime mold macrocysts (Erdos *et al.,* 1973). The precystic ameba generally rounds up and exhibits a concomitant increase in cytoplasmic density, owing probably to loss of water from the cytoplasm. Wall material, produced in the cytoplasm, deposits uniformly about the plasma membrane of the rounded ameba (Plate IV, Fig-

---

Plate IV: Figure 11. *Naegleria gruberi,* early cyst stage. Outer wall component (ectocyst) has formed, along with a pore filled with mucoid material. The peripheral cytoplasm of an encysting ameba contains round, electron-dense granules probably with wall material. Magnification: ×25,000.

Figure 12. Early cyst stage of *T. rostratus.* An encysted ameba surrounded by a simple wall with no pores. The nucleus of the ameba is seen at N. The slightly oval shape of the cyst is caused by compression during sectioning; typically the cyst is round. Magnification: ×19,000.

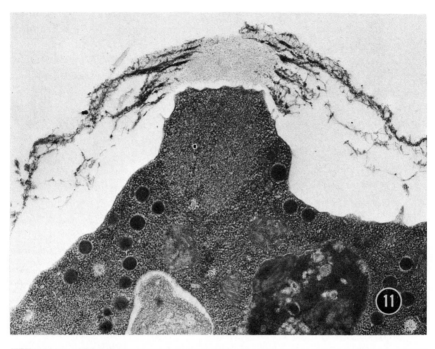

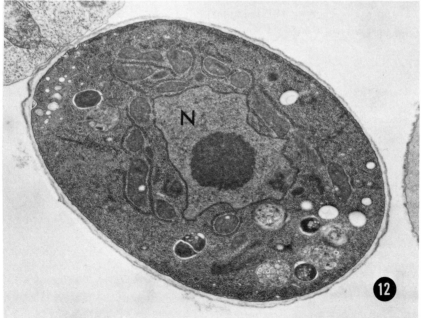

ure 11). The origin of the wall material is not clear, either from light or electron microscope studies. It is apparent, however, that the buildup of wall material increases with the age of the cyst, leading to thickening of the wall and, perhaps, a change in its nature. The walls of mature cysts prevent infiltration of embedding resin for transmission electron microscope studies, and thus most such studies have been done on young cysts (Bowers and Korn, 1969; Schuster, 1963b, 1975a). For this reason, scanning electron microscopy has been of great use in describing the morphology of the mature cyst and the surface changes occurring at the time of encystment (Pasternak et al., 1970).

Depending upon the species, the wall may be single or double (Plate IV, Figure 12). In forms with a double wall, the outer component is the thinner of the two and often erodes as the cyst ages. The inner and outer walls are termed the endo- and ectocyst wall, respectively. In species whose walls contain pores, these pores each become filled with an electron-lucent mucoid plug that effectively seals them. As encystment progresses, the cytoplasm of the ameba increases in density, as indicated by the close packing of ribosomes. Endoplasmic reticulum is often seen in close association with mitochondria. Food vacuoles, an obvious feature of the trophic ameba, are in most cases eliminated prior to formation of the wall. Latex spheres ingested by trophic A. castellanii are, like the contents of food vacuoles, eliminated by exocytosis at the time of encystment (Stewart and Weisman, 1972). The expelled beads are sometimes incorporated into the forming cyst wall.

A. Encystment

The organism of choice for studies in this area has been Acanthamoeba. It can be grown in large quantities and encysts spontaneously in axenic medium, and, more importantly for research, encystment can be induced by the transfer of amebas to nonnutrient media (Band, 1963; Band and Mohrlok, 1969b; Griffiths and Hughes, 1968, 1969; Neff et al., 1964b). Differences may exist among the various species and strains of Acanthamoeba with regard to response to various types of encystment media employed (Griffiths and Hughes, 1968). Encystment media vary in specifics but are similar in the absence of a carbon or nitrogen source and the presence of divalent cations (Ca^{2+} and/or Mg^{2+}); otherwise they may or may not be buffered, their osmolarities vary, and additional ions may be present. Prestarvation of cells in growth medium with a 10-fold reduction in glucose enhances encystation (Band and Mohrlok, 1969b). Three phases have been recognized leading to formation of the mature cyst: (1) induction, characterized by rounding of amebas, (2) synthesis of the protein and cellulose components of the cyst wall, and (3) dormancy, with reduced metabolic activity (Neff and Neff, 1969).

At the ultrastructural level, transmission electron microscopy has been used by Bowers and Korn (1969) to describe changes accompanying encystment of *Acanthamoeba*. Bauer (1967) has carried out freeze–fracture of encysting amebas. Wall formation starts at the time the ameba begins to round up, with the laminar ectocyst depositing as a fibrous layer ca. 0.5 μm thick. The fibrogranular endocyst forms inside this, the two layers separated by a space. Ostioles are circular openings (ca. 1 μm in diameter) present in both walls, which are used as exits at excystment. The ostiole is sealed by an operculum or lid which resembles the two wall components. Within the cytoplasm, there is an increase in density, budding of vesicles from the Golgi complex, formation of acid phosphatase–positive autolysosomes, development of intracristal concretions within the mitochondria, localization of cytoplasmic microfilaments (actin?) in the hyaline cortical region, and reduction in both nuclear and nucleolar volume, the former accompanied by vesicle blebbing into the cytoplasm. Chemical analyses reveal an increase in neutral lipid and reductions in phospholipid and glycogen. At the time the cells are rounding, a decrease in intracellular actin is detected (Rubin and Maher, 1976). Lasman (1977) noted that both ecto- and endocyst walls of *A. palestinensis* had the same densely granular ultrastructural appearance (cf. *A. castellanii* of Bowers and Korn, 1969). Nucleolarlike bodies appeared in the nucleoplasm, and there was a suggestion of blebbing at the periphery of the nucleus. Dense concretions developed within the mitochondria. Pasternak *et al.* (1970) followed surface changes with the scanning electron microscope, and Spies *et al.* (1975a) employed freeze–fracture to trace plasma membrane changes in encysting *Acanthamoeba*. Cytochemistry was used to identify diffuse material at the surface of the rounded trophozoite as phosphoprotein (Neff and Neff, 1969).

Fewer studies have been carried out on other genera of soil amebas, the main reason for this being the lack of a technique for producing synchronous encystment in axenic populations. Furthermore, many of these other genera, unlike *Acanthamoeba*, either encyst erratically or not at all in axenic culture. Most other studies, therefore, were of necessity done on amebas grown in the presence of bacteria.

Encystment can be readily followed in *Naegleria* grown on agar with bacteria as a food source. As the plate surface is cleared of bacteria, the amebas form large clusters of cysts carpeting the agar surface. Encystment is more or less synchronous in that cells on a single plate are all at roughly the same stage of encystment: rounding up, wall formation, and maturation of the newly formed cysts. *Naegleria* spp. have been studied by transmission electron microscopy (Schuster, 1975a), where an increase in cytoplasmic density was noted, autolysosome formation occurred, ecto- and endocyst walls developed on the surface of the rounded ameba, and pores were formed, each containing a mucoid plug. In *Naegleria*, as in *Acanthamoeba*, the cytoplasmic contents of the mature cyst were difficult to observe because of the tremendous increase in density (Chiovetti, 1978). Unlike *Acanthamoeba* where wall material—at least the ectocyst component—appears to originate from Golgi vesicles (Bowers and Korn, 1969), no organelles of the rounded *Naegleria* ameba appear to be directly involved in wall formation. Both the wall material and mucoid plug appear at the surface, though the latter can be seen in the cytoplasm in the region of the forming cyst pore. Cytoplasmic vesicles may be the source of wall material, since an increased number of them is found at the periphery of the ameba during the period of wall formation (Plate IV, Figure 11). An alkali-insoluble component in the *Naegleria* cyst wall was identified by Werth and Kahn (1967) as cellulose. Schuster and Svilha (1968) employed uv microscopy, electron microscope autoradiography, and acridine orange staining to demonstrate the presence of ribonucleoprotein vesicles in the cytoplasm of encysting *Naegleria*. Traces of mitochondrial profiles within these vesicles suggest that they are similar to the autolysosomes of *Acanthamoeba* observed by Bowers and Korn (1969). On the basis of a decrease in electron density and a reduction in acridine orange fluorescence, there appeared to be a reduction in nucleolar material at the time of encystment. An unusual pattern of reencystment has been described in *N. gruberi* in which an apparently normal cyst wall encloses a smaller encysted ameba (Chiovetti, 1976). This double encystment might occur when the original wall is in some way disrupted, as might happen during the collection of cysts from an agar substrate with a bacteriological loop.

In another soil ameba, *V. russelli,* Maitra *et al.* (1974b) found a tripartite wall consisting of an inner electron-dense fibrillar layer, an outer electron-dense granular layer, and between them an electron-lucent granular region or layer. "Black" bodies measuring 70–100 nm observed in the cytoplasm were equated with the Golgi vesicles observed by Bowers and Korn in *A. castellanii* and probably transported wall precursor material to the plasma membrane. They noted also a close association between endoplasmic reticulum and mitochondria, and ejection of cytoplasmic material during encystment. Elimination of vacuolar contents upon rounding of the trophozoites, or even after wall formation has occurred, is not uncommon (Bowers and Korn, 1969; Maitra *et al.,* 1974b). This debris may represent undigested residues from food vacuoles or the contents of autolysosomes; it may be incorporated into the forming wall, or may come to lie between the wall and the plasma membrane of the ameba.

A biochemical approach has been employed to define factors controlling encystment. Concerning encystment of *Acanthamoeba,* Neff and Neff (1969) have generalized that it (1) is initiated by depletion (or absence) of some factor in the medium, (2) occurs in the absence of an exogenous energy supply and/or metabolites, and (3) is an example of single-cell differentiation.

Based on the ability of mitomycin C, fluorodeoxyuridine (FUdR), and Trenimon to induce encystation, Neff and Neff (1969, 1972) suggested that blockage of DNA synthesis serves as an encystment signal in *Acanthamoeba.* Fluorodeoxyuridine is postulated to act by blocking synthesis of thymidylate synthetase; added thymidine serves to bypass the FUdR-induced inhibition. Mitomycin C (and Trenimon) were thought to induce encystment through an alteration in chromatin conformation, activating genes responsible for the initiation of dormancy. This hypothesis was supported by induction of encystment through inhibition of DNA synthesis with hydroxyurea (Rudick, 1971). RNA synthesis was also necessary—both rRNA and 4 S RNAs—and encystment was reduced if actinomycin D was added to precystic cultures before the RNA level had reached a maximum. Once this level was realized, the actinomycin D effect was neutralized. A 50% decrease in RNA occurs, according to Stevens and Pachler (1973), leading them to conclude that RNA buildup is not an encystment prerequisite. A rise in RNA and a decrease in DNA and glycogen were detected in late log-phase *Acanthamoeba* by Byers *et al.* (1969), and a 40% decline in protein was reported by Griffiths and Hughes (1969).

Bromodeoxyuridine (BUdR), which substitutes for thymidine in DNA, when added to *Acanthamoeba* cultures did not halt growth, nor did it block encystment. Bromodeoxyuridine-treated cells, however, responded by excysting at about one-third the level of those in control cultures (Roti Roti and Stevens, 1974). Roti Roti and Stevens (1975) found that FUdR or hydroxyurea, added to cultures of *Acanthamoeba* in the exponential growth phase, caused a delay in encystment, but they suggested that the delay caused by the inhibitors was due to their effect upon RNA, rather than DNA, synthesis. Roti Roti and Stevens (1975) also reported DNA synthesis, measured by incorporation of [^{3}H]thymidine, through the young cyst stage, another point not compatible with the Neffs' hypothesis. Chagla and Griffiths (1974) noted an overall increase in DNA content in mature cysts and a rise in the proportion of binucleate amebas. Ethidium bromide, an inhibitor of mitochondrial DNA synthesis, induced encystment of *A. castellanii* (Gessat and Jantzen, 1974).

An increase in specific activity of DNA-dependent RNA polymerase was observed by Rudick and Weisman (1973), following induction of encystment in *Acanthamoeba*. This period of increase overlapped with a period of maximal uptake of [³H]uridine. The two events may be related to the synthesis of protein needed for cyst maturation. Detke and Paule (1975) reported a doubling of RNA polymerase activity in encysting *Acanthamoeba* during the first 10 hr. Rubin *et al.* (1976) detected qualitative and quantitative changes in the electrophoretic profiles of phenol-soluble nucleolar proteins; some of these acidic proteins increase, while others decrease or disappear. These changes occurred at the time when trophic forms were rounding up and the outer cyst wall was making its initial appearance. They postulated that these alterations may be related to gene regulatory functions, particularly coding for RNA.

Cellulose was found to be a component of *Acanthamoeba* cyst walls by Tomlinson and Jones (1962) but was masked *in vivo* by a protein. The cellulose component represents approximately one-third of the dry weight of the cyst wall. Neff *et al.* (1964a) confirmed the presence of cellulose and found that protein constituted another third of the cyst wall dry weight. These two components are separated spatially and temporally from one another as the proteinaceous outer cyst wall (synthesized early in encystment) and the cellulose inner cyst wall (synthesized late in encystment). Other constituents include inorganics (7–8%) and a lipoprotein that may be bound to the cellulose inner wall. The cellulose wall component was traced in encysting *A. castellanii* by Stewart and Weisman (1974) by fractionation of isotopically labeled amebas and by electron microscope autoradiography. Their results indicated that cellular glycogen served as the precursor of cyst wall cellulose. This is consistent with a drop in glycogen reported during the induction stage (Byers *et al.,* 1969; Neff and Neff, 1969). Glycogen degradation may be associated with a two- to threefold increase in cyclic adenosine-3′,5′-monophosphate (cAMP) in preencysting populations of *A. castellanii,* probably through increased adenyl cyclase conversion of ATP to cAMP. Stewart and Weisman were unsuccessful in localizing developing wall material within the ameba, either at the periphery of the rounded ameba or in association with any of the cellular organelles (e.g., the Golgi complex) that have been implicated in the synthesis of wall material. They suggest that the Golgi vesicles might contribute protein or oligosaccharide components to the developing wall.

Coincident with the onset of encystment, a 30-fold increase in β-glucan synthetase was detected *in vitro* in populations of *Acanthamoeba* (Potter and Weisman, 1971, 1972, 1976) and was involved in the synthesis *in vivo* of cyst wall cellulose. Enzyme synthesis and encystment were found to be sensitive to the inhibitors actinomycin D and cycloheximide but only for ca. 14 hr in the encystment medium; beyond this point immature cysts transformed into mature cysts despite the presence of inhibitors. Presumably any RNA and/or protein required for the completion of encystment was in sufficient supply by the eighth hour after induction. Deichmann and Jantzen (1977) detected a cellulose-degrading system consisting of three different enzyme components, one of which was β-glucosidase, in growing cultures of *A. castellanii*. Activity of the system declines to 10% of its original level with the onset of encystation.

Less is known about the composition and synthesis of the outer cyst wall. In the course of studying nucleolar protein changes during encystment of *Acanthamoeba*, Rubin *et al.* (1976) characterized the proteins of the outer cyst wall by electrophoresis. They found nine protein bands, only one being a major protein component with a molecular weight of about 70,000.

Measurement of the respiratory metabolism of encysting *A. castellanii* in the presence of glutamate and succinate as substrates revealed elevated oxygen consumption (to about 10 hr) followed by a decrease upon completion of encystment (Griffiths and Hughes, 1969). This presumably reflects an increased energy requirement for encysting amebas, at least during the initial stage. Band and Mohrlok (1969b) reported an initial rise (1–6 hr) in respiratory rate for *A. castellanii*, strain HR, during encystment, followed by a drop upon formation of the mature cyst. Mitochondria isolated from encysting amebas respired at the same rate as mitochondria from vegetative cells. The respiratory rate of cyst mitochondria, however, declined markedly. Ultrastructural changes in mitochondria of encysting *Acanthamoeba* have been reported by Vickerman (1960) and Bowers and Korn (1969). The change involves the formation of large, electron-dense, intracristal masses (Plate II, Figure 6) which are either absent in mitochondria of trophic amebas or infrequently observed in the mitochondrial tubules. Vickerman suggested that formation of these bodies may be related to a shift in respiratory rate through a change in the volume of the mitochondrial lumen. Mitochondria of encysted *Naegleria*, though they may appear smaller in size than those from the ameba, have no intracristal material (Plate II, Figure 8).

Band (1963) found that malonate, chloramphenicol, and dinitrophenol inhibited encystment of *A. castellanii*, strain HR, but Griffiths and Hughes (1969) found no effect of malonate, dinitrophenol, or sodium fluoride on encystment of *A. castellanii*. Tetracycline was found to be the strongest inhibitor of encystment; of the other compounds tried by Griffiths and Hughes, iodoacetate, arsenate, and sodium azide were inhibitory to varying degrees.

An increase in the acid phosphatase activity of encysting *Acanthamoeba* spp. has been reported by Lasman (1967), Band and Mohrlok (1969b), and Griffiths and Bowen (1969). Activity of this lysosomal enzyme may have significance in regard to the degradative changes occurring at encystment, particularly the turnover of reserves in the synthesis of cyst wall materials. Band (1959), in addition, found an increase in the peroxisomal enzyme catalase in encysting populations of amebas.

Encystment of *A. culbertsoni* was induced on a nonnutrient agar containing Mg^{2+} and the biogenic amine taurine (Raizada and Krishna Murti, 1971). Loss of dehydrogenase activity coupled with glucose uptake denoted a shift from aerobic metabolism to cyst wall synthesis. Actinomycin D, cycloheximide, and mitomycin C were all inhibitory. Cyclic adenosine 3',5'-monophosphate induced ca. 60% encystment within 72 hr, paralleling results obtained in the Mg^{2+} plus taurine encystment medium (Raizada and Krishna Murti, 1972). Magnesium ion and taurine are assumed to stimulate cAMP synthesis by activation of adenyl cyclase. Thus, encystment is regulated by the relative concentrations of cAMP and phosphodiesterase, which is responsible for its conversion to noncyclic AMP. Actinomycin D and cycloheximide block the effect of taurine in promoting encystment (Krishna Murti, 1975). Based upon its regulatory role in carbohydrate metabolism, cAMP is believed to control the synthesis of cyst wall material from polysaccharide reserves. The precise mode of control remains conjectural (Verma and Raizada, 1975).

A cellulose component was reported present in the cyst wall of *V. russelli* by Rastogi *et al.* (1971b), sandwiched between the two electron-dense layers. In addition to the carbohydrate moiety that comprises about 38% of the wall, protein (27%) and lipid (15%) were also detected. The lipoprotein complex was associated with the thick inner layer of the wall and may be a factor in the control of cyst wall permeability. This conclusion is supported by evidence obtained from ultrasonication of mature cysts, which results in removal of the outer wall (Rastogi *et al.*, 1970). Despite the presence of only a single wall, the cysts respond normally to excystment and encystment agents. Encystment of *V. russelli* has been followed in an *Escherichia coli*-fed population of amebas (Rastogi *et al.*, 1977). [^{14}C]Glucose incorporation into amebas rose sharply at time of encystment, and was correlated with formation of the cellulose cyst wall. Berenil, an inhibitor of DNA synthesis (in mitochondria?), did not block the encystment of trophozoites, but did block the reencystment of excysted amebas. It is suggested that a phase of cell division sensitive to Berenil may be necessary before encystment can be initiated.

B. Excystment

Excystment of amebas can be triggered by a variety of factors. In some species, suspension of cysts in distilled water is sufficient to initiate resumption of vegetative activity, and the stimulus is probably osmotic. Cysts of other species may, however, be refractory to osmotic shock but respond to organic substances in the excystment medium, provided as bacterial food organisms, bacterial extracts, sterile complete growth medium, or component parts of the medium. Through careful attention to variables such as cyst age, degree of desiccation, and the like, it is possible to obtain a high degree of synchrony in excysting populations.

Chambers and Thompson (1972) employed scanning electron microscopy to study excystment of *A. castellanii*. After transfer to growth medium, the process of excystment began with the projection of a "bud" through a pore or ostiole in the cyst wall. The operculum, which formed over the ostiole at encystment, was apparently digested prior to emergence, though other aspects of the cyst wall remained unchanged (e.g., the surface ridges), and no collapse, as might be caused by digestion, was observed. The absence of generalized enzymatic digestion is also supported by the light microscope observations of Mattar and Byers (1971). Once excysted, the ameba develops its characteristic acanthopodia.

Naegleria spp. were studied during excystment using transmission electron microscopy (Schuster, 1963b, 1975a). Washing mature cysts with distilled water was sufficient to activate the dormant organisms and, once activated, excystment occurred within a couple of hours. The ameba apparently digests the mucoid plug sealing the cyst pore and then proceeds to squeeze through the small hole. Excysting *N. gruberi* leaves the cyst wall unaltered, though *N. fowleri* apparently digests some of the wall during the process of excystment. Excystment in *Naegleria* has also been induced by high hydrostatic pressure (Todd and Kitching, 1973).

In a study that examined excystment in *N. gruberi,* Averner and Fulton (1966) reported that CO_2 served as a signal for activation of the dormant ameba. Working with cysts formed by amebas in bacterized cultures, they found that contaminating bacteria in the washed cyst experimental populations were responsible for CO_2 production which over a period of ca. 20 min induced morphological changes associated with excystment. Proline, which appeared to stimulate excystment, served as a substrate for bacterial metabolism leading to elevation of the CO_2 level in the excystment medium. Increasing cyst density also enhanced excystment, presumably by supplying more adherent bacteria for CO_2 production. The nature of the CO_2-sensitive trigger in the cyst is unknown, but the mechanism may be important as a signal for renewal of trophic growth in *Naegleria* populations in soil and water habitats.

Following excystment in *V. russelli,* Rastogi *et al.* (1971a) noted a decrease in cellulose content, caused by disappearance of the cyst wall, but not glycogen of excysting amebas. In activated cysts, a doubling of DNA occurred, but no increase in numbers of organisms; RNA and protein also increased. Over a 4-hr period of excystment, these workers detected a marked increase in RNase activity, but drops in aldolase and glutamate oxaloacetate transaminase activity. Exposure of cysts to uv radiation retarded excystment, an effect that could be neutralized by dark incubation of irradiated cysts at 28°C (Rastogi *et al.*, 1972). It was suggested that dark incubation allowed repair of DNA lesions caused by uv light. Activation of resting cysts prior to uv treatment, however, reduced their sensitivity to radiation retardation of excystment.

Though growth in BUdR had no effect upon encystment of *A. castellanii,* the incorporation of this compound into DNA caused a significant reduction in the percentage of excysting amebas (Roti Roti and Stevens, 1974). Synthesis of RNA and protein was indicated in excysting *Acanthamoeba* through a sensitivity to actinomycin D and cycloheximide, respectively (Mattar and Byers, 1971). Hydroxyurea, which blocks DNA synthesis, does not inhibit excystment. Jehan and Dutta (1977) were unable, however, to block excystment of *A. castellanii* with actinomycin D.

Chambers and Thompson (1974) reported an obligatory period of aging of cysts of *A. castellanii* before amebas excysted in significant numbers. Cysts were stored in a saline medium for various time periods prior to transfer to growth medium. Cysts ≤ 2 days of age exhibit ca. 10% excystment, while ca. 90% excystment is obtained with cysts 1 month of age. Furthermore, excystment in aged cysts exhibits a more synchronous response than is seen in newly formed cysts. Thus, even though a cyst acquires all its morphological complexity within a short period after an ameba rounds up, physiological maturation of an unknown nature must occur before dormancy can end. Dubes and Jensen (1964) found that *Acanthamoeba* trophozoites, but not cysts, produced an excystment inhibitor insensitive to proteolytic and nuclease digestions, and 60°C for 10 min. Acid and base treatments, like heating at 100°C for 10 min, however, destroyed the activity of the inhibitor.

Cysts of several species of soil amebas (including *N. gruberi, T. rostratus,* and *V. russelli*) cleansed of bacteria by an HCl–trypsin treatment, gave 92–100% excystment in the presence of a heat-stable *E. coli* extract (Singh *et al.,* 1971). In *V. russelli,* replacement of the bacterial extract by an amino acid mixture gave a comparable percentage of excystment. The investigation also suggested that the pH of the medium may play a role in controlling cyst wall permeability and thus influence excystment. Rastogi *et al.* (1973) reported that an amino acid mixture plus riboflavin was effective in promoting excystment of *V. russelli,* leading them to conclude that trace contamination of the excystment medium—possibly from bacterial debris—with the vitamin riboflavin may account for previous successes.

Live *Aerobacter aerogenes,* bacterial extract, heat-killed bacteria, and a mixture of amino acids extracted from *A. aerogenes* were all found to stimulate excystment of a hartmannellid ameba (Drożański, 1961). Excystment was initiated in *A. culbertsoni* by heat-stable, cell-free extracts of several species of bacteria and fungi, as well as peptone, tryptone, and amino acids (Kaushal and Shukla, 1977).

Using aqueous extracts of bacteria and fungi, Jehan *et al.* (1975) obtained ca. 95% excystment of *V. russelli, T. rostratus,* and *A. castellanii,* strain HR. Encystment was completely blocked by cycloheximide, dinitrophenol, and sodium arsenite, but not by a variety of other inhibitors tested (mitomycin C, actinomycin D, chloromycetin, and sodium azide). Excystment was viewed as a two-stage process: activation by extracts followed by initiation of macromolecular synthesis, particularly of enzymes involved in cyst wall digestion. Chromatographic analyses of fungal extracts suggested that their constituent amino acids were responsible for excystment (Datta and Kaur, 1978). Jehan and Dutta (1977) found that pretreatment of cysts with sodium lauryl sulfate to increase the permeability of the cyst wall did not alter the response to mitomycin C, actinomycin D, or hydroxyurea. Normal levels of excystment occurred, presumably ruling out DNA and mRNA synthesis as preliminaries to excystment. They suggest that a stable mRNA is present in the dormant cyst, which initiates protein synthesis upon activation.

IV. THE FLAGELLATE

Cyst formation is one aspect of differentiation exhibited by all soil amebas as a necessary prerequisite for existence. Ameboflagellates have, in addition to a cyst stage of the life cycle, another pattern of differentiation, transformation from an ameba to a flagellate. The ease with which flagellation can be produced *in vitro* makes this an ideal system for the analysis of morphological and biochemical events leading to induction of the flagellate phenotype and synthesis of the flagellar apparatus. The same may be said for the process of encystment but, unlike the mature cyst which is the dormant end product, the flagellate is a dynamic, metabolically active stage of the life cycle. Though transformation has been studied in all ameboflagellate species, most of the research beyond the basic morphological description of flagellate formation has been done using *Naegleria,* which can be grown axenically and which responds reproducibly to induction. Other ameboflagellate species may exhibit erratic flagellation responses or require live bacteria as food, or both. Detailed treatments of flagellation with reviews of methodology and results of recent studies have been prepared by Fulton (1970a, 1977a).

A. Comparative Flagellate Morphology

Naegleria represents one of the least complex of the flagellate stages from a morphological viewpoint. Typically, the kinetic apparatus (Plate V, Figure 13) consists of two flagella, their kinetosomes, a fibrous rhizoplast and, radiating into the cytoplasm and under the plasma membrane, a microtubular cytoskeleton (Dingle and Fulton, 1966; Satir, 1965; Schuster, 1963a). Some strains of *Naegleria* are reported to produce regularly four flagella (Willmer, 1956), though this appears to be a variable property. Hyperflagellation has been demonstrated (2 to 18 flagella) experimentally through heat shock at 38°C during flagellate induction (Dingle, 1970; Okubo and Inoki, 1973). The additional kinetids that formed were structurally normal at the electron microscope level, and reversion to the typical number of flagella occurred in isolated clones upon subsequent retransformation at routine temperatures (Dingle, 1970). Okubo and Inoki (1973) produced a multiflagellate mutant by ultraviolet irradiation. The striated rhizoplast (Simpson and Dingle, 1971) consists of a single protein with a molecular weight of ca. 165,000 (unpublished work, cited in Fulton, 1977a). Attaching at its anterior end to the kinetosomes, the rhizoplast tapers distally to the level of the nucleus and beyond. Though often found closely associated with mitochondria, its main function is probably to stabilize the kinetid during swimming of the flagellate. The cytoskeleton with its microtubules is responsible for the semirigid shape of the flagellate.

The flagellate stage of *T. rostratus* consists of four flagella and associated kinetosomes, a rhizoplast, and radiating microtubules (Plate V, Figure 14). The *Tetramitus* rhizoplast, at the ultrastructural level, exhibits an unusual periodicity not seen in *Naegleria* (Pitelka, 1969, Figure 4). A greater number of radiating microtubules than found in the *Naegleria* cytoskeleton is responsible for the well-defined profile of the flagellate; part of the skeleton supports a prominent rostrum at the anterior end of the organism and a gullet associated with food vacuole formation leading

---

Plate V: Comparison of the flagellar apparatus of two ameboflagellates. Magnification: ×42,500.

Figure 13. Flagellate stage of *N. gruberi*. This organism probably was quadriflagellate, though only three of the four kinetosomes are seen. The rhizoplast (R) extends posteriorly along the surface of the nucleus (N). Note the attachment of the rhizoplast to the kinetosomes in the form of a microtubular band (arrow).

Figure 14. Similar view of *T. rostratus* flagellar apparatus. Three of the four kinetosomes are seen, along with the striated rhizoplast (R) and a microtubular band (arrow). The nucleus (N) is in close proximity to the rhizoplast. A portion of the gullet (G) can be seen, the wall of which is supported by microtubules.

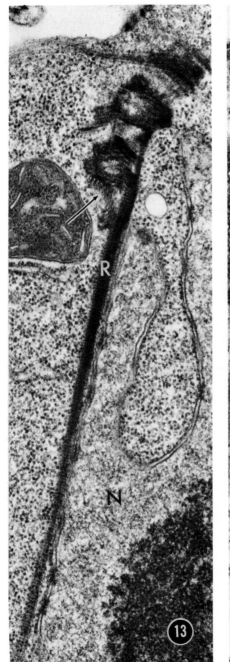

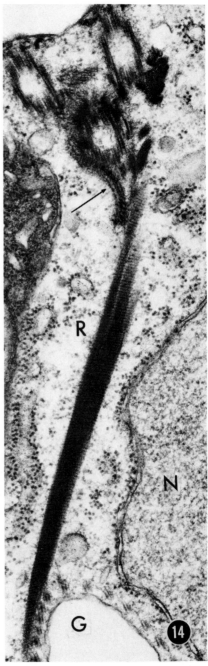

deep into the cytoplasm. *Tetramitus* both feeds and divides as a flagellate and can remain permanently in this stage in culture without reversion to the ameboid stage (Balamuth and Rowe, 1955; Outka, 1965). Brent (1957) was successful in inducing reversion by exposing flagellates to nitrogen mustard; the amebas produced were able to transform into flagellates again, indicating the absence of permanent injury to the cells.

A flagellate similar in appearance to *Tetramitus* is produced by *Paratetramitus* (Darbyshire *et al.*, 1976). Though biflagellate cells are typical, flagellar number ranges from one to six or more. The shape of the organism, seen in the scanning electron microscope, is reminiscent of *Tetramitus,* with a well-defined rostrum and a rigid outline. The pellicle is characterized by the presence of fine striations. Silver staining suggests the presence of a rhizoplast. A groove is present in the anterior third of the organism, which looks like it might be the opening to a gullet; flagellates, however, are not observed to ingest food particles. Flagellates divide by longitudinal fission.

Singh and Hanumaiah (1977a) have recently described a new genus of ameboflagellate, *Tetramastigamoeba,* with a flagellate stage that divides but does not feed. Repeated fission is responsible for the reduction in size of the flagellate until it either transforms to become a small ameba or dies out. A basal complex is seen from which four flagella originate. No gullet is present, nor has a rhizoplast been observed. Flagellate stages are also produced by other ameboflagellate species: *Adelphamoeba* (Napolitano *et al.*, 1970), *Heteramoeba* (Droop, 1962), and *Trimastigamoeba* (Bovee, 1959).

B. Flagellate Induction and Reversion

Transformation of *Naegleria* can be achieved by washing a population of amebas in distilled water or, for better control, buffer. This simulates a condition which in nature would lead to flagellation, as when the soil substrate in which amebas are found is diluted by rainwater. Within a period of 1 hr, $\geq 90\%$ of a population of amebas develop flagella (Fulton and Dingle, 1967). The flagellate stage is transitory, lasting ca. 2 hr; reversion to the ameboid stage eventually occurs, though as many as three additional cycles of retransformation might take place in the population before the amebas begin to encyst in large numbers. At the level of the organism, the ameba in the induction medium exhibits its normal polarity, moving over the substrate with well-defined anterior (ectoplasmic pseudopod) and posterior (sticky uroid) regions. Though there exists some difference of opinion about what happens next, flagella either appear first at the uroid followed by a rounding of the ameba (Willmer, 1956), or the ameba rounds up first and then forms flagella (Fulton, 1977a). In either case, during the period of rounding, the final growth of flagella and streamlining of the body occur. When the flagellate finally breaks free from the substrate—a flagellate is often attached to the substrate by a single sticky filopod—it swims off into the medium. Reversion can be leisurely or rapid. Ameboid processes are extended by the flagellate, which then settles upon the substrate; flagella are rapidly pulled into the organism which once again looks and behaves like an ameba.

Deuterium oxide (50–75%) induces a synchronous and rapid reversion to the ameboid stage, though the mechanism is not known (Preston and O'Dell, 1973). Resumption of pseudopod formation and cytoplasmic streaming occurs, indicative of fundamental alterations in plasma membrane and cytoplasmic organization, respectively. After flagellar resorption, naked axonemes have been seen in the cytoplasm of reversion stages (Dingle and Fulton, 1966; Preston and O'Dell, 1973). In living organisms, the still-active flagellum can be seen within the endoplasm of recently reverted amebas (Pittam, 1963; Preston and O'Dell, 1973).

The transformation of an ameba to a flagellate has been extensively described at both the light and electron microscope levels for *Naegleria* and *Tetramitus* (Fulton, 1970a). No evidence in the ameba of any part of the kinetic apparatus of the flagellate has ever been found, and the consensus is that the entire unit arises *de novo* following induction. Centrioles, which in many systems function either as kinetosomes or kinetosome precursors, are not present in *Naegleria*. They are not found at the initiation of transformation (Fulton and Dingle, 1971), nor are they seen in the mitotic figure of the ameba (Schuster, 1975b). *De novo* origin of the flagellar apparatus is also true for *Tetramitus* (Outka and Kluss, 1967). Kinetosomes, which are probably the first kinetid elements to form, give rise to the flagellar axonemes with their characteristic microtubular pattern. Early developmental stages in kinetosome assembly have been difficult to find (or recognize) in transforming amebas, partly because they are few in number in a relatively large organism, and partly because the period of kinetosome formation occupies a small portion of the time scale from activation to the swimming off of a well-developed flagellate. The same may be said for the rhizoplast which, when it is first seen in transforming individuals, is already a totally recognizable entity.

Willmer (1956, 1958) was the first to appreciate the utility of the *Naegleria* transformation as a means of studying differentiation. The effects of media of different qualitative and quantitative ionic composition in either enhancing or inhibiting flagellation were evaluated. Distilled water or phosphate buffer was used as the basic induction medium, and osmotic pressure, pH, and concentrations of K^+, Na^+, Li^+, Ca^{2+}, and Mg^{2+} (all as chlorides) were varied. Willmer concluded that changes in osmotic pressure or hydrogen ion concentration were not in themselves responsible for flagellation, but rather the loss of essential ions from the ameba. K^+, Na^+, and Ca^{2+} acted similarly at concentrations greater than 50 mM to suppress flagellation. Magnesium ion was more potent, suppressing flagellation at concentrations greater than 6 mM. Lithium behaved like Mg^{2+}, also inducing a change in the shape of amebas that Willmer (1961) compared to the effect of Li^+ in causing vegetalization of developing embryos. Employing growth conditions (monoxenic or axenic populations) more defined than those used by Willmer, Fulton (1972) and Jeffery and Hawkins (1976b) tested a variety of electrolytes and found 50–90 mM to be the critical concentration range above which flagellation was prevented, the actual inhibitory concentration differing for each salt tested. Nonelectrolytes such as glucose and sucrose blocked transformation, the effect being osmotic (Fulton, 1972). In examining complete growth medium to determine which components, when removed singly, would permit flagellation to occur, Fulton (1972) found that yeast extract was inhibitory. Upon fractionation of yeast extract, a small organic molecule, termed *Factor,* was responsible for the block. *Factor* by itself was a potent inhibitor at an estimated concentration of 1 mM, but it also acted synergistically when employed in combination with electrolytes. Both pH (Yuyama, 1971) and temperature (Fulton and Dingle, 1967; Corff and Yuyama, 1978) shifts have been used for controlling transformation. High hydrostatic pressure applied for a short time also induces transformation (Todd and Kitching, 1973). Population density is not a factor controlling transformation (Napolitano *et al.,* 1967), as long as the population is not of such magnitude that gaseous exchange becomes a problem in the medium (Fulton, 1977a). Amebas from bacterized culture have a shorter latent period and higher flagellation response than amebas from axenic culture (Gong and Balamuth, 1973).

Upon induction of flagellation in *Tetramitus,* flagella arise at the posterior end of the polarized ameba and then, in the ensuing morphogenesis, the posterior end of the ameba becomes the anterior end of the flagellate (Hollande, 1942). Outka (1965) examined the transformation of *T. rostratus* in bacterized populations of amebas in depression chambers and reported that the flagellation response was proportional to the number of amebas present in microdrops. Osmotic pressure and ionic balance appear to have no direct effect upon transformation, but might have a secondary role by lowering metabolic activity of the amebas. Outka postulated the elaboration of a flagellate-inducing metabolic by-product into the medium by transforming amebas. Population density was also found to be a factor in *Tetramitus* flagellation by Brent (1965) and Fulton (1970b). The addition

of cell-free filtrates from populations of amebas and flagellates to transforming ameba populations failed to enhance flagellation (Brent, 1965),
raising doubts about a flagellate-inducing metabolite. Fulton (1970b) concluded that the critical factor at induction was oxygen deficiency; maximal
numbers of flagellates were produced in an atmosphere of 0.3% oxygen in
nitrogen. Other things being equal, *Tetramitus* gives less reproducible
flagellation results than *Naegleria*, and is the more difficult of the two
morphogenetic systems to work with. Axenic populations of amebas do
not transform (Outka, 1965), necessitating the use of bacteria-fed populations. Under the conditions employed by Fulton (1970b), flagellates were
not produced until ca. 3 hr (as compared to 1 hr for *Naegleria*) after
induction, and in somewhat lesser numbers (80%, as compared to $\geq 95\%$
for *Naegleria*). Others were less successful, with 2–24 hr required for
flagellation and 3–80% of the ameba population transforming (Brent, 1965;
Outka, 1965).

The precise mechanism by which *Naegleria* amebas transform remains
elusive. An ameba suspended in a nonnutrient, dilute induction medium is
faced with water uptake and cation loss (Willmer, 1961, 1970). A system in
the flagellate for pumping cations back into the cell was suggested. According to Willmer, the induction step might produce a change in membrane phospholipids, leading sequentially to surface alterations, a shift in
the orientation of adsorbed proteins and, finally, the synthesis of protein
fibers. Elevating cation concentration of course restores the ameboid
stage. Willmer's efforts were directed toward analysis of the role of
steroid hormones, their control of ion movement across the plasma membrane, and their effects on the integrity of the membrane and the transformation process (Willmer, 1970). The results of these experiments, however, were not conclusive (Pearson and Willmer, 1963). Fulton (1977b)
formulated an hypothesis to explain morphogenetic changes in *Naegleria*
based upon a substance, ψ, that Fuller and Fulton (1974) extracted
from amebas and flagellates. ψ is a small, heat-stable molecule re-
plete with a heat-labile ψ-ase enzyme that destroys it, and two inhibitors,
one heat-stable and the other heat-labile. In conjunction with intracellular
Ca^{2+}, a Ca^{2+}-dependent actin motility mechanism, and a Ca^{2+}-sensitive,
tubulin-based cytoskeleton, ψ is postulated to regulate intracellular Ca^{2+}
release from hypothetical Ca^{2+} reservoirs. Release of ψ causes intracellular Ca^{2+} release, necessary for maintenance of the ameboid shape. Compartmentalization of ψ at the onset of differentiation leads to a reduction in
Ca^{2+}, alteration in cell shape, assembly of the microtubular cytoskeleton,
and flagellate formation. Reversal of the sequence of events leads to reversion to the ameboid stage. The location of ψ in *Naegleria*, its continued
release in the ameboid stage, and its sequestration at the onset of differ-

entiation all remain to be demonstrated. Calcium ion concentration is also a factor in another theory explaining flagellation. Perkins and Jahn (1970) have postulated that modification of the Gibbs–Donnan equilibrium through alteration of the ionic environment causes flagellation and/or reversion in *Naegleria*. Dilution of the ionic environment produces a change in surface cation concentration, particularly of Ca^{2+} ions, and this in some undetermined way initiates flagellation. They suggest that a low Ca^{2+} concentration at the cell surface favors the ameboid stage and, by increasing surface Ca^{2+}, the flagellate stage is favored. The critical control mechanism in their theory is the Gibbs–Donnan ratio. Fulton (1977b) and Perkins and Jahn (1970) cite the large body of literature that deals with Ca^{2+}-induced changes in protoplasm (sol–gel interconversions, assembly of tubulin into microtubules), and the cell surface (adhesive properties).

Various classes of inhibitors are capable of blocking flagellation in *Naegleria*. Dinitrophenol and cyanide, which block oxidative phosphorylation, inhibit flagellation (Yuyama, 1971). Actinomycin D also blocks flagellation (Balamuth, 1965; Fulton, 1977a; Preston and O'Dell, 1971; Yuyama, 1971), as do the protein synthesis inhibitors cycloheximide and puromycin (Preston and O'Dell, 1971; Yuyama, 1971). Acriflavin, which binds to nucleic acid, is also inhibitory (Napolitano and Persico, 1966), as is mitomycin C (Yuyama, 1971). These compounds may act indirectly, however, since other inhibitors of DNA synthesis (FUdR and hydroxyurea) do not block transformation, and DNA synthesis is not required for transformation (Corff and Yuyama, 1976). Other inhibitory compounds include α-amanitin, cordycepin, and daunomycin (Preston and O'Dell, 1974), and the sulfhydryl compounds mercaptoethanol (Wade and Satir, 1968) and dithiodiglycol (Yuyama, 1971). A temporal factor is involved, since inhibitors are effective blocking agents only if applied before a critical transition point (Fulton, 1977a; Gong and Balamuth, 1973; Jeffery and Hawkins, 1977). The system is sensitive to inhibition of RNA synthesis halfway, and to inhibition of protein synthesis three-quarters of the way to assembly of the flagellar apparatus. Transforming amebas are insensitive to colchicine (Yuyama, 1971), which is unexpected since this compound is an effective inhibitor of microtubule assembly. This may reflect a permeability problem rather than an actual insensitivity, since this difficulty has been encountered in previous studies using *Naegleria* (Fulton, 1970a). Preston and O'Dell (1971) reported that uptake of inhibitor is facilitated by the presence of polymyxin B. In general, the concentrations of inhibitors such as actinomycin D employed are unphysiologically high, and *Naegleria* amebas can remain active for prolonged time periods at inhibitor levels that are lethal for other cell systems (Bols *et al.*, 1977).

Inhibitor studies indicate that synthesis of RNA and protein occurs during transformation, which is expected since the ameba forms a kinetic apparatus during this period. A qualitative shift in RNA occurs in differentiating cells, with more mRNA and less rRNA as a result (Walsh and Fulton, 1973). A rapidly labeled, messengerlike RNA has been detected in transforming populations, with the RNA species first appearing in the nucleoplasm and then shifting to the cytoplasm (Walsh and Fulton, 1973). The change in type of RNA occurs without a change in the amounts or types of the three detectable RNA polymerases produced by *Naegleria* (Soll and Fulton, 1974). Suspension of amebas in 80 mM NaCl inhibited flagellation but not RNA synthesis associated with morphogenesis (Jeffery and Hawkins, 1977). Flagellar tubulin has been extracted and purified from the outer doublets of *Naegleria* axonemes (Kowit and Fulton, 1974a). It has a molecular weight of 55,000, which is similar to that of tubulins from other sources, but differs from other tubulins in the greater amount of guanine nucleotide present. Purified tubulin was used to detect, by immunochemical assay, the relative amounts of tubulin antigen present in amebas and flagellates. Kowit and Fulton (1974a) conclude that a 35- to 55-fold increase in tubulin antigen occurs during the period of transformation, which is consistent with *de novo* synthesis of the kinetid and not merely a change in antigenicity of preexisting tubulin (Kowit and Fulton, 1974b). Characterization and comparison of RNAs from amebas and flagellates indicated that they were the same, with molecular weights of 1.25 and 0.75 × 10^6 for large and small RNA molecules (Jeffery and Hawkins, 1976a). There appears, therefore, to be no change in the ribosomal genes transcribed at differentiation.

At the time of transformation the synthetic machinery of the ameba is mobilized for RNA and protein production, bringing to a halt DNA synthesis (Corff and Yuyama, 1976). Levels of thymidine kinase, a thymidine-phosphorylating enzyme, have been examined in differentiating populations (Bols *et al.*, 1977). Thymidine kinase activity declines, though a second phosphorylating enzyme, nucleoside phosphotransferase, remains constant. Although the significance of the thymidine kinase decline is not known, the change in enzyme level denotes a qualitative shift in the type of protein synthesized upon induction. The decline can also be demonstrated in populations of transforming *Naegleria* in which synthesis of mRNA and protein has been blocked by inhibitors. Nor is the thymidine kinase decline induced by washing amebas free of growth medium to induce flagellation. The same drop in enzyme level was observed when flagellation was accomplished by temperature shift (32° to 20°C) as a means of inducing transformation (Corff and Yuyama, 1978).

The lectin Con A inhibits flagellation in *Naegleria* (Fulton, 1977a; Perkins and Morris, 1974; Preston and O'Dell, 1974). Some of the effects produced include rounding and agglutination of cells, adhesion and immobilization of flagella, and initiation of cytoplasmic streaming. Fulton (1977a) associates Con A inhibition of flagellation with cell damage, though replacement of the Con A medium by buffer reverses the inhibition and flagellates form (Perkins and Morris, 1974). Also, Preston *et al.* (1975) report resumption of ameboid mobility after shedding of a Con A–label complex in the ameboid stage. Perkins and Morris (1974) suggest that Con A functions as a depolarizing agent, favoring reversion to the ameboid stage.

V. ECOLOGY

The ecology of soil amebas is in reality the study of two habitats, soil and water. The two are not mutually exclusive, and no sharp barrier exists between them where soil amebas are concerned, as might be true for large, ciliated protozoa, which require a fluid environment for swimming. Furthermore, for ameboflagellates, addition of water to the soil habitat is probably essential to flagellation and dissemination of the flagellate stage. Sufficient water, bound to soil particles, may allow flagellation and swimming to occur in the absence of actual flooding of the soil substrate. Singh (1975) has recently reviewed the literature on soil protozoology as it affects the small amebas. The ecology of the pathogenic strains of soil amebas is developing as a separate area, one whose main focus is epidemiology, and is treated in Section VII,D. These amebas lend themselves well to various ecological analyses: quantitation of organisms, differential counting of cysts versus trophozoites, testing of food preference, predator–prey interactions, and the use of small amebas as bioassay organisms for soil and/or water pollution.

A. Sampling

Singh (1975) emphasizes two considerations in sampling soil: (1) use of a nonnutrient agar substrate to retard growth of indigenous soil bacteria and fungi, and (2) provision of an edible bacterial species as food for amebas. The first criterion is satisfied by using water or saline agar as a substrate, and the second by supplying bacteria such as *A. aerogenes* (variously called *Enterobacter aerogenes* or *Klebsiella pneumoniae*) or *E. coli*. Antibiotics are generally not necessary to limit growth of undesirable bacteria and fungi. Hydrated soil samples can be plated directly onto the agar with or without dilution. Menapace *et al.* (1975) employed an agar overlay technique that restricts ameboid movement and enhances plaque formation as the amebas feed on bacteria. Once grown out, amebas can be freed from contaminating bacteria and fungi by migration techniques, with or without a streak of edible bacteria as a food source to encourage movement (Harrison, 1957; Neff, 1957, 1958). Testing of isolates for the ability to flagellate is accomplished by transferring small blocks of agar containing amebas to distilled water. Clonal populations can be established by dilution, migration, or any of several micromanipulative procedures.

Sampling from aqueous habitats is performed by direct plating of water samples or, to ensure a denser starting population, filtering of the water and plating of the filter (Jamieson and Anderson, 1973). It is virtually impossible *not* to get amebas from either soil or water samples, the most common isolates being *Acanthamoeba, Naegleria,* and *Vahlkampfia.* Menapace *et al.* (1975) found no *Naegleria* in their study and attributed its absence to lack of moisture in the sampled soils.

The typical spectrum of soil amebas is not found in seawater habitats; but there are exceptions, notably *Acanthamoeba* (Davis *et al.,* 1978; Sawyer *et al.,* 1977). Unlike *Naegleria* and its immediate allies from soil, *Acanthamoeba* is relatively tolerant of high osmolarities, which accounts for its presence in seawater and its ability to overgrow cells in tissue culture systems (Jahnes, *et al.,* 1957).

B. Food and Feeding

Soil amebas are voracious consumers of bacteria, provided the food is of the right type. Bacterial types range from those that are readily eaten to inedible forms (Singh, 1975). Given a choice between bacterial species, pigmented types are generally among the less preferred food organisms, owing probably to the toxicity of the pigments. Menapace *et al.* (1975) found that gram-positive forms were consumed more readily than gram-negative forms, but Upadhyay (1968) found the reverse to be true. A variety of gram-negative bacilli were used successfully by Anderson and Jamieson (1974) for isolation and growth of *Naegleria,* in addition to heat-killed *E. coli* (Jamieson and Anderson, 1973). Drożański and Drożańska (1961) found that heat-killed gram-negative bacteria—but not gram-positive bacteria—were unsuitable for hartmannellid amebas.

Feeding upon bacteria is facilitated by ameba enzymes that have been reported to lyse food organisms. A constitutive bacteriolytic enzyme, capable of lysing *Micrococcus lysodeikticus* and its cell wall, was released into the culture supernatant by *A. castellanii.* Subsequent studies demonstrated that it was a hexosaminidase with a pH optimum of 5 (Drożański,

1972). Upadhyay *et al.* (1977) characterized a bacteriolytic enzyme from *A. glebae* as a muramidase (lysozyme). Tracey (1955) detected chitinase and cellulase in soil ameba strains.

Although amebas are capable of attacking plant and animal polysaccharides by virtue of their enzymes, the reverse is also true. Culture filtrates of the fungus *Alternaria* sp. contain cellulase, chitinase, and protease activities which, in concert, were found capable of degrading cysts of *A. culbertsoni* over 2–3 days (Verma *et al.*, 1974).

C. Population Dynamics

Glucose or hay-water added to soil samples enhanced the growth of amebas and bacteria (Menapace *et al.*, 1975). Of the two additives tested, glucose caused a more rapid stimulation of growth which decreased with time. The response to hay-water was slower, but larger ameba populations were ultimately produced.

The predator–prey relationship between soil amebas (predators) and bacteria (prey) has been examined by Danso and Alexander (1975). With abundant prey organisms present, the ameba population increased proportionally. As the number of prey decreased through feeding, the likelihood of predator encountering prey decreased and the ameba population stabilized. Increasing the number of food organisms in the system led to another increase in amebas. Prey density appears, therefore, to set a limit on continued increase in the ameba population. Amebas are well adapted to fluctuations of prey population since, when food is scarce, amebas are likely to encyst. Another type of predator–prey relationship was postulated by Nielson *et al.* (1978), in which soil amebas function as a mechanism of selection. They reported that feeding by amebas may limit populations of the pathogenic yeast *Cryptococcus neoformans* in soil, a conclusion based on selective feeding of *Acanthamoeba polyphaga* on virulent yeast cells *in vitro,* leaving avirulent pseudohyphal colonies.

D. Soil Amebas As Bioassay Organisms

Pesticides of different chemical types, polychlorinated biphenyls, and various metals have been tested *in vitro* on *A. castellanii.* The results suggest that this organism, and perhaps other soil amebas, may be useful in evaluating the effects of these and other chemicals on soil or water fauna (Prescott and Olson, 1972; Prescott *et al.*, 1977).

E. Intracellular Symbioses

Drożański (1963) reported the presence of motile gram-negative bacteria in association with amebas from soil. Amebas ingested the bacteria, which multiplied within food vacuoles, causing enlargement of the vacuole, rounding up of the ameba, and its eventual lysis. The bacterium could not be cultivated independently of the ameba.

Another sort of relationship has been described in the EG_s strain of *N. gruberi* involving the ameba and viruslike particles (Schuster and Dunnebacke, 1971, 1976b). Viruslike elements develop within the nuclear compartment of amebas, pass across the envelope, and enter the cytoplasm where a distinct structure forms about the particles using membranes synthesized *de novo*. The particle-containing membranous body transforms into a sphere which is ejected by the ameba into the medium. Spheres are picked up by other amebas, either directly or indirectly, and this is the presumed mode of transmission of the viruslike particles. Amebas containing viruslike elements do not encyst or flagellate, suggesting that the presence of the particles interferes with normal synthetic pathways. Infected cells ultimately undergo lysis.

Viruslike particles were not observed in amebas grown axenically but were induced by growth in the presence of living bacteria (Schuster, 1969; Schuster and Dunnebacke, 1974), though the bacteria were not the source of the particles. It was subsequently shown that BUdR could induce particle formation in axenic populations of the ameba (Schuster and Clemente, 1977). Evidence, therefore, suggests that the particles are present in a latent or temperate form within the nuclei of the EG_s strain of *N. gruberi*. How particles are induced by food organisms is unknown, but may be related to DNA synthesis as indicated by the efficacy of BUdR in causing particle development. The presence of latent particles may be more widespread in soil organisms than the relatively few reports that have appeared in the literature suggest (Lemke, 1976).

In the pathogenic CJ strain of *N. fowleri*, Phillips (1974) reported an apparent intracellular diphtheroid. This symbiont could be eliminated by antibiotic treatment of the amebas, but growth of the amebas in culture was better with the diphtheroid present. Bacterialike endosymbionts were demonstrated using electron microscopy in the cytoplasm of an *Acanthamoeba* strain by Proca-Ciobanu *et al.* (1975), who suggested that the pathogenicity of the isolate may be related to the presence of the endosymbionts. Jadin (1976) has postulated that soil amebas may serve as vectors of microorganisms, based on a study demonstrating intracellular multiplication of mycobacteria, including *Mycobacterium leprae*. An apparently stable intracellular relationship, persisting over a period of 15 days, between *Acanthamoeba castellanii* and several species of mycobacteria developed when the bacteria were fed to axenically grown amebas (Krishna Prasad and Gupta, 1978).

VI. NUTRITION AND GROWTH

No problem exists in obtaining amebas from soil or water samples, and the use of antibiotics has made it reasonably simple to rid isolates of contaminating bacteria, establishing them in monoxenic or axenic cultures. Monoxenic cultures have certain advantages in that large numbers of amebas can be grown out with little effort within a short period of time. Establishing some amebas in axenic culture can be difficult—in some cases, seemingly impossible. Axenic cultures may not produce as heavy populations as bacteria-fed cultures, but they are nevertheless essential for critical studies of growth, metabolism, differentiation, and so on. The ultimate goal of nutrition studies is the formulation of defined media for soil amebas; so far, this has been achieved only for *Acanthamoeba*. *Naegleria* is more exacting in its nutritional requirements, and *Tetramitus* even more so.

A. *Acanthamoeba*

Neff *et al.* (1958) studied metabolism and defined growth conditions for *Acanthamoeba*. Proteose–peptone was a satisfactory nitrogen source and, of the various carbon sources tested, glucose supported the heaviest growth. Based on chemical analysis of reducing material in the culture supernatant, glucose was found to be utilized at a slow rate. Working with *A. castellanii,* strain HR, and labeled glucose, Band (1959) demonstrated, however, that glucose was actively assimilated and served as an energy source during growth. Subsequent efforts led to the formulation of defined media for *Acanthamoeba*. A medium consisting of 18 amino acids, acetate as a carbon source, and the vitamins B_{12} and thiamine was used by Adam (1959) for the cultivation of *A. castellanii*. Band (1961, 1962) demonstrated a requirement for biotin and devised a medium composed of seven amino acids, glucose as a carbon source, B_{12}, thiamine, and biotin. Adam (1964a) confirmed the requirement for biotin, and Adam and Blewett (1967) further refined the *Acanthamoeba* medium to 5 of the 10 essential amino acids plus glycine, and an energy source. The various strains of *Acanthamoeba* differed in their ability to utilize the carbohydrate sources added. The type of carbon source added to the medium had a modifying effect upon the amino acid requirements (Dolphin, 1976); with acetate (instead of glucose) as the sole carbon source, glycine was required. Dense populations of *A. castellanii* (3×10^7 amebas/ml, with a 6-hr generation time) were produced by Jensen *et al.* (1970), who used a peptone–yeast extract–glucose medium and shaking; amebas, however, did not encyst in this medium.

Thermolabile factors are produced in axenic cultures of *Acanthamoeba* that limit cell number (Pigon, 1970). Contacts between cells in culture may lead to an exchange of material. The introduction of conditioned medium to cell cultures reduces exchanges and shortens periods of contact between amebas (Pigon and Morita, 1973). Pigon (1978) found that amebas in culture exude material into the medium from their cell surface. *Acanthamoeba* transferred into medium containing exudate bind the material to their surface.

B. *Naegleria*

Attempts to grow *Naegleria* axenically met with more difficulties than those leading to cultivation of *Acanthamoeba*. Chang (1958), working with an isolate of *N. gruberi* from river water, eliminated contaminating bacteria but could not grow amebas without live food organisms. Heat-killed bacteria (60°C for 60 min) produced poor growth of amebas. *Aerobacter aerogenes* subjected to a shorter heat treatment (100°C for 1 min) supported growth of *N. gruberi* (Schuster, 1961). Heat-killed bacteria were added to a rich nutrient medium, as used by Brent (1954) for the growth of *Tetramitus,* consisting of yeast extract, peptone, and liver concentrate. Though these results suggested that a heat-labile factor in bacteria was needed by *Naegleria,* O'Dell and Brent (1974) obtained growth of *Naegleria* on *E. coli* subjected to 121°C for 10 min. Alcohol-killed bacteria have also been used successfully to support growth of *Naegleria* in short-term cultures (Napolitano *et al.,* 1977; Napolitano and Gamble, 1978). Schuster and Svilha (1968) used chick embryo extract as a supplement in place of dead bacteria.

A distinct improvement in cultivation of *Naegleria* was the use of serum or a serum fraction (IV-4) added as a supplement to a basic medium of proteose–peptone, yeast extract, liver concentrate, and glucose (Balamuth, 1964). Band and Balamuth (1974) found that hemin (1 μg/ml) could replace the serum fraction or fetal calf serum. They obtained yields of 5×10^6 amebas/ml, with a generation time of 8 hr. Hemin, however, did not support growth unless proteose–peptone was used in the basic medium (Fulton, 1977a). A variant strain of *N. gruberi* was cultivated by Fulton (1974) in a simplified medium in which proteose–peptone was replaced by a single amino acid, L-methionine; a serum fraction was still required.

O'Dell and Stevens (1973) employed a Casitone-based medium supplemented with serum and red blood cell lysate to grow a strain of *Naegleria* isolated as a tissue culture contaminant. They obtained yields of 5×10^6 amebas/ml with a 9-hr generation time. The same yields were reported for *N. gruberi* by Weik and John (1977b) with an improved generation time of 7 hr, using a hemin-supplemented medium. Agitation of cultures growing in fernbach flasks, using a gyrorotary shaker, produced two phases of logarithmic increase. The growth-promoting action of shaking suggested that oxygen deficiency and not waste accumulation in the medium limited logarithmic multiplication. These investigators observed various culture changes during the second log phase: a reduction in initial growth rate (from 7- to 19-hr generation times), a decrease in oxygen tension, a doubling of mean cell size, an increase in cell dry mass, and a DNA increase. On the basis of their finding that amebas did not metabolize glucose during log phase, they concluded that amino acids served as carbon and energy sources. This is also supported by the gradual increase in pH, presumably from the deamination of amino acids.

In contrast to *N. gruberi*, pathogenic *N. fowleri* strains grow in less rich media. Červa (1969, 1977) grew *N. fowleri* in a Casitone-based medium with serum as an additive, and Nelson has formulated a basal medium (unpublished) using glucose plus liver infusion (cited in Wong *et al.*, 1975a) or glucose plus Panmede liver digest (cited in Weik and John, 1977a), both media containing serum at 9 and 2%, respectively. Using the Casitone–serum medium, Červa (1977) obtained generation times of ≥ 7 hr, depending upon the growth temperature. The optimal growth temperature was dependent upon culture medium composition. Weik and John (1977a) used Nelson's medium for mass cultivation of *N. fowleri*, obtaining yields of 3×10^6 amebas/ml with a 5.5-hr generation time. They found a predominantly aerobic metabolism but only slight utilization of glucose. As in *N. gruberi*, amino acids appear to serve as carbon and energy sources.

C. *Tetramitus*

Still more difficult to grow axenically than *Naegleria* is *Tetramitus*. Brent (1954) established the ameboid stage of *Tetramitus* in a yeast extract–peptone–liver concentrate medium with autoclaved bacteria. In a later study, Brent and Paxton (1974) reported growth of the flagellate stage in a trypticase–yeast extract medium to which autoclaved *E. coli* was added. The growth-promoting factor present in the bacteria appeared to reside in the bacterial ribosomes. Generation time in this medium was 17 hr. Yeast extract could be replaced by elevation of the trypticase level and addition of a mixture of B vitamins. The heat-killed bacteria could be replaced by purified ribosomes of *E. coli,* though it is suggested that their importance may be less nutritional and more as a stimulant for engulfment of particles and formation of food vacuoles. Outka (personal communication) has maintained the flagellate stage of *Tetramitus* in a liver-based axenic medium with autoclaved bacteria as a food supplement.

VII. SOIL AMEBAS AS PATHOGENS

One of the more striking shifts in research on this group of organisms followed the recognition of soil amebas as lethal pathogens. Unlike *Entamoeba histolytica* and other assorted species of amebas found in humans and other vertebrates, soil amebas are not obligate parasites. They are free-living forms, with a wide distribution in soil and water, that invade and cause disease in humans and other organisms when conditions are right. For this reason, they might best be regarded as opportunists rather than parasites, since they exploit an as yet largely undefined set of circumstances in becoming established in human hosts. The ability to occupy free-living as well as endozoic ecological niches is reflected in the term amphizoic, which Page (1974b) has proposed to characterize the catholic tastes in habitats exhibited by the small amebas. The severity of the disease that some of these soil amebas cause may be indicative of poor adaptation to a parasitic existence on the part of these protozoa. Appreciation of the disease potential of pathogenic soil amebas has engendered widespread interest in their identification, cultivation, nutrition, or ecology, paralleling research efforts on the biology of their nonpathogenic counterparts. This section concentrates on the features associated with, or unique to, the pathogenic variants and their disease-producing capacity. In the relatively short time since their role as pathogens was recognized, a large body of reports has appeared in the literature, including several comprehensive reviews (Carter, 1972; Červa, 1975; Chang, 1974; Culbertson, 1971; Griffin, 1978; Jadin, 1974; Lockey, 1978; Singh, 1973, 1975; Willaert, 1977b).

A. Tissue Culture Contaminants

Prior to the recognition of their role as pathogens, amebas were found as contaminants of tissue cultures both in routine maintenance (Jahnes *et al.,* 1957) and in safety tests of polio vaccine (Culbertson *et al.,* 1958, 1959), except that at the time they often went unrecognized. They were described as transformed tissue culture cells; or the devastation caused as the amebas ate their way across tissue culture monolayers was attributed to viruses (lipovirus or Ryan virus). Identification, however, was not long in coming, and in most of these situations the contaminant was *Acanthamoeba* (Armstrong and Pereira, 1967; Casemore, 1969; Dunnebacke and Williams, 1967; Moore and Hlinka, 1968; Peloux *et al.,* 1974; Warhurst and Armstrong, 1968). *Naegleria,* while it can grow in tissue culture media (Cursons and Brown, 1978; Chang, 1974; De Jonckheere and Van De Voorde, 1977a; Visvesvara and Callaway, 1974), does not adapt as readily to the relatively high osmolarity of the media employed as *Acanthamoeba* does (Wong *et al.,* 1977). The logical continuation of this research was demonstration of the pathogenicity of these contaminants for laboratory animals (Culbertson *et al.,* 1958). The study of Culbertson and co-workers (1958), in addition to revealing an unsuspected ability of soil amebas, alerted others to their potential as pathogens and, within a short period of time, reports implicating them as the causal agent of a newly described disease of humans, primary amebic meningoencephalitis (PAM), appeared (Butt, 1966; Červa and Novak, 1968; Fowler and Carter, 1965).

B. Primary Amebic Meningoencephalitis

The early reports of PAM were not specific about the identity of the causal agent, except that it was an ameba. *Acanthamoeba* was already known as a pathogen and not easily distinguished from *Naegleria.* With continued descriptions of new cases and the isolation of causal agents, it appeared that *Naegleria,* not *Acanthamoeba,* was responsible for PAM (Culbertson, 1971). Now, with careful analyses of cases, retrospective studies of autopsied tissues, and the application of immunochemical methods for identification, *Acanthamoeba* has also been identified as a causal agent of meningoencephalitis (Martinez *et al.,* 1977a; Willaert *et al.,* 1978a). Differences exist between meningoencephalitis caused by the two amebas. Cases in which *Naegleria* is the causal agent are rapidly fatal, the victims almost invariably having had a history of swimming or bathing in water that presumably contained the pathogen. Victims are generally young—children, teenagers, or young adults—and in good health (Duma *et al.,* 1969, 1971). The portal of entry of the ameba is probably across the nasal mucosa, amebas then migrating along the olfactory nerves to the olfactory lobes of the brain where, within a matter of days, they proliferate and do extensive damage to tissue. In contrast, *Acanthamoeba* infections are chronic and not necessarily the result of exposure to swimming or bathing. The portal of entry is uncertain but need not be limited to the nasal passages and the olfactory mucosa. The invasion of brain tissue is probably secondary to establishment of amebas at some other point in the body. Victims are of all ages and often reported to be in poor health. A feature useful in distinguishing the two different amebas in brain tissue is that *Acanthamoeba* encysts, but *Naegleria* does not (Martinez, 1977; Willaert, 1977b).

A detailed discussion of the disease as a medical phenomenon is beyond the scope of this chapter. Several points concerning PAM are relevant to the biology of the pathogens and are examined here, with emphasis on laboratory studies. Invasion of the central nervous system by *Naegleria* occurs most likely by a nasal route. This has been substantiated by laboratory studies in which mice have been infected by intranasal instillation of amebas. Introduced into the nasal passages, amebas cross the nasal mucosa, invade (or are phagocytosed by) sustentacular cells that make up part of the olfactory epithelium, and ultimately, after migrating along olfactory nerves, cross the cribriform plate to enter the brain cavity (Martinez *et al.*, 1973). This mode of entry in human infections is supported by the extensive damage done to the olfactory lobes of human PAM victims. Once in the central nervous system, amebas reproduce, causing extensive damage to host tissue (Martinez *et al.*, 1971; Schuster and Dunnebacke, 1977; Visvesvara and Callaway, 1974). It is not clear whether the amebas penetrate by direct pseudopodial activity, literally shoving their way through the neuropil, or whether their progress is aided by enzymes that act upon the cells, causing lysis. Food vacuoles seen within trophic amebas in the brain contain host tissue in various stages of digestion. Large numbers of amebas come to occupy perivascular spaces and may use them as channels for deeper penetration into the brain. An inflammatory response is suggested by association between amebas and host phagocytic cells (Figures 15 and 16). Schuster and Dunnebacke (1977) reported the presence of intranuclear particles of an unknown nature in both amebas and brain cells of mice dying from PAM. Myocarditis of unknown etiology has been reported in several human cases of PAM (Markowitz *et al.*, 1974).

Adams *et al.* (1976) showed that mice which survived intravenous or intraperitoneal infection with *Naegleria* were resistant to intravenous challenge with pathogens. Formalin-treated or freeze-thawed pathogenic or live nonpathogenic *Naegleria* administered to mice intravenously protected animals from challenge with pathogens via intranasal or intravenous routes (John *et al.*, 1977). Live *N. fowleri* administered intraperitoneally protected mice from subsequent intranasal instillation of pathogens, with a 27% survival rate (Thong *et al.*, 1978c). Ameba dose and age, sex, and strain of mouse were variables found to affect mouse mortality following intravenous inoculation with *N. fowleri* (Haggerty and John, 1978). Culbertson *et al.* (1972) used subcutaneous inoculation of *Naegleria*

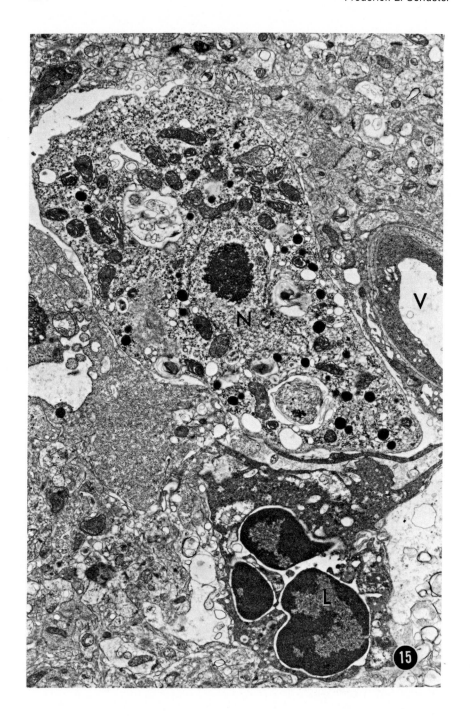

amebas into guinea pigs, producing loss of strength and weight and, in some cases, death. Some deaths were also reported by Diffley *et al.* (1976) in subcutaneously inoculated guinea pigs. In neither of these studies was the central nervous system involved. Germ-free guinea pigs were used by Phillips (1974) in a study in which pathogens were inoculated intranasally, intraorally, into skin lesions, and into the conjunctival sac of the eye; the intranasal route caused death, and *Naegleria* was found in the brain. Wong *et al.* (1975a) successfully inoculated monkeys with *N. fowleri*. The variability encountered in that study was attributed to the culture history and virulence of the amebas used, as well as to age and other host factors.

Variability has also been encountered in producing mouse mortality using pathogenic amebas with long culture histories (De Jonckheere *et al.*, 1975; Visvesvara and Callaway, 1974; Wong *et al.* 1977). This is most likely due to loss of virulence upon continued *in vitro* cultivation, which can be restored by repeated animal passage. Intracerebral inoculation is more effective than intranasal instillation, since with the latter technique there is some question about the number of amebas actually introduced. Singh and Das (1972b) have produced PAM in mice using the flagellate stage of pathogenic *Naegleria* which, they reason, is more likely to be encountered by humans in pools, lakes, etc.

Human PAM infections have resulted from contact with amebas in swimming pools (Červa and Novak, 1968; Jadin *et al.*, 1971), lakes (Butt *et al.*, 1968; Callicott *et al.*, 1968; Duma *et al.*, 1969, 1971), a mineral hot pool (Cursons and Brown, 1975; Mandal *et al.*, 1970), thermal streams (Cursons *et al.*, 1976; Van Den Driessche *et al.*, 1973), mud puddles (Apley *et al.*, 1970), and a household water supply (Anderson and Jamieson, 1972b). Since the first report of PAM little more than a decade ago, there have been more than 90 cases reported in the literature.

Pathogenic *Naegleria* isolated from PAM victims were initially classified as *N. gruberi*, the familiar and widespread species. Červa (1970) compared three strains of pathogenic *Naegleria* from Czechoslovakia and the United States and found them to be relatively similar. Differences between the pathogenic isolates and *N. gruberi* were, however, apparent, which led to the creation of a new species by several different individuals to include the pathogenic strains. The new species was described by Carter (1970) as *N. fowleri* on the basis of a comparative study of the pathogenic and non-

---

Plate VI: Figure 15.   Trophic *N. fowleri* as seen in the brain of an experimentally infected mouse. An ameba with a characteristic vesicular nucleus (N) is seen alongside a blood vessel (V). Note the presence of food vacuoles within the ameba containing cellular debris in various stages of digestion. A leukocyte (L) is in close proximity to the ameba. Magnification: × 10,500.

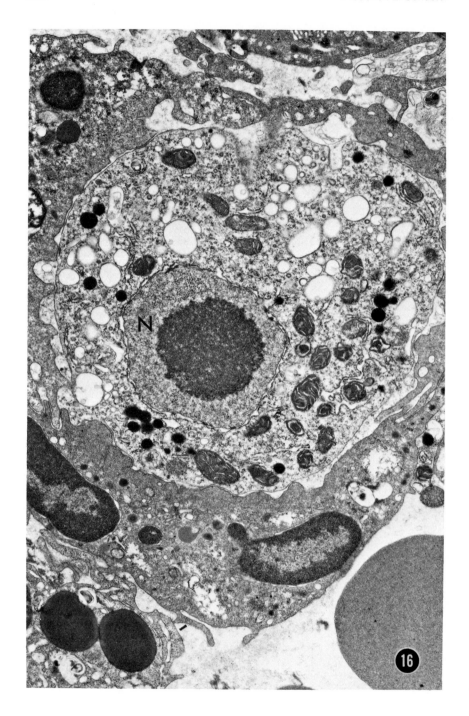

pathogenic variants. Shortly thereafter, the pathogenic variant was described in the literature as *Naegleria aerobia* and *N. invades* following the studies of Singh and Das (1970) and Chang (1971), respectively. Carter's report and species designation have priority, and *N. fowleri* is generally the name used for the pathogenic *Naegleria* strains.

C. Comparison of *Naegleria gruberi* and *Naegleria fowleri*

The major differences recognized by Carter (1970) between the two species, aside from the most obvious difference in virulence, are in cyst structure, growth temperature optima, electron microscope appearance of the mitochondria, and conditions for bacterized culture. Subsequent studies by others confirmed most of these distinctions, adding differences in axenic culture conditions and antigenicity. These differences are described in greater detail:

1. Pathogenicity

Isolation of strains of *N. fowleri* from the cerebrospinal fluid of PAM victims is ipso facto proof of their virulence. Numerous studies, however, have compared *Naegleria* spp. with regard to their ability to cause PAM in experimental animals. *Naegleria gruberi* strains do not cause PAM; *N. fowleri* isolates either from PAM victims or from nature cause death of mice within a period of 5–15 days (Culbertson *et al.,* 1968). Ability to cause death of mice is subject to a number of variables, including age of mice, culture history of the ameba strain, ameba dose, and method of inoculation.

2. Cyst Structure

Evidence of differences in cyst structure is based on light and electron microscope observations of the two species. Initial reports claimed that, while *N. gruberi* had distinct pores in the cyst wall, *N. fowleri* cysts contained no such pores, and indeed the wall was shown to have a smooth outline (Lastovica, 1974; Page, 1975). Comparative studies of cysts of the two species, using the scanning electron microscope, demonstrated the presence of pores in *N. fowleri* cysts that were flush with the surface of the cyst wall and not readily obvious at the light microscope level (Cursons and Brown, 1976; Jadin *et al.,* 1974). *Naegleria fowleri* cysts had an average of 1.6 pores per cyst, while *N. gruberi* cysts averaged 7.2 pores per

---

Plate VII: Figure 16.   Trophic *N. fowleri* from the brain of a mouse. The nucleus of the ameba is seen at N. The ameba is completely encircled in this section by host leukocytes. Magnification: ×15,000.

cyst (Jadin *et al.*, 1974). Van Saanen (1977) reported an average of 1.7 pores per cyst for *N. fowleri*. Transmission electron microscopy confirmed the presence of pores (Schuster, 1975a); the collar around the cyst pore of *N. gruberi,* making it obvious at the light microscope level, was not present around *N. fowleri* pores. *Naegleria fowleri,* upon excystment, appears to digest the cyst wall partially, as suggested by a reduction in thickness of the wall during and following excystation (Schuster, 1975a).

3. Temperature Tolerance

Pathogenic *N. fowleri* amebas grow well at 37°C, as might be expected, but also continue to grow at 42°C (Singh and Hanumaiah, 1977b) and 45°–46°C (Griffin, 1972). Strains of *N. gruberi* grow at temperatures up to 37°C, but generally not beyond. Singh and Hanumaiah (1977b), however, isolated nonpathogenic *N. gruberi* from soil which grew at 42°C, and Stevens *et al.* (1977b) also found nonpathogenic, thermophilic amebas. Thus, temperature tolerance is not a foolproof method of distinguishing between the two species, nor does the ability to tolerate mammalian body temperatures signify that an ameba strain is pathogenic.

4. Morphological Differences

Carter (1970) reported that mitochondria of *N. fowleri* had a cup or dumbbell shape in contrast to the oval organelles of *N. gruberi*. The distinction is not, however, consistent, and both types of mitochondria are evident in transmission electron micrographs of pathogenic variants (Carosi *et al.,* 1977; Maitra *et al.,* 1974a; Rondanelli *et al.,* 1976; Schuster and Dunnebacke, 1977; Visvesvara and Callaway, 1974). The greater number of mitochondria with unusual shapes in *N. fowleri* may reflect some aspect of growth conditions, such as oxygen availability.

5. Nutritional Differences

Both species grow on nonnutrient agar with living bacteria (*E. coli* is suitable) as a food source. In axenic culture, a nutritional difference between pathogenic and nonpathogenic amebas is evident and not unexpected. The direction of the difference is surprising. Pathogenic and/or parasitic species generally have more exotic nutritional requirements than their free-living counterparts. With *Naegleria* spp., however, pathogenic strains are less discriminating than the nonpathogenic variants. Strains of *N. gruberi* require a rich nutrient medium supplemented with calf serum or a serum fraction (Balamuth, 1964; Fulton, 1970a) for axenic growth. Strains of *N. fowleri* grow in a serum-supplemented, relatively low-nutrient medium (Červa, 1969, 1977; De Jonckheere, 1977; Nelson, unpublished observations, cited in Wong *et al.*, 1975a; also cited in Weik and John, 1977b), which does *not* support growth of *N. gruberi* or non-pathogenic variants of *N. fowleri* (De Jonckheere, 1977). Červa (1978) reported that *N. gruberi* favored phagotrophy and *N. fowleri* osmotrophy as means of nutrient uptake, which is confirmed by the observation of Thong *et al.* (1978a) that *N. gruberi*, but not *N. fowleri*, actively ingested baker's yeast.

6. Antigenic Differences

A number of studies have revealed antigenic differences between the two species using agglutination testing (Anderson and Jamieson, 1972a), a fluorescent antibody technique (De Jonckheere *et al.*, 1974; Van Dijck *et al.*, 1974), and gel diffusion and immunoelectrophoresis (Visvesvara and Healy, 1975b; Willaert, 1977a). Protein patterns resulting from disc electrophoresis of the two species are different (Hadas *et al.*, 1977).

7. Miscellaneous Distinctions

Several other differences have been reported. Incorporation of 0.5% NaCl into the medium inhibits pathogenic *N. fowleri* but not *N. gruberi* (Singh and Das, 1970), though this has been reexamined and disputed by Červa (1978). Cysts of *N. fowleri* are more rapidly destroyed by chlorine than those of *N. gruberi*—1 hr as compared to 3 hr at a chlorine concentration of 0.5 μg/ml (De Jonckheere and Van De Voorde, 1976). Optimal pH for *N. gruberi* growth was 6.0–6.5, and 6.5 for *N. fowleri* (Červa, 1978). *Naegleria fowleri* tolerated higher viscosity, produced with 0.5% methylcellulose, than *N. gruberi* (Červa, 1978). Concanavalin A agglutinates *N. gruberi* but not *N. fowleri* amebas (Josephson *et al.*, 1977; Stevens and Stein, 1977). Microfilaments, possibly representing contractile proteins, are more likely to be found in *N. fowleri* amebas (Lastovica, 1976). Pathogenic strains appear more sensitive to the antibiotic amphotericin B (Am B) (Schuster and Rechthand, 1975). A general comparison of macromolecular properties of the two species in culture was made by Weik and John (1978).

D. Ecology of Pathogenic Amebas

Considerable interest has developed in the physical and biological factors influencing the ecology of pathogenic *Naegleria* amebas. This concern is not simply academic, since there is strong incentive to define conditions under which amebas might proliferate, which in turn relates to the epidemiology of PAM. The tolerance of pathogenic strains for temperatures in excess of 37°C, the occurrence of PAM during warm months of the year, and the ability of these amebas to thrive in warm water, particularly where bacteria might be present as a food source, provide the basis for understanding the ecology of the pathogens. Attempts to isolate pathogens from particular streams, pools, etc., where humans have contracted PAM have been largely unsuccessful, which Griffin (1972) attributes to the high temperatures required for their optimal growth. At room temperature, nonpathogenic *N. gruberi* can outgrow the high-temperature amebas. With a greater awareness of the use of temperature as a tool in screening for pathogens, many isolates of pathogenic *Naegleria* and *Acanthamoeba* have been obtained from a variety of sources.

Singh and Das (1972a) isolated both pathogenic *Naegleria* and *Acanthamoeba* from sewage sludge, using 37°C as the incubation temperature for nonnutrient agar plates with a bacterial lawn. Pathogenic strains of amebas were isolated from lakes in Poland (Kasprzak *et al.*, 1974). In association with cases of PAM in South Australia, pathogens were cultured from a domestic water supply, the implication being that the amebas had multiplied in warm water standing in the pipes (Anderson and Jamieson, 1972b; Anderson *et al.*, 1973). *Naegleria fowleri* was isolated from a thermally polluted canal in Belgium (De Jonckheere *et al.*, 1975), in which a case of PAM had been contracted (Van Den Driessche *et al.*, 1973).

Water samples from lakes thermally polluted with cooling discharge water (35°–38°C) from power stations in Florida and Texas were found positive for pathogenic *Naegleria*, while control lakes were negative (Willaert and Stevens, 1976). Stevens *et al.* (1977b) emphasize the role of thermal pollution, either solar or man-made, in enhancing the growth of pathogenic variants. A somewhat different conclusion was reached by Wellings *et al.* (1977) in a study of Florida lakes. Stressing the importance of large volumes of water (50–100 gal) in sampling, these workers found pathogenic strains to be relatively widespread. On the basis of a year-long study, Wellings *et al.* (1977) concluded that thermal pollution by cooling discharge water was of negligible importance in the distribution of pathogens, since they could be isolated from bottom sediment during the cold months. De Jonckheere and Van De Voorde (1977b) sampled thermal discharge water from various sources during winter (10°–37°C) and summer (20°–36°C) in Belgium. They found pathogenic amebas, more isolates being obtained during the hot-weather months. Pathogenic *Naegleria*, seeded into water samples, were found by Janitschke *et al.* (1978) to survive particularly well in tap water, though less well in pool and lake waters, at moderate temperatures (25° and/or 35°C) and periods of time up to 42 days.

Swimming areas have come under increasing scrutiny because of the association between PAM and water sports. Some comfort to epidemiologists is found in Chang's (1978) comprehensive study of the effects of environmental parameters on survival of trophic and cystic pathogenic *Naegleria*. Trophic amebas are sensitive to desiccation, low (<10°C) and high (>51°C) temperatures, and especially freezing. Cysts are also sensitive to drying, and survive poorly at subzero but tolerate high temperatures (>51°C). Disinfection of water with chlorine or iodine is successful in destroying amebas, provided that pH, temperature, and contact time are taken into consideration. Chang's results suggest that the danger of dust-borne transmission of *Naegleria* is negligible as is cyst survival in areas of the world where prolonged low temperatures and freezing of surface waters occur. Kadlec *et al.* (1978) studied the cause of one of the major recurrent outbreaks of PAM involving an indoor swimming pool and found that moderate temperatures (27°–30°C), abundant organic matter, and low concentration of disinfectant in a confined water layer behind a cracked wall of the pool supported a population of pathogenic *Naegleria*, which were periodically flushed out into the main swimming area. Pathogenic *Acanthamoeba* have been isolated from frozen swimming areas in Norway where, despite severe cold (0°–2°C), survival was possible (Brown and Cursons, 1977). Other studies have examined amebas in swimming pools in New York State (Lyons and Kapur, 1977) and northern France (Dive *et al.*, 1978); neither study detected pathogenic *Naegleria*.

In all the studies referred to, in addition to the pathogenic *Naegleria* and/or *Acanthamoeba* strains, nonpathogenic thermophilic amebas were isolated. These appeared to be members of several other genera, including *Naegleria* and *Acanthamoeba*. De Jonckheere and Van De Voorde (1977a) found isolates that bore a morphological and immunological resemblance to *N. fowleri,* but which were nonpathogenic. Another isolate described as a new species, *N. jadini,* appears to be intermediate between *N. gruberi* and *N. fowleri.* It is morphologically like *N. fowleri* (Carosi *et al.,* 1976) but nonpathogenic (Willaert and Le Ray, 1973). With continued isolations from nature, the genus *Naegleria* appears to be a continuum of species, with *N. fowleri* and *N. gruberi* at opposite ends of the spectrum and the new isolates fitting in somewhere between the two (Willaert, 1977a).

Naegleria was also isolated from the nasal mucosa of an apparently healthy child (Chang *et al.,* 1975; Shumaker *et al.,* 1971). This strain was subsequently characterized as a pathogenic *N. fowleri* (Visvesvara and Healy, 1975a). *Naegleria* and/or *Acanthamoeba* have been isolated from other organisms including reptiles (Bosch and Deichsel, 1972; Frank and Bosch, 1972), snails (Kingston and Taylor, 1976), and fish (Taylor, 1977).

Acanthamoeba which, unlike *Naegleria,* is tolerant of relatively high salt concentrations, has been found in seawater. Sawyer *et al.* (1977) isolated two species of *Acanthamoeba* from marine habitats, *A. culbertsoni* and *A. hatchetti,* both pathogenic for mice. A hartmannellid ameba was found in the tissues of dead or dying oysters, but it was thought to be a secondary invader of already weakened oysters rather than the cause of their sickness (Cheng, 1970).

Whatever the factors are that ultimately define the ecology of pathogenic amebas and the epidemiology of PAM, they probably extend beyond dependence upon warm waters. As is the case for nonpathogenic amebas, they probably include pH, O_2 and CO_2 levels, presence of edible bacterial flora, and competition from other protozoa and perhaps fungi, as well as inorganic and organic constituents of the fluid environment.

E. *Acanthamoeba* as a Pathogen

Attention has focused on *Naegleria* as the ameba involved in PAM, but more interest is developing in the pathogenic *Acanthamoeba* with confirmation of its role as a causal agent of amebic meningoencephalitis. Techniques are now available for distinguishing between the two types of pathogens in formalin-fixed tissue using antisera and fluorescein or peroxidase labeling (Culbertson, 1975; Stamm, 1974).

In *Acanthamoeba* meningoencephalitis, the entry of amebas into the host may be via the respiratory tract, urogenital system, or skin (Martinez *et al.,* 1977b). Amebas enter the central nervous system as a result of hematogenous spread. Experimental infection of mice with pathogenic *Acanthamoeba* suggests that there is an initial pulmonary involvement followed by dissemination of amebas, probably through the circulatory system, to the brain (Martinez *et al.,* 1975). In a scanning electron microscope study of *Acanthamoeba* in the subarachnoid space surrounding the nerve cord of dogs, Sarphie and Allen (1977) report that trophic amebas enter underlying nerve tissue by pseudopodial action rather than by enzyme-induce cytolysis. Intranasal inoculation of amebas into monkeys has also resulted in the production of meningoencephalitis (Wong *et al.,* 1975b). Several recent reports of *Acanthamoeba*-caused meningoencephalitis have been published (Bhagwandeen *et al.,* 1975; Hoffmann *et al.,* 1978; Ringsted *et al.,* 1976; Sotelo-Avila *et al.,* 1974). Interference with the host's immunological defense system by administration of corticosteroid (methyl prednisolone) resulted in increased mortality of mice infected with *A. castellanii* (Markowitz *et al.,* 1978).

Acanthamoeba polyphaga has been isolated from eye infections in humans (Jones *et al.,* 1975). Subsequent studies using this isolate revealed that it was of low virulence, requiring about 1 month before intranasally instilled amebas killed mice (Visvesvara *et al.*, 1975b). In addition, cellular slime mold myxamebas (*Polysphondylium*) were reported to have caused the death of mice upon intracerebral inoculation (Srivastava *et al.*, 1971). A case of fatal meningoencephalitis caused by an ameba that was neither *Naegleria* nor *Acanthamoeba* was reported by Duma *et al.* (1978), emphasizing the potential for pathogenicity of other genera of small amebas.

F. Mechanisms of Pathogenesis

Nothing is known of a specific mechanism by which pathogenic *Naegleria* or *Acanthamoeba* causes meningoencephalitis. The simplest explanation is that amebas have a voracious appetite for nerve tissue, which has been well documented at the electron microscope level (Martinez *et al.*, 1971; Schuster and Dunnebacke, 1977; Visvesvara and Callaway, 1974). Ingestion, however, may be facilitated by enzymatic cytolysis of host cells or by a cytopathic agent released from amebas that causes cytolysis. Virulence may be related to the ability of amebas to cause a cytopathic effect (CPE) in tissue culture systems (Cursons and Brown, 1978; Visvesvara and Balamuth, 1975); thus an understanding of the mechanism of CPE may explain amebic pathogenicity.

Acanthamoeba spp. produce phospholipases that might explain their destructive potential (Hax *et al.*, 1974; Victoria and Korn, 1975; Visvesvara and Balamuth, 1975). Pathogenic *A. culbertsoni* produced more phospholipase and CPE than other *Acanthamoeba* species (Visvesvara and Balamuth, 1975). Enzyme was released by intact amebas into the medium, and crude phospholipase prepared from 6-day-old ameba culture supernatants initiated CPE in monkey kidney cells within 24 hr of addition. Enzyme from nonpathogenic *A. castellanii* was not as effective in producing CPE, in that higher concentrations of crude material were required to demonstrate lysis. A lecithinase was isolated from a culture supernatant of an *Acanthamoeba* initially described as a lipogenic virus in tissue cultures (Chang *et al.*, 1966). The enzyme was heat stable, with a pH optimum of 8.6 (Elson *et al.*, 1970). Upon addition to mammalian cell cultures, the enzyme caused rounding of cells and accumulation of fat droplets, possibly because of its effects upon cell phospholipid.

Filtrates from cultures of pathogenic *Naegleria* produced CPE in monkey kidney cell cultures (Chang, 1971). Filtrates caused rounding of cells (>75%) after 24 hr. The CPE agent was unstable at room temperature and inactivated by heat (50°–60°C) and by storage at 3°–5°C. Not all pathogenic strain filtrates produced CPE, nor did filtrates from cultures of nonpathogenic *Naegleria*. Cursons and Brown (1978) obtained results similar to those of Chang; nonpathogenic *Naegleria* did not produce CPE.

When amebas are added directly to tissue cell cultures, they, in addition to producing CPE, ingest cells in the monolayer, ultimately leading to a plate surface covered with amebas. The conclusion to be drawn from these studies is that amebas release extracellular lipolytic enzymes into the culture supernatant that cause lysis of tissue culture monolayers. If, in addition to lytic enzymes, amebas are present in the tissue culture, feeding on dead or dying cells occurs. Such feeding has been demonstrated by Visvesvara and Callaway (1974). Phospholipases A and B were detected by Cursons *et al.* (1978) in *Naegleria* culture supernatants, with pathogenic strains producing greater amounts of phospholipase A than nonpathogenic strains. This is suggested as the basis for the CPE in cell cultures, and the invasivness and virulence of pathogens in host tissues. Employing cinephotomicrography to study pathogenic *Naegleria* in secondary mouse embryo cell cultures, Brown (1978) concluded that cell destruction resulted from phagocytic behavior of the amebas. No evidence of cytopathic, cytotoxic, or cytolytic agents was detected.

The work of Dunnebacke and Schuster (1971, 1974) is the basis for another explanation of CPE. An agent, *Naegleria* ameba cytopathic material (NACM), has been found to be peculiar to pathogenic and nonpathogenic *Naegleria* spp. Cell-free filtrates of lysed *Naegleria* cause CPE in a variety of tissue culture systems: secondary cultures of chick embryo fibroblasts, rat glioma cells, human embryonic brain cells, and HeLa cells (Schuster and Dunnebacke, 197oa). Unlike the previously described enzyme-mediated CPE which occurs within 24 hr, NACM-induced CPE requires 4–6 days (depending upon the cell line used). Prior to the occurrence of CPE the cell sheet is normal in appearance; then, within less than 24 hr, complete destruction of the monolayer occurs with extensive cell blebbing and lysis. Amebic contamination is ruled out by sterility testing for amebas and by passage of ameba lysate through 0.22-μm and molecular filters. No evidence is seen of particles, virus or otherwise, when cytopathic cells are examined in the electron microscope (Schuster and Dunnebacke, 1974).

NACM has been partially purified by column chromatography and characterized (Dunnebacke and Schuster, 1977). It has a molecular weight of ca. 50,000 and is uv-insensitive ($\leq \times 10^5$ ergs/mm^2). Crude NACM resists digestion by proteolytic enzymes, DNase, and RNase; upon column purification, papain and Pronase inactivate it. Cold storage does not destroy its activity, either 12 months at 2°C or 18 months at −20°C. Complete inactivation occurs at 100°C after 5 min; a decrease in activity (ca. 50%) occurs after 24 hr at 37°C. Ultracentrifugation (100,000g for 4 hr) fails to sediment biologically active NACM. Isoelectric focusing indicates that NACM behaves as an acid protein (Dunnebacke and Schuster, 1978). Serial passage of NACM five times in mammalian cell lines is routine, and 23 passages have been carried out in chick embryo cells. Ability to passage NACM and produce CPE, in conjunction with its partial inactivation at 37°C, is taken as evidence of replication of the agent in the tissue culture systems. The possibility that NACM is an enzyme or a toxin is ruled out, because of its specificity for certain lines of tissue culture cells and not others (e.g., rabbit kidney), the several-day latent period before the onset of CPE in cultures, and its apparent ability to replicate in tissue culture cells prior to the occurrence of CPE.

G. Chemotherapy

The search for chemotherapeutic agents effective against pathogenic strains of *Naegleria* has led to extensive testing of amphotericin B (Am B) *in vitro* and *in vivo* (reviewed in Schuster and Rechthand, 1975). Am B is a polyene antibiotic which acts by binding to the plasma membrane and disrupting its selective permeability (Hamilton-Miller, 1973; Kobayashi and Medoff, 1977). With few exceptions, *in vitro* studies have reported Am B to be inhibitory at concentrations of ≤ 1 μg/ml; amebicidal levels may be higher, but are influenced by several variables such as age of culture, duration of exposure to the drug, size of inoculum, and whether the amebas were grown in bacterized or axenic cultures. Upon exposure to Am B, amebas round up and cease forming pseudopods. Pronounced membrane-related changes are evident at the electron microscope level, including enhanced nuclear plasticity, an increase in both smooth and rough endoplasmic reticulum, a decrease in food vacuole formation, and blebbing of the plasma membrane. Mitochondria are seen in various stages of degeneration, but this may be a secondary effect of the drug. Pathogenic *N. fowleri* strains appeared more sensitive to Am B than non-pathogenic *N. gruberi* amebas. A less toxic, water-soluble derivative of Am B, called amphotericin methyl ester, also exhibited amebicidal activity (Visvesvara *et al.*, 1975a). In addition to testing Am B, Duma and Finley (1976) tested clotrimazole and miconazole *in vitro* and found the latter two drugs effective against *N. fowleri;* ameba response, however, was slower and more variable than to Am B. Though clotrimazole and miconazole were inhibitory, poor motility inhibition suggested that rapid killing of amebas did not occur. Jamieson (1975) reported clotrimazole to be effective *in vitro,* and established minimal inhibitory and amebicidal levels for the drug at 0.03–0.215 μg/ml and 0.125–0.25 μg/ml, respectively. The drug was ineffective *in vivo* in preventing PAM in mice. Significant *in vitro* growth inhibition by miconazole, tetracycline, and rifamycin were reported by Thong *et al.* (1977). A protective synergistic effect between Am B and tetracycline was recognized in mice by Thong *et al.* (1978b). Few clinical studies of drug efficacy are available chiefly because of the rapid progress of PAM as caused by *Naegleria,* though Am B has been the drug of choice (Apley *et al.,* 1970; Carter, 1969; Duma *et al.,* 1971). A recent case of PAM in California was successfully treated with intravenous and intrathecal Am B, miconazole, and oral rifampin (Anonymous, 1978).

Acanthamoeba was relatively insensitive to Am B, clotrimazole, and miconazole (Duma and Finley, 1976). Polymyxin B and pentamidine isethionate were slightly effective but at nonphysiological levels. 5-Fluorocytosine, reported effective against *Acanthamoeba* (Casemore, 1970), was found to be ineffective by Duma and Finley (1976). Administration of tetracycline to mice subsequently infected with *Acanthamoeba* produced 60% mortality, compared to 10% for controls not treated with tetracycline (Markowitz *et al.,* 1978). The tetracycline may increase mortality by modifying the oropharyngeal bacterial flora or by suppression of the host's neutrophil response to the amebas.

REFERENCES

Adam, K. M. G. (1959). *J. Gen. Microbiol.* **21,** 519–529.
Adam, K. M. G. (1964a). *J. Protozool.* **11,** 98–100.
Adam, K. M. G. (1964b). *J. Protozool.* **11,** 423–430.
Adam, K. M. G., and Blewett, D. A. (1967). *J. Protozool.* **14,** 277–282.
Adam, K. M. G., and Blewett, D. A. (1974). *Ann. Soc. Belge Med. Trop.* **54,** 387–393.
Adam, K. M. G., Blewett, D. A., and Flamm, W. G. (1969). *J. Protozool.* **16,** 6–12.
Adams, A. C., John, D. T., and Bradley, S. G. (1976). *Infect. Immun.* **13,** 1387–1391.
Anderson, K., and Jamieson, A. (1972a). *Pathology* **4,** 273–278.
Anderson, K., and Jamieson, A. (1972b). *Lancet* **1,** 902–903.
Anderson, K., and Jamieson, A. (1974). *Pathology* **6,** 79–84.
Anderson, K., Jamieson, A., and Willaert, E. (1973). *Lancet* **1,** 672.
Anonymous. (1978). *Morbid. Mortal. Weekly Rep.* **27,** 343–344.
Apley, J., Clarke, S. K. R., Roome, A. P. C. H., Sandry, S. A., Saygi, G., Silk, B., and
    Warhurst, D. C. (1970). *Br. Med. J.* **1,** 596–599.
Armstrong, J. A., and Pereira, M. S. (1967). *Br. Med. J.* **1,** 212–214.
Averner, M., and Fulton, C. (1966). *J. Gen. Microbiol.* **42,** 245–255.
Balamuth, W. (1964). *J. Protozool.* **11,** Suppl., 19–20.
Balamuth, W. (1965). *Prog. Protozool., Abstr. Int. Conf. Protozool., 2nd, London,* p. 40.
Balamuth, W., and Rowe, M. B. (1955). *J. Protozool.* **2,** Suppl., 10.
Balamuth, W., and Thompson, P. E. (1955). *In* "Biochemistry and Physiology of Protozoa"
    (S. H. Hutner and A. Lwoff, eds.), Vol. II, pp. 277–345. Academic Press, New York.
Band, R. N. (1959). *J. Gen. Microbiol.* **21,** 80–95.
Band, R. N. (1961). *Nature (London)* **192,** 674.

Band, R. N. (1962). *J. Protozool.* **9**, 377–379.

Band, R. N. (1963). *J. Protozool.* **10**, 101–107.

Band, R. N., and Balamuth, W. (1974). *Appl. Microbiol.* **28**, 64–65.

Band, R. N., and Irvine, B. (1965). *Exp. Cell Res.* **39**, 121–128.

Band, R. N., and Mohrlok, S. N. (1969a). *J. Protozool.* **16**, 35–44.

Band, R. N., and Mohrlok, S. N. (1969b). *J. Gen. Microbiol.* **59**, 351–358.

Band, R. N., and Mohrlok, S. N. (1973a). *Exp. Cell Res.* **79**, 327–337.

Band, R. N., and Mohrlok, S. N. (1973b). *J. Protozool.* **20**, 654–657.

Batzri, S., and Korn, E. D. (1975). *J. Cell Biol.* **66**, 621–634.

Bauer, H. (1967). *Vierteljahresschr. Naturforsch. Ges. Zürich* **112**, 173–197.

Bhagwandeen, S. B., Carter, R. F., Naik, K. G., and Levitt, D. (1975). *Am. J. Clin. Pathol.* **63**, 483–492.

Bohnert, H. J. (1973). *Biochim. Biophys. Acta* **324**, 199–205.

Bols, N. C., Corff, S., and Yuyama, S. (1977). *J. Cell. Physiol.* **90**, 271–280.

Bosch, I., and Deichsel, G. (1972). *Z. Parasitenkd.* **40**, 107–129.

Bovee, E. (1959). *J. Protozool.* **6**, 69–75.

Bowers, B. (1977). *Exp. Cell Res.* **110**, 409–417.

Bowers, B., and Korn, E. D. (1968). *J. Cell Biol.* **39**, 95–111.

Bowers, B., and Korn, E. D. (1969). *J. Cell Biol.* **41**, 786–805.

Bowers, B., and Korn, E. D. (1973). *J. Cell Biol.* **59**, 784–791.

Bowers, B., and Korn, E. D. (1974). *J. Cell Biol.* **62**, 533–540.

Bowers, B., and Olszewski, T. E. (1972). *J. Cell Biol.* **53**, 681–694.

Brent, M. M. (1954). *Biol. Bull. (Woods Hole, Mass.)* **106**, 269–278.

Brent, M. M. (1957). *Nature (London)* **179**, 1029.

Brent, M. M. (1965). *Can. J. Microbiol.* **11**, 441–446.

Brent, M. M., and Paxton, H. (1974). *Can. J. Microbiol.* **20**, 1183–1185.

Brown, T. (1978). *J. Med. Microbiol.* **11**, 249–259.

Brown, T. J., and Cursons, R. T. M. (1977). *Scand. J. Infect. Dis.* **9**, 237–240.

Butt, C. G. (1966). *N. Engl. J. Med.* **274**, 1473–1476.

Butt, C. G., Baro, C., and Knorr, R. W. (1968). *Am. J. Clin. Pathol.* **50**, 568–574.

Byers, T. J., Rudick, V. L., and Rudick, M. J. (1969). *J. Protozool.* **16**, 693–699.

Callicott, J. H., Jr., Nelson, E. C., Jones, M. M., Dos Santos, J. G., Utz, J. P., Duma, R. J., and Morrison, J. V., Jr. (1968). *J. Am. Med. Assoc.* **206**, 579–582.

Carosi, G., Scaglia, M., Filice, G., and Willaert, E. (1976). *Protistologica* **12**, 31–36.

Carosi, G., Scaglia, M., Filice, G., and Willaert, E. (1977). *Arch. Protistenkd.* **119**, 264–273.

Carter, R. F. (1969). *J. Clin. Pathol.* **22**, 470–474.

Carter, R. F. (1970). *J. Pathol.* **100**, 217–244.

Carter, R. F. (1972). *Trans. R. Soc. Trop. Med. Hyg.* **66**, 193–213.

Casemore, D. P. (1969). *J. Clin. Pathol.* **22**, 254–257.

Casemore, D. P. (1970). *J. Clin. Pathol.* **23**, 649–652.

Červa, L. (1969). *Science* **163**, 576.

Červa, L. (1970). *Folia Parasitol. (Prague)* **17**, 127–133.

Červa, L. (1975). *Angew. Parasitol.* **16**, 1–12.

Červa, L. (1977). *Folia Parasitol. (Prague)* **24**, 221–228.

Červa, L. (1978). *Folia Parasitol. (Prague)* **25**, 1–8.

Červa, L., and Novak, K. (1968). *Science* **160**, 92.

Chagla, A. H., and Griffiths, A. J. (1974). *J. Gen. Microbiol.* **85**, 139–145.

Chambers, J. A., and Thompson, J. E. (1972). *Exp. Cell Res.* **73**, 415–421.

Chambers, J. A., and Thompson, J. E. (1974). *J. Gen. Microbiol.* **80**, 375–380.

Chambers, J. A., and Thompson, J. E. (1976). *J. Gen. Microbiol.* **92**, 246–250.

Chang, R. S., Pan, I.-H., and Rosenau, B. J. (1966). *J. Exp. Med.* **124**, 1153–1166.

Chang, S. L. (1958). *J. Gen. Microbiol.* **18**, 565–578.

Chang, S. L. (1971). *Curr. Top. Comp. Pathobiol.* **1**, 201–254.

Chang, S. L. (1974). *Crit. Rev. Microbiol.* **3**, 135–159.

Chang, S. L. (1978). *Appl. Environ. Microbiol.* **35**, 368–375.

Chang, S. L., Healy, G. R., McCabe, L., Shumaker, J. B., and Schultz, M. G. (1975). *Health Lab. Sci.* **12**, 1–7.

Chatton, E. (1953). *In* "Traité de Zoologie" (P.-P. Grassé, ed.), Vol. I, Part 2, pp. 5–91. Masson, Paris.

Cheng, T. C. (1970). *J. Invertebr. Pathol.* **15**, 405–419.

Childs, G. E. (1973). *Exp. Parasitol.* **34**, 44–55.

Chiovetti, R., Jr. (1976). *Trans. Am. Microsc. Soc.* **95**, 122–124.

Chiovetti, R., Jr. (1978). *Trans. Am. Microsc. Soc.* **97**, 245–249.

Chlapowski, F. J., and Band, R. N. (1971a). *J. Cell Biol.* **50**, 625–633.

Chlapowski, F. J., and Band, R. N. (1971b). *J. Cell Biol.* **50**, 634–651.

Corff, S., and Yuyama, S. (1976). *J. Protozool.* **23**, 587–593.

Corff, S., and Yuyama, S. (1978). *Exp. Cell Res.* **114**, 175–183.

Culbertson, C. G. (1971). *Annu. Rev. Microbiol.* **25**, 231–254.

Culbertson, C. G. (1975). *Am. J. Clin. Pathol.* **63**, 475–482.

Culbertson, C. G., Smith, J. W., and Minner, J. R. (1958). *Science* **127**, 1506.

Culbertson, C. G., Smith, J. W., Cohen, H. K., and Minner, J. R. (1959). *Am. J. Pathol.* **35**, 185–197.

Culbertson, C. G., Ensminger, P. W., and Overton, W. M. (1968). *J. Protozool.* **15**, 353–363.

Culbertson, C. G., Ensminger, P. W., and Overton, W. M. (1972). *Am. J. Clin. Pathol.* **57**, 375–386.

Cursons, R. T. M., and Brown, T. J. (1975). *N.Z. Med. J.* **82**, 123–125.

Cursons, R. T. M., and Brown, T. J. (1976). *N.Z. J. Mar. Freshwater Res.* **10**, 245–262.

Cursons, R. T. M., and Brown, T. J. (1978). *J. Clin. Pathol.* **31**, 1–11.

Cursons, R. T. M., Brown, T. J., Bruns, B. J., and Taylor, D. E. M. (1976). *N.Z. Med. J.* **84**, 479–481.

Cursons, R. T. M., Brown, T. J., and Keys, E. A. (1978). *J. Parasitol.* **64**, 744–745.

Danso, S. K. A., and Alexander, M. (1975). *Appl. Microbiol.* **29**, 515–521.

Darbyshire, J. F., Page, F. C., and Goodfellow, L. P. (1976). *Protistologica* **12**, 375–387.

Datta, T., and Kaur, J. (1978). *Protistologica* **14**, 121–123.

Davis, P. G., Caron, D. A., and Siebarth, J. McN. (1978). *Trans. Am. Microsc. Soc.* **97**, 73–88.

Dearborn, D. G., Smith, S., and Korn, E. D. (1976). *J. Biol. Chem.* **251**, 2976–2982.

Deichmann, U., and Jantzen, H. (1977). *Arch. Microbiol.* **113**, 309–313.

De Jonckheere, J. (1977). *Appl. Environ. Microbiol.* **33**, 751–757.

De Jonckheere, J., and Van De Voorde, H. (1976). *Appl. Environ. Microbiol.* **31**, 294–297.

De Jonckheere, J., and Van De Voorde, H. (1977a). *J. Protozool.* **24**, 304–309.

De Jonckheere, J., and Van De Voorde, H. (1977b). *Am. J. Trop. Med. Hyg.* **26**, 10–15.

De Jonckheere, J., Van Dijck, P., and Van De Voorde, H. (1974). *Appl. Microbiol.* **28**, 159–164.

De Jonckheere, J., Van Dijck, P., and Van De Voorde, H. (1975). *J. Hyg.* **75**, 7–13.

Detke, S., and Paule, M. R. (1975). *Biochim. Biophys. Acta* **383**, 67–77.

Diffley, P., Skeels, M. R., and Sogandares-Bernal, F. (1976). *Z. Parasitenkd.* **49**, 133–137.

Dingle, A. D. (1970). *J. Cell Sci.* **7**, 463–482.

Dingle, A. D., and Fulton, C. (1966). *J. Cell Biol.* **31**, 43–54.

Dive, D., Leclerc, H., Picard, J. P., Telliez, E., and Vangrevelinghe, R. (1978). *Ann. Microbiol.* (*Paris*) **129B**, 225–244.

Dolphin, W. D. (1976). *J. Protozool.* **23**, 455–457.

Drainville, G., and Gagnon, A. (1973). *Comp. Biochem. Physiol. A* **45**, 379–388.

Droop, M. R. (1962). *Arch. Protistenkd.* **42**, 254–266.

Drożański, W. (1961). *Acta Microbiol. Pol.* **10**, 147–153.

Drożański, W. (1963). *Acta Microbiol. Pol.* **12**, 9–24.

Drożański, W. (1972). *Acta Microbiol. Pol.* **21**, 33–52.

Drożański, W., and Drożańska, D. (1961). *Acta Microbiol. Pol.* **10**, 379–388.

Dubes, G. R., and Jensen, T. (1964). *J. Parasitol.* **50**, 380–385.

Duma, R. J., and Finley, R. (1976). *Antimicrob. Agents Chemother.* **10**, 370–376.

Duma, R. J., Ferrell, H. W., Nelson, E. C., and Jones, M. M. (1969). *N. Engl. J. Med.* **281**, 1315–1323.

Duma, R. J., Rosenblum, W. I., McGehee, R. F., Jones, M. M., and Nelson, E. C. (1971). *Ann. Intern. Med.* **74**, 861–869.

Duma, R. J., Helwig, W. B., and Martinez, A. J. (1978). *Ann. Intern. Med.* **88**, 468–473.

Dunnebacke, T. H., and Schuster, F. L. (1971). *Science* **174**, 516–518.

Dunnebacke, T. H., and Schuster, F. L. (1974). *J. Protozool.* **21**, 327–329.

Dunnebacke, T. H., and Schuster, F. L. (1977). *Am. J. Trop. Med. Hyg.* **26**, 412–421.

Dunnebacke, T. H., and Schuster, F. L. (1978). *J. Supramol. Struct., Suppl.* **2**, 280.

Dunnebacke, T. H., and Williams, R. C. (1967). *Proc. Natl. Acad. Sci. U.S.A.* **57**, 1363–1370.

Dykstra, M. J., and Aldrich, H. C. (1978). *J. Protozool.* **25**, 38–41.

Elson, C., Geyer, R. P., and Chang, R. S. (1970). *J. Protozool.* **17**, 440–445.

Erdos, G. W., Raper, K. B., and Vogen, L. K. (1973). *Proc. Natl. Acad. Sci. U.S.A.* **70**, 1828–1830.

Forrester, J. A., Gingell, D., and Korohoda, W. (1967). *Nature* (*London*) **215**, 1409–1410.

Fowler, M., and Carter, R. F. (1965). *Br. Med. J.* **ii**, 740–742.

Frank, W., and Bosch, I. (1972). *Z. Parasitenkd.* **40**, 139–150.

Fuller, M., and Fulton, C. (1974). *J. Gen. Physiol.* **64**, Suppl., 4a.

Fulton, C. (1970a). *Methods Cell Physiol.* **4**, 341–476.

Fulton, C. (1970b). *Science* **167**, 1269–1270.

Fulton, C. (1972). *Dev. Biol.* **28**, 603–619.

Fulton, C. (1974). *Exp. Cell Res.* **88**, 365–370.

Fulton, C. (1977a). *Annu. Rev. Microbiol.* **31**, 597–629.

Fulton, C. (1977b). *J. Supramol. Struct.* **6**, 13–43.

Fulton, C., and Dingle, A. (1967). *Dev. Biol.* **15**, 165–191.

Fulton, C., and Dingle, A. (1971). *J. Cell Biol.* **51**, 826–836.

Fulton, C., and Guerrini, A. M. (1969). *Exp. Cell Res.* **56**, 194–200.

Gessat, M., and Jantzen, H. (1974). *Arch. Microbiol.* **99**, 155–166.

Gong, T., and Balamuth, W. (1973). *J. Protozool.* **20**, 498.

Gray, W. D., and Alexopoulos, C. J. (1968). "Biology of Myxomycetes." Ronald Press, New York.

Grell, K. G. (1973). "Protozoology." Springer-Verlag, Berlin and New York.

Griffin, J. L. (1972). *Science* **178**, 869–870.

Griffin, J. L. (1978). *In* "Parasitic Protozoa" (J. P. Kreier, ed.), Vol. ?, pp. 507–549. Academic Press, New York.

Griffiths, A. J. (1970). *Adv. Microbial Physiol.* **4**, 105–129.

Griffiths, A. J., and Bowen, S. M. (1969). *J. Gen. Microbiol.* **59**, 239–245.

Griffiths, A. J., and Hughes, D. E. (1968). *J. Protozool.* **15**, 673–677.
Griffiths, A. J., and Hughes, D. E. (1969). *J. Protozool.* **16**, 93–99.
Griffiths, A. J., Curnick, L., Unitt, M. D., and Wilcox, S. L. (1978). *J. Gen. Microbiol.* **107**, 211–215.
Hadas, E., Kasprzak, W., and Mazur, T. (1977). *Tropenmed. Parasitol.* **28**, 35–43.
Haggerty, R. M., and John, D. T. (1978). *Infect. Immun.* **20**, 73–77.
Hamilton-Miller, J. M. T. (1973). *Bacteriol. Rev.* **37**, 166–196.
Harrison, D. (1957). *Nature (London)* **180**, 1301–1302.
Hax, V. M. A., Demel, R. A., Spies, F., Vossenberg, J. B. J., and Linnemans, W. A. M. (1974). *Exp. Cell Res.* **89**, 311–319.
Hoffmann, E. O., Garcia, C., Lunseth, J., McGarry, P., and Coover, J. (1978). *Am. J. Trop. Med. Hyg.* **27**, 29–38.
Holberton, D. (1969). *Nature (London)* **223**, 680–681.
Hollande, A. (1942). *Arch. Zool. Exp. Gen.* **83**, 1–268.
Hoover, R. L. (1974). *Exp. Cell Res.* **87**, 265–276.
Ito, S., Chang, R. S., and Pollard, T. D. (1969). *J. Protozool.* **16**, 638–645.
Jadin, J. B. (1974). *Bull. Acad. R. Med. Belg.* **129**, 439–466.
Jadin, J. B. (1976). *Pathol. Biol.* **24**, 171–173.
Jadin, J. B., Hermanne, J., Robyn, G., Willaert, E., Van Maercke, Y., and Stevens, W. (1971). *Ann. Soc. Belge Med. Trop.* **51**, 255–266.
Jadin, J. M., Eschbach, H. L., Verheyen, F., and Willaert, E. (1974). *Ann. Soc. Belge Med. Trop.* **54**, 259–264.
Jahnes, W. G., Fullmer, H. M., and Li, C. P. (1957). *Proc. Soc. Exp. Biol. Med.* **96**, 484–488.
Jamieson, A. (1975). *J. Clin. Pathol.* **28**, 446–449.
Jamieson, A., and Anderson, K. (1973). *Pathology* **5**, 55–58.
Janitschke, K., Werner, H., Lervy, M., and Schmitt, I. (1978). *Zentralbl. Bakteriol., Parasitenkd., Infektionskr. Hyg.* **166**, 244–249.
Jeffery, S., and Hawkins, S. E. (1976a). *Microbios Lett.* **2**, 153–155.
Jeffery, S., and Hawkins, S. E. (1976b). *Microbios* **15**, 27–36.
Jeffery, S., and Hawkins, S. E. (1977). *Microbios* **18**, 35–49.
Jehan, M., and Dutta, G. P. (1977). *Protistologica* **13**, 181–186.
Jehan, M., Dutta, G. P., Kaushal, D. C., and Shukla, O. P. (1975). *Indian J. Exp. Biol.* **13**, 375–377.
Jensen, T., Barnes, W. G., and Myers, D. (1970). *J. Parasitol.* **56**, 904–906.
John, D. T., Weik, R. R., and Adams, A. C. (1977). *Infect. Immun.* **16**, 817–820.
Jones, D. B., Visvesvara, G. S., and Robinson, N. M. (1975). *Trans. Ophthalmol. Soc. U.K.* **95**, 221–232.
Josephson, S. L., Weik, R. R., and John, D. T. (1977). *Am. J. Trop. Med. Hyg.* **26**, 856–858.
Kadlec, V., Červa, L., and Škvařová, J. (1978). *Science* **201**, 1025.
Kasprzak, W., Mazur, T., and Rucka, A. (1974). *Ann. Soc. Belge Med. Trop.* **54**, 351–357.
Kaushal, D. C., and Shukla, O. P. (1977). *J. Gen. Microbiol.* **98**, 117–123.
King, C. A., and Preston, T. M. (1977). *J. Cell Sci.* **28**, 133–149.
Kingston, N., and Taylor, P. C. (1976). *Proc. Helminthol. Soc. Wash.* **43**, 227–229.
Klein, R. L. (1961). *Exp. Cell Res.* **25**, 571–584.
Klein, R. L. (1964). *Exp. Cell Res.* **34**, 231–238.
Klein, R. L., and Neff, R. J. (1960). *Exp. Cell Res.* **19**, 133–155.
Kobayashi, G. S., and Medoff, G. (1977). *Annu. Rev. Microbiol.* **31**, 291–308.
Korn, E. (1978). *Proc. Natl. Acad. Sci. U.S.A.* **75**, 588–599.
Korn, E. D., and Olivecrona, T. (1971). *Biochem. Biophys. Res. Commun.* **45**, 90–97.
Korn, E. D., and Wright, P. L. (1973). *J. Biol. Chem.* **248**, 439–497.

Kowit, J. D., and Fulton, C. (1974a). *J. Biol. Chem.* **249,** 3638–3646.

Kowit, J. D., and Fulton, C. (1974b). *Proc. Natl. Acad. Sci. U.S.A.* **71,** 2877–2881.

Krishna Murti, C. R. (1975). *Indian J. Med. Res.* **63,** 757–767.

Krishna Prasad, B. N., and Gupta, S. K. (1978). *Curr. Sci.* **47,** 245–247.

Larochelle, J., and Gagnon, A. (1976). *Comp. Biochem. Physiol. A.* **54,** 275–279.

Larochelle, J., and Gagnon, A. (1978). *Comp. Biochem. Physiol. A.* **59,** 119–123.

Lasman, M. (1967). *J. Cell. Physiol.* **69,** 151–154.

Lasman, M. (1975). *J. Protozool.* **22,** 435–437.

Lasman, M. (1977). *J. Protozool.* **24,** 244–248.

Lastovica, A. J. (1974). *Int. J. Parasitol.* **4,** 139–142.

Lastovica, A. J. (1976). *Z. Parasitenkd.* **50,** 245–250.

Lastovica, A. J., and Dingle, A. D. (1971). *Exp. Cell Res.* **66,** 337–345.

Lemke, P. A. (1976). *Annu. Rev. Microbiol.* **30,** 105–145.

Lockey, M. W. (1978). *Laryngoscope* **88,** 484–503.

Lyons, T. B., III, and Kapur, R. (1977). *Appl. Environ. Microbiol.* **33,** 551–555.

McIntosh, A. H., and Chang, R. S. (1971). *J. Protozool.* **18,** 632–636.

Maitra, S. C., Krishna Prasad, B. N., Das, S. R., and Agarwala, S. C. (1974a). *Trans. R. Soc. Trop. Med. Hyg.* **68,** 56–60.

Maitra, S. C., Sagar, P., and Agarwala, S. C. (1974b). *J. Protozool.* **21,** 507–511.

Mandal, B. N., Gudex, D. J., Fitchett, M. R., Pullon, D. H. H., Malloch, J. A., David, C. M., and Apthorp, J. (1970). *N.Z. Med. J.* **71,** 16–23.

Markowitz, S. M., Martinez, A. J., Duma, R. J., and Shiel, F. O. M. (1974). *Am. J. Clin. Pathol.* **62,** 619–628.

Markowitz, S. M., Sobieski, T., Martinez, A. J., and Duma, R. J. (1978). *Am. J. Pathol.* **92,** 733–744.

Martinez, A. J. (1977). *Proc. Int. Conf. Amebiasis, Mexico City, 1975* pp. 64–81.

Martinez, A. J., Nelson, E. C., Jones, M. M., Duma, R. J., and Rosenblum, W. I. (1971). *Lab. Invest.* **25,** 465–475.

Martinez, A. J., Duma, R. J., Nelson, E. C., and Moretta, F. L. (1973). *Lab. Invest.* **29,** 121–133.

Martinez, A. J., Markowitz, S. M., and Duma, R. J. (1975). *J. Infect. Dis.* **131,** 692–699.

Martinez, A. J., Dos Santos Neto, J. G., Nelson, E. C., Stamm, W. P., and Willaert, E. (1977a). *In* "Pathology Annual" (S. C. Sommers and P. P. Rosen, eds.), Vol. 12, pp. 225–250. Appleton, New York.

Martinez, A. J., Sotelo-Avila, C., Garcia-Tamayo, J., Morón, J. T., Willaert, E., and Stamm, W. P. (1977b). *Acta Neuropathol.* **37,** 183–191.

Maruta, H., and Korn, E. D. (1977a). *J. Biol. Chem.* **252,** 399–402.

Maruta, H., and Korn, E. D. (1977b). *J. Biol. Chem.* **252,** 6501–6509.

Maruta, H., and Korn, E. D. (1977c). *J. Biol. Chem.* **252,** 8329–8332.

Marzzoco, A., and Colli, W. (1974). *Biochim. Biophys. Acta* **374,** 292–303.

Marzzoco, A., and Colli, W. (1975). *Biochim. Biophys. Acta* **395,** 525–534.

Mattar, F. E., and Byers, T. J. (1971). *J. Cell Biol.* **49,** 507–519.

Menapace, D., Klein, D. A., McClellan, J. F., and Mayeux, J. V. (1975). *J. Protozool.* **22,** 405–410.

Moore, A. E., and Hlinka, J. (1968). *J. Natl. Cancer Inst.* **40,** 569–581.

Müller, M. (1969). *J. Protozool.* **16,** 428–431.

Napolitano, J. J., and Gamble, H. R. (1978). *Protistologica* **14,** 183–187.

Napolitano, J. J., and Persico, F. J. (1966). *J. Protozool.* **13,** 202–203.

Napolitano, J. J., and Smith, B. H. (1975). *J. Protozool.* **22,** 196–199.

Napolitano, J. J., Smith, B. H., and Persico, F. J. (1967). *J Protozool.* **14,** 108–109.

Napolitano, J. J., Wall, M. E., and Ganz, C. S. (1970). *J. Protozool.* **17,** 158–161.

Napolitano, J. J., LaVerde, A. V., and Gamble, H. R. (1977). *Acta Protozool.* **16,** 207–217.

Neff, R. J. (1957). *J. Protozool.* **4,** 176–182.

Neff, R. J. (1958). *J. Protozool.* **5,** 226–231.

Neff, R. J., and Neff, R. H. (1969). *Symp. Soc. Exp. Biol.* **23,** 51–81.

Neff, R. J., and Neff, R. H. (1972). *C.R. Trav. Lab. Carlsberg* **39,** 111–168.

Neff, R. J., Neff, R. H., and Taylor, R. E. (1958). *Physiol. Zool.* **31,** 73–91.

Neff, R. J., Benton, W. F., and Neff, R. H. (1964a). *J. Cell Biol.* **23,** Suppl., 66A.

Neff, R. J., Ray, S. A., Benton, W. F., and Wilborn, M. (1964b). *Methods Cell Physiol.* **1,** 55–83.

Nielson, J. B., Ivey, M. H., and Bulmer, G. S. (1978). *Infect. Immun.* **20,** 262–266.

O'Dell, D. S., Preston, T. M., King, C. A., and Gardiner, P. R. (1976). *Biochem. Soc. Trans.* **4,** 124–125.

O'Dell, W. D., and Brent, M. M. (1974). *J. Protozool.* **21,** 129–133.

O'Dell, W. D., and Stevens, A. R. (1973). *Appl. Microbiol.* **25,** 621–627.

Okubo, S., and Inoki, S. (1973). *Biken J.* **16,** 181–184.

Olive, L. S. (1975). "The Mycetozoans." Academic Press, New York.

Outka, D. E. (1965). *J. Protozool.* **12,** 85–93.

Outka, D. E., and Kluss, B. C. (1967). *J. Cell Biol.* **35,** 323–346.

Page, F. C. (1967a). *J. Protozool.* **14,** 499–521.

Page, F. C. (1967b). *J. Protozool.* **14,** 709–724.

Page, F. C. (1974a). *Arch. Protistenkd.* **116,** 149–184.

Page, F. C. (1974b). *Acta Protozool.* **13,** 143–154.

Page, F. C. (1975). *Protistologica* **11,** 195–204.

Page, F. C. (1976). "An Illustrated Key to Freshwater and Soil Amoebae." Freshwater Biolog. Assoc., Ambleside, Cumbria.

Pal, R. A. (1972). *J. Exp. Biol.* **57,** 55–76.

Pasternak, J. J., Thompson, J. E., Schultz, T. M. G., and Zachariah, K. (1970). *Exp. Cell Res.* **60,** 290–298.

Pearson, J. L., and Willmer, E. N. (1963). *J. Exp. Biol.* **40,** 493–515.

Peloux, Y., Nicolas, A., Peyron, L., and Beurlet, J. (1974). *Pathol. Biol.* **22,** 587–592.

Perkins, D. L., and Jahn, T. (1970). *J. Protozool.* **17,** 168–172.

Perkins, D. L., and Morris, G. S. (1974). *J. Protozool.* **21,** Suppl., 445.

Phillips, B. P. (1974). *Am. J. Trop. Med. Hyg.* **23,** 850–855.

Pigon, A. (1970). *Protoplasma* **70,** 405–414.

Pigon, A. (1978). *Cytobiologie* **16,** 259–267.

Pigon, A., and Morita, M. (1973). *Cytobiologie* **8,** 76–88.

Piteika, D. R. (1969). *In* "Research in Protozoology" (T.-T. Chen, ed.), Vol. 3, pp. 278–388. Academic Press, New York.

Pittam, M. D. (1963). *Q. J. Microsc. Sci.* **104,** 513–529.

Pollard, T. D. (1976). *J. Cell Biol.* **68,** 579–601.

Pollard, T. D., and Korn, E. D. (1973a). *J. Biol. Chem.* **248,** 4682–4690.

Pollard, T. D., and Korn, E. D. (1973b). *J. Biol. Chem.* **248,** 4691–4697.

Pollard, T. D., and Korn, E. D. (1973c). *Cold Spring Harbor Symp. Quant. Biol.* **37,** 573–583.

Pollard, T. D., Shelton, E., Weihung, R. R., and Korn, E. D. (1970). *J. Mol. Biol.* **50,** 91–97.

Potter, J. L., and Weisman, R. A. (1971). *Biochim. Biophys. Acta* **237,** 64–74.

Potter, J. L., and Weisman, R. A. (1972). *Dev. Biol.* **28,** 472–479.

Potter, J. L., and Weisman, R. A. (1976). *Biochim. Biophys. Acta* **428,** 240–252.

Prescott, L. M., and Olson, D. L. (1972). *Proc. S.D. Acad. Sci.* **51,** 136–141.

Prescott, L. M., Kubovec, M. K., and Tryggestad, D. (1977). *Environ. Contam. Toxicol.* **18**, 29–34.

Preston, T. M., and King, C. A. (1978a). *J. Gen. Microbiol.* **104**, 347–351.

Preston, T. M., and King, C. A. (1978b). *J. Cell Sci.* **34**, 145–158.

Preston, T. M., and O'Dell, D. S. (1971). *Exp. Cell Biol.* **68**, 465–466.

Preston, T. M., and O'Dell, D. S. (1973). *J. Gen. Microbiol.* **75**, 351–361.

Preston, T. M., and O'Dell, D. S. (1974). *Ann. Soc. Belge Med. Trop.* **54**, 279–286.

Preston, T. M., O'Dell, D. S., and King, C. A. (1975). *Cytobios* **13**, 207–216.

Proca-Ciobanu, M., Lupascu, Gh., Petrovici, Al., and Ionescu, M. D. (1975). *Int. J. Parasitol.* **5**, 49–56.

Pussard, M. (1972). *J. Protozool.* **19**, 557–563.

Pussard, M. (1973). *Protistologica* **9**, 163–173.

Rafalko, J. S. (1946). *Stain Technol.* **21**, 91–93.

Rafalko, J. S. (1947). *J. Morphol.* **81**, 1–44.

Rafalko, J. S. (1951). *J. Morphol.* **89**, 71–90.

Raizada, M. K., and Krishna Murti, C. R. (1971). *J. Protozool.* **18**, 115–119.

Raizada, M. K., and Krishna Murti, C. R. (1972). *J. Cell Biol.* **52**, 743–748.

Rastogi, A. K., Sagar, P., and Agarwala, S. C. (1970). *J. Gen. Microbiol.* **60**, 387–392.

Rastogi, A. K., Sagar, P., and Agarwala, S. C. (1971a). *J. Protozool.* **18**, 506–509.

Rastogi, A. K., Shipstone, A. C., and Agarwala, S. C. (1971b). *J. Protozool.* **18**, 176–179.

Rastogi, A. K., Sagar, P., Kapoor, S. C., and Agarwala, S. C. (1972). *J. Protozool.* **19**, 363–365.

Rastogi, A. K., Sagar, P., and Agarwala, S. C. (1973). *J. Protozool.* **20**, 453–455.

Rastogi, A. K., Sagar, P., and Agarwala, S. C. (1977). *J. Protozool.* **24**, 294–296.

Ringsted, J., Val Jager, B., Suk, D., and Visvesvara, G. S. (1976). *Am. J. Clin. Pathol.* **66**, 723–730.

Rondanelli, E. G., Carosi, G., Scaglia, M., and Dei Cas, A. (1976). *Protistologica* **12**, 25–30.

Roti Roti, L. W., and Stevens, A. R. (1974). *J. Cell Biol.* **61**, 233–237.

Roti Roti, L. W., and Stevens, A. R. (1975). *J. Cell Sci.* **17**, 503–515.

Rubin, R. W., and Maher, M. (1976). *Exp. Cell Res.* **103**, 159–168.

Rubin, R. W., Hill, M. C., Hepworth, P., and Boehmer, J. (1976). *J. Cell Biol.* **68**, 740–751.

Rudick, V. L. (1971). *J. Cell Biol.* **49**, 498–506.

Rudick, V. L., and Weisman, R. A. (1973). *Biochim. Biophys. Acta* **299**, 91–102.

Ryter, A., and Bowers, B. (1976). *J. Ultrastruct. Res.* **57**, 309–321.

Sarphie, T. G., and Allen, D. J. (1977). *Am. J. Clin. Pathol.* **68**, 485–492.

Satir, P. (1965). *Protoplasmatologia* **3**, 1–52.

Sawyer, T. K., and Griffin, J. L. (1975). *Trans. Am. Microsc. Soc.* **94**, 93–98.

Sawyer, T. K., Visvesvara, G. S., and Harker, B. A. (1977). *Science* **196**, 1324–1325.

Schultz, T. M. G., and Thompson, J. E. (1969). *Biochim. Biophys. Acta* **193**, 203–211.

Schuster, F. L. (1961). *J. Protozool.* **8**, Suppl., 19.

Schuster, F. L. (1963a). *J. Protozool.* **10**, 297–313.

Schuster, F. L. (1963b). *J. Protozool.* **10**, 313–320.

Schuster, F. L. (1969). *J. Protozool.* **16**, 724–727.

Schuster, F. L. (1975a). *J. Protozool.* **22**, 352–359.

Schuster, F. L. (1975b). *Tissue Cell* **7**, 1–12.

Schuster, F. L., and Clemente, J. S. (1977). *J. Cell Sci.* **26**, 359–371.

Schuster, F. L., and Dunnebacke, T. H. (1971). *J. Ultrastruct. Res.* **36**, 659–668.

Schuster, F. L., and Dunnebacke, T. H. (1974). *Ann. Soc. Belge Med. Trop.* **54**, 359–370.

Schuster, F. L., and Dunnebacke, T. H. (1976a). *J. Protozool.* **23**, Suppl., 7A.

Schuster, F. L., and Dunnebacke, T. H. (1976b). *Cytobiologie* **14**, 131–147.

Schuster, F. L., and Dunnebacke, T. H. (1977). *J. Protozool.* **24**, 489–497.
Schuster, F. L., and Rechthand, E. (1975). *Antimicrob. Agents Chemother.* **8**, 591–605.
Schuster, F. L., and Svilha, G. (1968). *J. Protozool.* **15**, 752–758.
Shumaker, J. B., Healy, G. R., English, D., Schultz, M., and Page, F. C. (1971). *Lancet* **ii**, 602–603.
Simpson, P. A., and Dingle, A. D. (1971). *J. Cell Biol.* **51**, 323–328.
Singh, B. N. (1952). *Philos. Trans. R. Soc. London, Ser. B* **236**, 405–461.
Singh, B. N. (1973). *J. Sci. Ind. Res.* **32**, 399–432.
Singh, B. N. (1975). "Pathogenic and Non-pathogenic Amoebae." Wiley, New York.
Singh, B. N., and Das, S. R. (1970). *Philos. Trans. R. Soc. London, Ser. B* **259**, 435–476.
Singh, B. N., and Das, S. R. (1972a). *Curr. Sci.* **41**, 277–281.
Singh, B. N., and Das, S. R. (1972b). *Curr. Sci.* **41**, 625–628.
Singh, B. N., and Hanumaiah, V. (1977a). *Protozoology* **3**, 183–191.
Singh, B. N., and Hanumaiah, V. (1977b). *Indian J. Parasitol.* **1**, 71–73.
Singh, B. N., Datta, T., and Dutta, G. P. (1971). *Indian J. Exp. Biol.* **9**, 350–357.
Sobota, A., Hrebenda, B., and Przelecka, A. (1977). *Eur. J. Cell Biol.* **15**, 259–268.
Soll, D. R., and Fulton, C. (1974). *Dev. Biol.* **36**, 236–244.
Sotelo-Avila, C., Taylor, F. M., and Ewing, C. W. (1974). *J. Pediatr.* **85**, 131–136.
Spies, F., Elbers, P. F., and Linnemans, W. A. M. (1972). *Cytobiologie* **6**, 327–341.
Spies, F., Linnemans, W. A. M., Ververgaert, P. H. J. T., Leunissen, J. L. M., and Elbers, P. F. (1975a). *Cytobiologie* **11**, 50–64.
Spies, F., Linnemans, W. A. M., and De Ruyter De Wildt, T. M. (1975b). *Cytobiologie* **11**, 65–86.
Srivastava, A., Gupta, H. P., Mathur, I. S., and Gupta, S. K. (1971). *J. Gen. Appl. Microbiol.* **17**, 251–257.
Stamm, W. P. (1974). *Ann. Soc. Belge Med. Trop.* **54**, 321–325.
Stevens, A. R., and Kaufman, A. E. (1974). *Nature (London)* **252**, 43–45.
Stevens, A. R., and Pachler, P. F. (1973). *J. Cell Biol.* **57**, 525–537.
Stevens, A. R., and Stein, S. (1977). *J. Parasitol.* **63**, 151–152.
Stevens, A. R., Kilpatrick, T., Willaert, E., and Capron, A. (1977a). *J. Protozool.* **24**, 316–324.
Stevens, A. R., Tyndall, R. L., Coutant, C. C., and Willaert, E. (1977b). *Appl. Environ. Microbiol.* **34**, 701–705.
Stewart, J. R., and Weisman, R. A. (1972). *J. Cell Biol.* **52**, 117–130.
Stewart, J. R., and Weisman, R. A. (1974). *Arch. Biochem. Biophys.* **161**, 488–498.
Taylor, P. W. (1977). *J. Parasitol.* **63**, 232–237.
Thong, Y. H., Rowan-Kelly, B., Shepherd, C., and Ferrante, A. (1977). *Lancet* **ii**, 876.
Thong, Y. H., Ferrante, A., Shephard, C. (1978a). *Trans. R. Soc. Trop. Med. Hyg.* **72**, 207–209.
Thong, Y. H., Rowan-Kelly, B., Ferrante, A., and Shephard, C. (1978b). *Med. J. Aust.* **1**, 663–664.
Thong, Y. H., Shephard, C., Ferrante, A., and Rowan-Kelly, B. (1978c). *Am. J. Trop. Med. Hyg.* **27**, 238–240.
Todd, S. R., and Kitching, J. A. (1973). *J. Protozool.* **20**, 421–424.
Tomlinson, G., and Jones, E. (1962). *Biochim. Biophys. Acta* **63**, 194–200.
Tracey, M. V. (1955). *Nature (London)* **175**, 815.
Ulsamer, A. G., Wright, P. L., Wetzel, M. G., and Korn, E. D. (1971). *J. Cell Biol.* **51**, 193–215.
Upadhyay, J. M. (1968). *J. Bacteriol.* **95**, 771–774.

Upadhyay, J. M., Mares, B. A., Hemelt, D. M., and Rivet, P. G. (1977). *Appl. Environ. Microbiol.* **33**, 1–5.

Van Den Driessche, E., Vandepitte, J., Van Dijck, P. J., De Jonckheere, J., and Van De Voorde, H. (1973). *Lancet* **ii**, 971.

Van Dijck, P., De Jonckheere, J., Reybrouck, G., and Van De Voorde, H. (1974). *Zentralbl. Bakteriol., Parasitenkd., Infektionskr. Hyg.* **158**, 541–551.

Van Saanen, M. (1977). *Experientia* **33**, 1680–1681

Verma, A. K., and Raizada, M. K. (1975). *Cell Differ.* **4**, 167–177.

Verma, A. K., Raizada, M. K., Shukla, O. P., and Krishna Murti, C. R. (1974). *J. Gen. Microbiol.* **80**, 307–309.

Vickerman, K. (1960). *Nature (London)* **188**, 248–249.

Vickerman, K. (1962). *Exp. Cell Res.* **26**, 497–519.

Victoria, E. J., and Korn, E. D. (1975). *J. Lipid Res.* **16**, 54–60.

Visvesvara, G. S., and Balamuth, W. (1975). *J. Protozool.* **22**, 245–256.

Visvesvara, G. S., and Callaway, C. S. (1974). *J. Protozool.* **21**, 239–250.

Visvesvara, G. S., and Healy, G. R. (1975a). *Health Lab. Sci.* **12**, 8–11.

Visvesvara, G. S., and Healy, G. R. (1975b). *Infect. Immun.* **11**, 95–108.

Visvesvara, G. S., Healy, G. R., and Jones, D. B. (1975a). *J. Protozool.* **22**, 26A.

Visvesvara, G. S., Jones, D. B., and Robinson, N. M. (1975b). *Am. J. Trop. Med. Hyg.* **24**, 784–790.

Wade, J., and Satir, P. (1968). *Exp. Cell Res.* **50**, 81–92.

Walsh, C., and Fulton, C. (1973). *Biochim. Biophys. Acta* **312**, 52–71.

Warhurst, D. C., and Armstrong, J. A. (1968). *J. Gen. Microbiol.* **50**, 207–215.

Weihing, R. R., and Korn, E. D. (1971). *Biochemistry* **10**, 590–600.

Weik, R. R., and John, D. T. (1977a). *J. Parasitol.* **63**, 868–871.

Weik, R. R., and John, D. T. (1977b). *J. Protozool.* **24**, 196–200.

Weik, R. R., and John, D. T. (1978). *J. Parasitol.* **64**, 746–747.

Weisman, R. A. (1976). *Annu. Rev. Microbiol.* **30**, 189–219.

Weisman, R. A., and Korn, E. D. (1967). *Biochemistry* **6**, 485–497.

Wellings, F. M., Amuso, P. T., Chang, S. L., and Lewis, A. L. (1977). *Appl. Environ. Microbiol.* **34**, 661–667.

Werth, J. M., and Kahn, A. J. (1967). *J. Bacteriol.* **94**, 1272–1274.

Wetzel, M. G., and Korn, E. D. (1969). *J. Cell Biol.* **43**, 90–104.

Willaert, E. (1977a). *Acta Zool. Pathol. Antverp.* **65**, 1–239.

Willaert, E. (1977b). *Rev. Assoc. Belge Technol. Lab.* **4**, 101–122.

Willaert, E., and Le Ray, D. (1973). *Protistologica* **9**, 417–426.

Willaert, E., and Stevens, A. R. (1976). *Lancet* **ii**, 741.

Willaert, E., Stevens, A. R., and Healy, G. R. (1978a). *J. Clin. Pathol.* **31**, 717–720.

Willaert, E., Stevens, A. R., and Tyndall, R. L. (1978b). *J. Protozool.* **25**, 1–14.

Willmer, E. N. (1956). *J. Exp. Biol.* **33**, 583–603.

Willmer, E. N. (1958). *J. Embryol. Exp. Morphol.* **6**, 187–214.

Willmer, E. N. (1961). *Exp. Cell Res. Suppl.* **8**, 32–46.

Willmer, E. N. (1970). "Cytology and Evolution," 2nd Ed. Academic Press, New York.

Wong, M. M., Karr, S. L., Jr., and Balamuth, W. (1975a). *J. Parasitol.* **61**, 199–208.

Wong, M. M., Karr, S. L., Jr., and Balamuth, W. (1975b). *J. Parasitol.* **61**, 682–690.

Wong, M. M., Karr, S. L., Jr., and Chow, C. K. (1977). *J. Parasitol.* **63**, 872–878.

Yuyama, S. (1971). *J. Protozool.* **18**, 337–343.

The Bioluminescence of Dinoflagellates

9

BEATRICE M. SWEENEY

> About, about, in reel and rout
> The death-fires danced at night;
> The water, like a witch's oils,
> Burnt green, and blue, and white.
> *The Rime of the Ancient Mariner*
> Samuel Taylor Coleridge

287

BIOCHEMISTRY AND PHYSIOLOGY OF PROTOZOA
SECOND EDITION, VOL. 1

I. INTRODUCTION

A. Widespread Occurrence of Bioluminescence in the Sea

Few of us land creatures realize that every cubic meter of the sea probably contains at least one luminous cell, in all likelihood a dinoflagellate. Sometimes tiny flashes from dinoflagellates are so numerous that, looking down into the dark water at night, one sees points of light more brilliant and numerous than the stars. The marine world is the home of the vast majority of luminous creatures. In the surface layers, and perhaps deeper, marine dinoflagellates, both photosynthetic and heterotrophic species, occur in vast numbers everywhere, even in arctic and antarctic seas. Many of these are capable of emitting a bioluminescent flash when stimulated. Their populations vary, both in space and time, as is the case with all planktonic organisms, and when conditions are favorable can reach densities of thousands of cells per milliliter. When a luminous dinoflagellate species is present in such numbers, every disturbance of the water, every breaking wave or darting fish, produces a brilliant blue-green flash. Raindrops can excite spreading rings of bioluminescence. Along the shore, as the outermost waves curl and crash, an electric blue light flashes along their length.

Although bioluminescence has been recorded from every ocean (Staples, 1966; Tett, 1971; Tett and Kelly, 1973), there are certain localities where large populations of dinoflagellates often allow these spectacular displays. Along the western coasts of continents where upwelling of nutrients occurs and plankton in general are present in higher numbers than in the open sea, dinoflagellate blooms are also encountered more frequently. Other common sites for bioluminescent displays are the waters close to Japan and the eastern coast of Borneo, the Banda Sea, and the Gulf of Akaba. In the Caribbean and in New Guinea, there are a number of small bays ringed with mangroves and with narrow access to the sea where populations of the bioluminescent *Pyrodinium bahamense* persist throughout the whole year. So reliably bioluminescent are some of these bays in Jamaica and Puerto Rico that they have become tourist attractions.

B. Historical Views of the Source of Bioluminescence

Mariners from earlier times have marveled at the displays of bioluminescence that accompany large populations of dinoflagellates. The "burning" seas were at first thought to be of supernatural origin, omens of the pleasure or displeasure of the gods. As science began to usurp the explanation of natural phenomena from religion, the light emitted from friction between salt molecules or from phosphorus burning in the water was invoked as an explanation for bioluminescence. In fact, the term "phosphorescence" still survives today. The biological origin was overlooked, because the cells responsible are for the most part quite small, beyond the resolving power of the human eye or even early microscopes. By 1800, living cells were suspected, but the last arguments were not settled in favor of a biological origin of bioluminescence until 1830, when Molisch and Ehrenberg summarized the evidence (Harvey, 1952).

Table I Bioluminescent and Nonbioluminescent Dinoflagellates[a]

Bioluminescent species	Nonbioluminescent species
Ceratium fusus	Ceratium dens
Dissodinium lunula	Ceratium furca
Fragilidium heterolobum	Ceratium lineatum
Gonyaulax catenata	Ceratium macroceros*
Gonyaulax catenella	Ceratocorys horrida*
Gonyaulax digitata	Dinophysis acuminata
Gonyaulax excavata (some strains)	Dinophysis caudata
Gonyaulax hyalina	Dinophysis tripos
Gonyaulax monilata	Diplopeltopsis minor
Gonyaulax polyedra	Glenodinium lenticulatum
Gonyaulax polygramma*	Gonyaulax excavata (some strains)
Gonyaulax sphaeroidea	Gymnodinium breve
Gonyaulax spinifera	Gymnodinium splendens
Gonyaulax tamarensis (some strains)	Ornithocercus steinii*
Gymnodinium flavum	Oxyrrhis marina
Gymnodinium sanguineum	Noctiluca miliaris, small strain
Noctiluca miliaris	Peridinium claudicans
Peridinium brochi	Peridinium subsalsum
Peridinium claudicans	Peridinium trochoideum
Peridinium conicum	Prorocentrum micans
Peridinium depressum	Scrippsiella sweeneyae
Peridinium elegans*	
Peridinium ovatum	
Peridinium pentagonum	
Peridinium steinii	
Pyrocystis fusiformis	
Pyrocystis noctiluca	
Pyrodinium bahamense	
Polykrikos schwartzii	

[a] From tables in Sweeney (1963), Tett (1971), and Tett and Kelly (1973), where the original references can be found. Asterisks indicate new additions to the list from Sweeney (unpublished observations).

C. Species of Dinoflagellates Which Are Bioluminescent

Not all dinoflagellates are capable of bioluminescence. No freshwater forms can emit light. When a very large population of a single dinoflagellate species, as in a "red tide," is accompanied by bioluminescence, it is a simple matter to assign this property to the species involved. Thus *Noctiluca miliaris*, *P. bahamense*, *Gonyaulax polyedra*, *G. excavata*, and *G. monilata* are known to be luminous species, while it is clear that *Gymnodinium breve* of the extensive Florida red tides and *Prorocentrum micans* are not bioluminescent. Positive identification of the ability to emit light in other species requires isolation of the organism and detection of light emitted either spontaneously or on stimulation. Usually a photomultiplier photometer is used for this purpose, although the light is often bright enough to be seen by the dark-adapted eye. In this way, a number of additional luminescent dinoflagellates have been identified (Table I). The isolation and subsequent testing of cells for bioluminescence have pitfalls. Time must be allowed for recovery after the manipulations which stimulate flashing. Cells should be tested at night because light inhibits luminescence, and in many species light emission is under the control of circadian rhythm (see Section IV). Furthermore, consistently luminous and non-luminous strains of the same species of dinoflagellate are known, for example, in *N. miliaris* (Eckert and Findlay, 1962) and *G. excavata* (Schmidt *et al.*, 1978). Therefore, negative results in tests for bioluminescence may be inconclusive. The effects of nutrient deficiency on the capacity to emit light have not been thoroughly investigated and should be considered. In several parts of the world, the bioluminescence of water samples is closely correlated with the number of known bioluminescent species present (Sweeney, 1963; Kelly, 1968). Thus it is probable that there are only a few species of luminescent dinoflagellates which have escaped detection.

II. *IN VIVO* BIOLUMINESCENCE

A. Characteristics of the Bioluminescent Flash

Living dinoflagellates can emit light in three modes. They can flash when stimulated mechanically, chemically, or electrically; they can flash apparently spontaneously; and late at night they can glow very dimly. The bioluminescent flash is short, with a half rise time of about 8–9 msec and a half decay time slightly longer, 16–18 msec, as determined in single cells of both *G. polyedra* stimulated mechanically (Christianson and Sweeney, 1972) and *N. miliaris* stimulated either mechanically or electrically (Eckert,

1966). The maximum amount of light emitted in a flash differs widely among species, larger species emitting more light per flash than smaller ones (Table II). Among photosynthetic dinoflagellates, the maximum number of photons emitted in a flash is quite constant for a given species throughout the log and stationary phases of growth until cell senescence (Biggley *et al.*, 1969; Sweeney, unpublished observations). In heterotrophic dinoflagellates such as *Noctiluca,* the nutritional status of the cell influences the brightness of the flash. For example, in *Noctiluca* from Southeast Asia which contains green flagellate symbionts, the number of photons emitted per flash increases as the irradiance and consequently the photosynthesis of the symbionts increases, following a saturation curve. Feeding this *Noctiluca* also results in an increase in photon emission, as compared with *Noctiluca* depending on the photosynthesis of its symbionts (Sweeney, 1971). When *G. polyedra* is kept in darkness for several days, the maximum bioluminescence declines with each successive night (Sweeney and Hastings, 1957), but can be increased if the cells are irradiated for a short time during the preceding day. The action spectrum from this effect is that for photosynthesis (Sweeney *et al.*, 1959).

B. Mechanical Stimulation and Its Inhibition by Light

The mechanical stimulus which elicits a bioluminescent flash is a distortion of the cell by shear forces. That shear is involved is shown by the observation that cells flash when their surface is deformed by a mechanical pulse from a piezoelectric crystal (Reynolds, 1970), or by passing the cell suspension through a capillary tube (Christianson and Sweeney, 1972). When cells are accelerated, as in a centrifuge, they flash only on making contact with the bottom of the container. The threshold for a response to mechanical stimulation in *G. polyedra* is on the order of 5 dynes/cm² during the day phase when the cells are in darkness, but considerably lower during the night phase (Figure 1), a manifestation of circadian rhythmicity (see Section IV,A). Bright light (1000 μW/cm² or more for 1–2 hr) inhibits flashing almost completely in *G. polyedra* (Haxo and Sweeney, 1955). Under these conditions, only a prolonged light emission of very low intensity can be observed following mechanical stimulation. Light inhibits via a change in the reception or transmission of mechanical stimuli, leaving the biochemistry unaffected. Stimulation of bioluminescence by the addition of hydrogen ions (see Section II,C, and Sweeney, 1969b) is unaffected by exposing the cells to light. Larger, rather than smaller, yields of luciferin and luciferase can be obtained from brightly lighted cells. Light has no effect on the emission of bioluminescence from cell extracts or from purified luciferin and luciferase (Hastings and Sweeney, 1957a).

Table II The Brightness of a Flash from a Single Cell and the Wavelength of Maximum Emission in a Variety of Bioluminescent Dinoflagellates

Organism	Maximum photons per cell $\times 10^{-9}$	Reference	Wavelength of maximum emission (nm)	Reference
Ceratium fusus	0.53	Esaias *et al.* (1973)	—	
Dissodinium lunula	3.9	Biggley *et al.* (1969)	—	
D. lunula	5–12	Swift *et al.* (1973)	475	Swift *et al.* (1973)
D. lunula	4	Hamman and Seliger (1972), Seliger *et al.* (1970)	—	
D. lunula	7.3	Schmitter *et al.* (1976)	477.5	Swift and Taylor (1967)
Gonyaulax acatenella	0.091	Esaias *et al.* (1973)	—	
Gonyaulax catenella	0.006–0.008	Esaias *et al.* (1973)	479	Esaias *et al.* (1973)
Gonyaulax excavata	0.036	Esaias *et al.* (1973)	—	
	0.09	Schmitter *et al.* (1976)	—	
Gonyaulax monilata	0.012	Esaias *et al.* (1973)	—	
Gonyaulax polyedra	0.117	Biggley *et al.* (1969)	474	Bode and Hastings (1963)
G. excavata	0.12	Hamman and Seliger (1972), Seliger *et al.* (1970)	478	Hamman and Seliger (1972)
G. excavata	0.35	Schmitter *et al.* (1976)	—	
Noctiluca miliaris	21	Hamman and Seliger (1972)	474	Nicol (1958)
	2.7	Eckert (1967)	—	

Species	Value		Reference	
Noctiluca miliaris (green)	370	—	Sweeney (1971)	
Peridinium pentagonum	5	—	Esaias et al. (1973)	
Polykrikos schwartzii	7.8	—	Tett (1969)	
Pyrocystis acuta	3–6	475	Swift et al. (1973)	Swift et al. (1973)
Pyrocystis fusiformis	88	473	Hamman and Seliger (1972)	Swift et al. (1973)
P. fusiformis	40			
P. fusiformis	45	473	Swift et al. (1973) (different strains)	Swift et al. (1973)
P. fusiformis	62			
P. fusiformis	23			
Pyrocystis noctiluca	110	—	Hamman and Seliger (1972)	
P. noctiluca	66	—		
P. noctiluca	101	473	Swift et al. (1973) (different strains)	Swift et al. (1973)
P. noctiluca	79			
P. noctiluca	19			
P. noctiluca	290	—	Schmitter et al. (1976)	
Pyrodinium bahamense	0.1	475	Swift et al. (1973)	Taylor et al. (1966)
P. bahamense	0.105	—	Biggley et al. (1969)	
P. bahamense	0.335	—	Biggley et al. (1969)	
P. bahamense	0.28	—	Hamman and Seliger (1972), Seliger et al. (1970)	

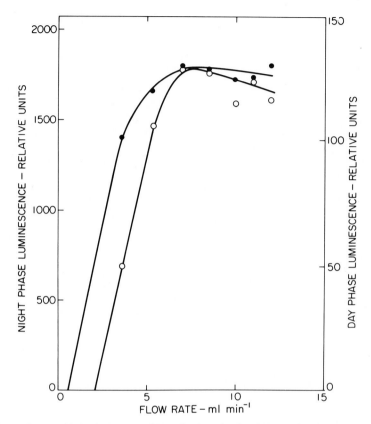

Figure 1. The bioluminescence of *G. polyedra* stimulated by passing through a capillary coil at different flow rates. The bioluminescence has been corrected to a constant number of cells passing through the capillary (9000/min). A flow rate of 1.8 ml/min gives a shear force of about 5 dynes/cm². All cell suspensions have been in constant darkness. Solid circles and left-hand ordinate refer to cells in the night phase; open circles and right-hand ordinate refer to cells in the day phase. Note the differences in luminescence capacity and threshold for stimulation of bioluminescence between cells in these opposite phases of their circadian rhythm. [Recalculated from the data of Christianson and Sweeney (1972).]

The bioluminescent flash is also photoinhibited in a number of other dinoflagellates besides *Gonyaulax polyedra,* including *Ceratium fusus, G. monilata* (Hamman and Seliger, 1972), *G. acatenella, G. catenella,* and *G. tamarensis* (*G. excavata*) (Esaias *et al.,* 1973) and *Pyrocystis acuta, P. noctiluca,* and *P. fusiformis* (Hamman and Seliger, 1972). Interestingly, no photoinhibition of bioluminescence is found in the nonphotosynthetic species *N. miliaris* (Harvey, 1926), *Peridinium pentagonum,* and *Polykrikos schwartzii* (Hamman and Seliger, 1972). Sensitivity to light inhibition varies greatly from species to species, *G. acatenella, G. catenella,* and *G. tamarensis* being 10^4 times more sensitive than *G. polyedra* (Esaias *et al.,* 1973). Action spectra for the first three species show a single maximum in the green region of the spectrum at 562 nm, whereas *G. polyedra* is sensitive to both blue and red but not to green wavelengths (Sweeney *et al.,* 1959). Action spectra for the photoinhibition of other species are yet to be determined.

C. Chemical Stimulation of Bioluminescence

The bioluminescence of all dinoflagellates can be stimulated by lowering the pH to 5. On the addition of acid, light is emitted very rapidly, and there is about the same number of photons in a flash as following mechanical stimulation of cells in darkness (Sweeney, 1972; McMurry and Hastings, 1972). Some other cations have also been reported to elicit a flash: Ca^{2+} and NH_4^+ in *G. polyedra* (Sweeney, 1972; Hamman and Seliger, 1972), *Pyrodinium bahamense, Pyrocystis fusiformis, Pyrocystis noctiluca,* and *Dissodinium (Pyrocystis) lunula* (Hamman and Seliger, 1972). Stimulation by K^+ was reported in *G. polyedra* and *P. bahamense* by Hamman and Seliger (1972), but Sweeney (1972) reported negative results with K^+ in *G. polyedra.* Final concentrations of these ions which were required to stimulate bioluminescence were on the order of 0.1–0.4 M. As mentioned above, no photoinhibition is observed when luminescence is stimulated by the addition of H^+, Ca^{2+}, NH_4^+, or K^+.

D. Electrical Stimulation of Bioluminescence

Electrical stimulation of a bioluminescent flash has only been successful with the large dinoflagellates *N. miliaris* and *P. fusiformis.* Eckert (1965) succeeded in inserting microelectrodes into the vacuole of *Noctiluca* and recording both a negative-going action potential (40–70 mV) and the flash which followed with a latency of about 3 msec. The all-or-none nature of the stimulation, as well as apparent summation and fatigue, could be demonstrated. Recently, the bioluminescence of *P. fusiformis,* held between two suction electrodes, has been stimulated electrically (Smith and Case, unpublished observations). Electrical stimulation of bioluminescence via a microelectrode inserted into the vacuole of *Pyrocystis* was preceded by a negative-going action potential as in *Noctiluca* (Smith and Case, unpublished observations).

In both these large dinoflagellates, bioluminescence can be seen by dark-field microscopy or image intensification techniques to originate from small, discrete regions in the cytoplasm called "microsources" (Eckert and Reynolds, 1967; Smith and Sweeney, unpublished observations). The structure of the microsources is still unknown, although it has been suggested that they may be identical with lipid-containing vesicles seen in the peripheral cytoplasm of both *Noctiluca* (Sweeney, 1978) and *P. fusiformis* (Sweeney, unpublished observations). Such microsources cannot be seen in *G. polyedra*.

E. Spontaneous Flashing

Flashes which arise apparently spontaneously have been observed both in nature and in cultures of a number of dinoflagellates. When cell suspensions are placed within the field of view of a photomultiplier tube in darkness in the absence of mechanical stimulation (Figure 2), occasional flashes are recorded from *G. polyedra* (Sweeney, 1969a) but not from *P. fusiformis* (Sweeney, unpublished observations). Since *Gonyaulax* is motile, while *P. fusiformis* is not, these flashes may be stimulated when the cells bump each other or the sides of the container. Similar spontaneous flashes can be detected in *P. bahamense* and *D. lunula* (Biggley *et al.,* 1969). By visual observation of the sea surface at night during a red tide of *G. polyedra,* individual flashes, also occurring apparently spontaneously, can be detected.

F. Bioluminescent Glow

A continuous dim light emission or glow can be detected when *G. polyedra* cell suspensions are placed into a photomultiplier photometer and left undisturbed during the night (Figure 2). This glow does not re-

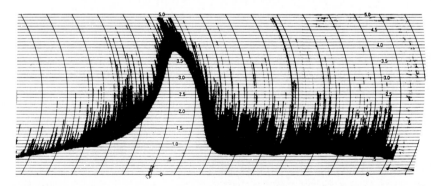

Figure 2. Spontaneous flashes (vertical lines) and the bioluminescent glow emitted by a cell suspension of *G. polyedra* in darkness. Recorded without stimulation by a photomultiplier photometer, starting at the beginning of a night. Chart speed was one division per hour, reading from right to left.

quire stimulation. It increases in intensity as the temperature is increased from 20° to 25°C (Sweeney and Hastings, 1958; Hastings and Sweeney, 1957b), while stimulated bioluminescence decreases over this range of temperature. The glow can be observed in stationary-phase cultures, and hence it is not the result of processes coincident to cell division. A similar glow occurs in *P. fusiformis* (Sweeney, unpublished observations), *P. bahamense,* and *D. lunula* (Biggley *et al.,* 1969). The cause of this mode of bioluminescence is not understood at present.

III. BIOCHEMISTRY OF DINOFLAGELLATE BIOLUMINESCENCE

A. The Bioluminescent Reaction

In luminescence, light is emitted from a molecule in an excited state, as it is in fluorescence or phosphorescence, but the excited state is attained by energy released in a chemical reaction rather than by previous absorption of light energy. If the reaction is biochemical, the resulting light is *bio*luminescence. In dinoflagellates, as in many other organisms, the emission of light is the result of the interaction of an enzyme, luciferase (a general name for all enzymes taking part in such a process) with its substrate, luciferin (again a general term which does not imply identity of structure). The reaction is usually the oxidation of luciferin by molecular oxygen, and the energy released excites the luciferin or the luciferin–enzyme complex. The details of this interaction in dinoflagellates are unknown at present. On decay of the excited state, a photon is emitted. This light is blue-green in dinoflagellates. The emission spectra of all the species examined so far are very similar and show a maximum at about 474 nm (Table II). Cofactors required in other bioluminescent systems, ATP in fireflies and FMNH in bacteria, have no effect on dinoflagellate bioluminescent extracts (Hastings and Sweeney, 1957a). Since light has been obtained from highly purified luciferin and luciferase, probably none are required (McMurry and Hastings, 1972). Neither the luciferin nor the luciferase extracted from dinoflagellates reacts with luciferases or luciferins from other systems, and both are thus assumed to be different and specific to dinoflagellates. Cross-reactions do take place between the luciferin from one dinoflagellate and the luciferase from another (Schmitter *et al.,* 1976); thus the biochemistry appears to be similar in the entire group. There are, however, some differences between different species with respect to the biochemistry of light emission, as noted in Sections III,C, and III,D.)

B. The Luciferin of Dinoflagellates

The luciferin of dinoflagellates is a small molecule, since it can diffuse through a dialysis membrane; it has a molecular weight of about 500 (Bode and Hastings, 1963). There is very little in each cell, and it is highly unstable at room temperature, its half-life being a matter of hours. It is relatively more stable at very cold temperatures (90°K). Only very small quantities have been obtained in pure form, and it has so far been impossible to get any idea of its chemical composition or structure. The luciferin molecule is fluorescent, with a maximum emission at 470 nm, a fairly good match with the maximum bioluminescence emission at 474 nm.

C. The Luciferases of Dinoflagellates

The luciferases of dinoflagellates are large proteins, 360,000 daltons in *G. polyedra*, 230,000 daltons in *G. tamarensis* (*G. excavata*), 200,000 daltons in *D. lunula*, and 310,000 daltons in *P. noctiluca* (Schmitter *et al.*, 1976). When luciferases are treated with 6 *M* guanidine–hydrochloric acid single chains are recovered which also vary in size somewhat: *G. polyedra*, 130,000 daltons; *G. tamarensis*, 130,000 daltons; *D. lunula* and *Pyrocystis*, 60,000 daltons (Schmitter *et al.*, 1976). Since these luciferases can be used interchangeably, the active sites must be identical. Fragments of about 35,000 daltons, which still contain the active site, have been obtained through the activity of endogenous proteases (Krieger *et al.*, 1974). The complete luciferases show sharp pH optima near pH 6.5, slightly different from one another. The pH optimum for the active proteolytic fragment from *G. polyedra* is much broader, so that there is some activity at pH 8 or above (Krieger and Hastings, 1968).

D. Luciferin-Binding Protein

In addition to luciferin and luciferase, some dinoflagellate extracts contain a third component, a protein of large size, 120,000 daltons, which binds to the luciferin at pH 8 and thereby inactivates it (Fogel and Hastings, 1971). When the pH is lowered to pH 6.5, the luciferin is released (Schmitter *et al.*, 1976). This luciferin-binding protein is found in *G. polyedra* and *G. tamarensis* (*G. excavata*) but is absent in *P. noctiluca, D. lunula,* and perhaps other species. The function of this protein *in vivo* is not known at present, but it could act in controlling light emission.

E. Soluble and Particulate Bioluminescent Fractions

The emission of light by extracts of dinoflagellates has some very unusual features. When cells are homogenized and the extract is centrifuged for biochemical experiments, both the supernatant and the sediment can give off light *in vitro*. Moreover, the requirements for light emission by the two fractions are not the same. The supernatant containing the soluble luciferin and luciferase emits light when salt is added and the pH is adjusted to pH 6.6 (Hastings and Sweeney, 1957a). Almost any salt can be used—sodium chloride, potassium chloride, or even ammonium sulfate. The molarity required is high, about 1 *M*, close to that of seawater. The kinetics of the flash produced depends on the salt added, ammonium sulfate producing a much longer light emission with slower kinetics than sodium chloride (Hastings and Sweeney, 1957a).

The sediment, on the other hand, emits light without the addition of salt. Only the addition of acid to bring the pH to about 5 is necessary for bioluminescence from this source (DeSa *et al.*, 1963). Dilution of the sediment does not lead to loss of activity. No soluble luciferin or luciferase is released from the sediment. Both are apparently bound firmly to some particle or membrane fragment in the sediment. Their presence can be shown because the emission spectrum from the sediment is the same as that from the soluble fraction (Hastings *et al.*, 1966), and if luciferin is added in excess, emission from the particulate fraction can be increased (Fogel and Hastings, 1972; Henry and Hastings, 1974). Experiments in two laboratories showed that spent particulate material could be "recharged" by the addition of luciferin from a soluble preparation, so that a small amount of additional light could be obtained (Fuller *et al.* 1972; Fogel and Hastings, 1972; Henry and Hastings, 1974). Luciferin-binding protein is also present in the particulate fraction from *G. polyedra*. Particulate material capable of bioluminescence has also been obtained from extracts of *Noctiluca* (Hastings *et al.*, 1966), *P. bahamense*, and *D. lunula* (Fuller *et al.*, 1972).

The nature of the particle to which the luciferin and luciferase are apparently so tightly bound is at present unknown. At first it was thought that a crystal of guanine or a similar substance with a typical chevron shape was the particle in question, and it was given the name "scintillon." Some bioluminescent dinoflagellates lack these crystals, for example, *P. bahamense* (Sweeney and Bouck, 1966) and *N. miliaris* (Sweeney, 1978), and they are present in nonluminous species such as *P. micans* and *Gymnodinium nelsoni* (Gold and Pokorny, 1973). There is evidence that in the latter species the crystals are nitrogen reserves (Gold and Pokorny, 1973). Although the nature of the particles is clearly different from that of the original scintillon, the name survives and is used to denote the unknown particulate material in the sediment of bioluminescent species (Fogel *et al.*, 1972).

IV. CIRCADIAN RHYTHMS IN BIOLUMINESCENCE IN DINOFLAGELLATES

A. *Gonyaulax polyedra*

One of the most interesting features of dinoflagellate bioluminescence is the rhythmicity observed in some species. This has been studied most extensively in *G. polyedra*. When the luminescence of samples from a culture of this organism growing in alternating light and darkness, each 12 hr long, is stimulated mechanically or by the addition of acid, the light emitted from cells during the dark part of the light–dark cycle is 50 to 100 times brighter than that emitted during the light part (Sweeney and Hastings, 1957). At the beginning of the night, the bioluminescent capacity increases rapidly, reaching a maximum after 6 hr, and then decreasing again (Figure 3). Bioluminescence remains low during the day. This time course is not merely the result of light inhibition of luminescence during the day. The luminescence of cells kept in darkness for several days continues to vary rhythmically, with a maximum occurring about every 24 hr. In continuous darkness, there is a progressive decline in the maximum bioluminescence in successive cycles because of the lack of light for

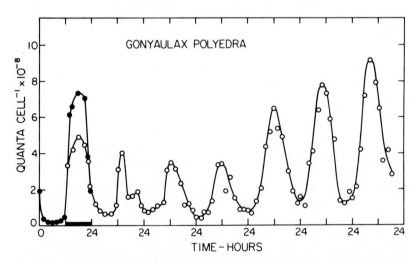

Figure 3. The circadian rhythm in the bioluminescence of *G. polyedra* during one light–dark cycle and then after transfer of the cell suspension to constant light of 450 μW/cm². Temperature, 20°C; mechanical stimulation for 1 min. Note the continuation of the circadian rhythm without damping for the duration of the experiment. The black bar on the abscissa represents the last 12-hr dark period of the light–dark cycle. (Unpublished observations of B. M. Sweeney.)

photosynthesis (Haxo and Sweeney, 1955). This loss can be prevented if the cells are kept in continuous light bright enough to be above the compensation point of photosynthesis. Under these conditions (Hastings and Sweeney, 1958, 1959), the rhythm continues to be evident for many days and even increases (Figure 3).

The oscillation in bioluminescence shares a number of properties with other circadian systems (Sweeney, 1976); under constant conditions (low irradiance) the period changes only a little with a change in the ambient temperature. The oscillation is temperature compensated but not perfectly, so that as the temperature is increased the period of the rhythm becomes slightly longer (Hastings and Sweeney, 1958). The rhythm can be phase-shifted by short exposures to bright light during the "night" part of the cycle. Light exposures during the first half of the night delay the next maximum in luminescence, while in the last half, light advances the next maximum. The bioluminescence can be entrained to a variety of light–dark cycles by this mechanism.

Like stimulated bioluminescence, the glow in *Gonyaulax* shows a strong circadian variation. In constant light of low irradiance, it is bright only during a short time each day, about 3 hr. The maximum glow appears about 4 hr after that in stimulated light emission. Since it is easier to automate the measurement of the glow, which does not require stimulation, than to automate other modes of luminescence, the glow has been used as an indicator for changes in the circadian rhythm following treatment with a variety of chemicals and conditions (Hastings and Bode, 1962).

The sensitivity to stimulation by mechanical means also has a circadian component (Christianson and Sweeney, 1972). The threshold for cells in the night phase is lower than that for cells in the day phase *Gonyaulax* (Figure 1).

When homogenates of *Gonyaulax* are prepared at different times in a circadian cycle and assayed for luciferin and luciferase, both enzyme and substrate vary in activity in a cyclic fashion (Hastings and Bode, 1962; Bode *et al.,* 1963). The maximum extractable luciferase activity coincides with the maximum *in vivo* luminescence. The greatest luciferin activity is found in extracts made somewhat later. The differences in activity do not seem to be due to changes in the extractability of the luminescent system with time, since different activities of luciferase are still observed when guanidine–hydrochloric acid is present during extraction (McMurry, 1971). No extra activator or inhibitor can be detected in the homogenate by mixing experiments (Sweeney, 1969b; McMurry, 1971), and none separates from the luciferase during purification by gradient centrifugation or column chromatography (McMurry and Hastings, 1972). Experiments to determine whether or not new synthesis of luciferase occurred each day gave at best equivocal results. An activator or inhibitor which binds tightly to the luciferase during part of each circadian cycle, of necessity by covalent linkage, has been postulated (McMurry, 1971) but never detected. Luciferin-binding protein could presumably play such a role in *G. polyedra*. However, its absence in *Pyrocystis* makes this unlikely.

Bioluminescence is only one of a number of physiological functions in *Gonyaulax* which show circadian rhythms. Photosynthesis (Hastings *et al.*, 1960; Prézelin *et al.*, 1977; Prézelin and Sweeney, 1977) and cell division (Sweeney and Hastings, 1958) are also strongly rhythmic. Other circadian processes, particularly membrane-related processes, follow a rhythmic pattern as well (Sweeney and Herz, 1977).

B. Circadian Rhythms in Bioluminescence in Other Dinoflagellates

Circadian rhythms in bioluminescence, like that in *G. polyedra*, have been demonstrated in *G. catenella*, *P. bahamense*, and *P. fusiformis* (Sweeney, unpublished observations). Under constant conditions, cycles

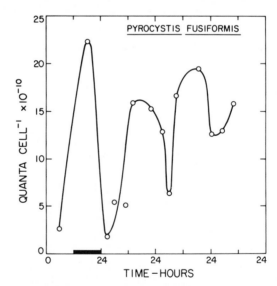

Figure 4. The circadian rhythm of bioluminescence in *P. fusiformis* during one light–dark cycle and then after transfer of the cell suspension to constant light of 130 μW/cm². Temperature, 22°C. The ordinate is the quanta emitted by one cell 3 min after the beginning of stimulation by the addition of acetic acid (final concentration, 5 m*M*). Note the continuation of the rhythm under constant conditions for at least two cycles. The black bar on the abscissa represents the last 12-hr dark period of the light–dark cycle. (Unpublished observations of N. Kittle and B. M. Sweeney.)

of bioluminescence do not persist for as long a time as in *G. polyedra*, but this may be because the most favorable conditions have not been used. In *P. fusiformis* is continuous light of 130 μW/cm^2, the rhythm can be followed for two cycles (Figure 4). Gradually the light emission becomes constant and very bright.

In the heterotrophic dinoflagellate *N. miliaris*, the bioluminescence shows no sign of circadian rhythm (Nicol, 1958). Even in the presence of photosynthetic flagellate symbionts, a constant level of bioluminescence capacity is observed (Sweeney, 1971).

V. THE ADAPTIVE VALUE OF BIOLUMINESCENCE IN DINOFLAGELLATES

There has been much speculation concerning the adaptive value of bioluminescence in dinoflagellates. The widespread occurrence of this ability among marine species implies that it must serve some function. The only experimental evidence suggests that this might be the inhibition of filter feeding in marine zooplankton. Making use of the circadian rhythm of bioluminescence in *G. polyedra* and the consequent availability of cell suspensions either capable of bright flashes or almost completely insensitive to stimulation, Esaias and Curl (1972) monitored the number of cells consumed by copepods and the amount of flashing during copepod feeding activity under both conditions. About half as many *Gonyaulax* disappeared from cell suspensions when they were capable of flashing than when they were not. Certainly further work along these lines would be desirable.

VI. PROSPECTS AND CONCLUSIONS

While much is known about bioluminescence in both living dinoflagellates and extracts made from them, there still remain a number of vexing problems which require investigation. Nothing is known concerning the intracellular location of light emission, although microsources are certainly present in *Noctiluca* and *Pyrocystis* and luminescence comes only from the central region in *D. lunula* (Swift and Reynolds, 1968). Microsources have not been seen in other species which do not have large vacuoles.

The sensor for mechanical stimulation is probably in the plasma membrane, but this cannot be said with any certainty. In *Noctiluca* and *Pyrocystis*, where action potentials precede the flash, the fact that they are negative-going, rather than positive-going, has led to the supposition that the vacuolar membrane is their source (Eckert and Sibaoka, 1968). The train of events by which the action potential actually generates a flash 2–3 msec later is completely unknown, as is the intracellular site of chemical stimulation. It has been postulated that ions act at a step later than the action potential (Hamman and Seliger, 1972). The observation that chemical stimulation is not photoinhibited, while mechanical stimulation shows photoinhibition, supports this hypothesis. While it seems probable that particulate luminescence may be the natural *in vivo* emitter and the soluble luminescence in extracts represents a pool of luciferin and luciferase, neither supposition is firmly established (Arrio and Lecuyer, 1975).

Questions with respect to the adaptive significance of bioluminescence in dinoflagellates remain. Many dinoflagellates are not bioluminescent, both in marine and freshwater environments, yet they survive. Among 10 clones isolated at the same time from a red tide, three were nonbioluminescent (Schmidt *et al.*, 1976; 1978).

Observations of bioluminescence in various parts of the world's oceans may disclose new bioluminescent species. The fascinating phenomenon of "phosphorescent wheels," widespread circling spokelike luminous bands or stripes seen at sea at rare intervals by mariners, has never been examined by scientists (Staples, 1966). The cause of such displays is completely unknown, but their existence suggests other modes for the stimulation of luminescence, possibly radar (Hilder, 1962).

The flashes emitted by luminous dinoflagellates will continue to excite the curiosity of scientists and layman alike.

REFERENCES

Arrio, B., and Lecuyer, B. (1975). *FEBS Lett.* **53**, 49–52.
Biggley, W. H., Swift, E., Buchanan, R. J., and Seliger, H. H. (1969). *J. Gen. Physiol.* **54**, 96–122.
Bode, V. C., and Hastings, J. W. (1963). *Arch. Biochem. Biophys.* **103**, 488–499.
Bode, V. C., DeSa, R., and Hastings, J. W. (1963). *Science* **141**, 913–915.
Christianson, R., and Sweeney, B. M. (1972). *Plant Physiol.* **49**, 994–997.
DeSa, R., Hastings, J. W., and Vatter, A. E. (1963). *Science* **141**, 1269–1270.
Eckert, R. (1965). *Science* **147**, 1140–1145.
Eckert, R. (1966). *Science* **151**, 349–352.
Eckert, R. (1967). *J. Gen. Physiol.* **50**, 2211–2237.
Eckert, R., and Findlay, M. (1962). *Biol. Bull.* (*Woods Hole, Mass.*) **123**, 494–495.
Eckert, R., and Reynolds, G. T. (1967). *J. Gen. Physiol.* **50**, 1429–1458.
Eckert, R., and Sibaoka, T. (1968). *J. Gen. Physiol.* **52**, 258–282.
Esaias, W. E., and Curl, H. C. (1972). *Limnol. Oceanogr.* **17**, 901–906.
Esaias, W. E., Curl, H. C., Jr., and Seliger, H. H. (1973). *J. Cell. Physiol.* **82**, 363–372.

Fogel, M., and Hastings, J. W. (1971). *Arch. Biochem. Biophys.* **142**, 310–321.
Fogel, M., and Hastings, J. W. (1972). *Proc. Natl. Acad. Sci. U.S.A.* **69**, 690–693.
Fogel, M., Schmitter, R. E., and Hastings, J. W. (1972). *J. Cell Sci.* **11**, 305–317.
Fuller, C. W., Kreiss, P., and Seliger, H. H. (1972). *Science* **177**, 884–885.
Gold, K., and Pokorny, K. S. (1973). *J. Phycol.* **9**, 225–229.
Hamman, J. P., and Seliger, H. H. (1972). *J. Cell. Physiol.* **80**, 397–408.
Harvey, E. N. (1926). *Biol. Bull. (Woods Hole, Mass.)* **51**, 85–88.
Harvey, E. N. (1952). "Bioluminescence." Academic Press, New York.
Hastings, J. W., and Bode, V. C. (1962). *Ann. N.Y. Acad. Sci.* **98**, 876–889.
Hastings, J. W., and Sweeney, B. M. (1959a). *J. Cell. Comp. Physiol.* **49**, 209–225.
Hastings, J. W., and Sweeney, B. M. (1956b). *Proc. Natl. Acad. Sci. U.S.A.* **43**, 804–811.
Hastings, J. W., and Sweeney, B. M. (1958). *Biol. Bull. (Woods Hole, Mass.)* **115**, 440–458.
Hastings, J. W., and Sweeney, B. M. (1959). *In* "Photoperiodism and Related Processes in Plants and Animals" (R. B. Withrow, ed.), pp. 567–584. Am. Assoc. Adv. Sci., Washington, D.C.
Hastings, J. W., Astrachan, L., and Sweeney, B. M. (1960). *J. Gen. Physiol.* **45**, 69–76.
Hastings, J. W., Vergin, M., and DeSa, R. (1966). *In* "Bioluminescence in Progress" (F. H. Jonson and Y. Hameda, eds.), pp. 301–329. Princeton Univ. Press, Princeton, New Jersey.
Haxo, F. T., and Sweeney, B. M. (1955). *In* "The Luminescence of Biological Systems" (F. H. Johnson, ed.), pp. 415–420. Am. Assoc. Adv. Sci., Washington, D.C.
Henry, J. P., and Hastings, J. W. (1974). *Mol. Cell. Biochem.* **3**, 81–91.
Hilder, B. (1962). *J. Aust. Inst. Navigation* **1**, 43–60.
Kelley, M. G. (1968). *Biol. Bull. (Woods Hole, Mass.)* **135**, 279–295.
Krieger, N., and Hastings, J. W. (1968). *Science* **161**, 586–589.
Krieger, N., Njus, D., and Hastings, J. W. (1974). *Biochemistry* **13**, 2871–2877.
McMurry, L. (1971). Ph.D. Thesis, Harvard Univ., Cambridge, Massachusetts.
McMurry, L., and Hastings, J. W. (1972). *Biol. Bull. (Woods Hole, Mass.)* **143**, 196–206.
Nicol, J. A. C. (1958). *J. Mar. Biol. Assoc. U.K.* **37**, 535–549.
Prézelin, B. B., and Sweeney, B. M. (1977). *Plant Physiol.* **60**, 388–392.
Prézelin, B. B., Meeson, B. W., and Sweeney, B. M. (1977). *Plant Physiol.* **60**, 384–387.
Reynolds, G. T. (1970). *Biophys. J.* **10**, 132A.
Reynolds, G. T. (1972). *Q. Rev. Biophys.* **5**, 295–347.
Schmidt, R. J., Gooch, V. D., and Loeblich, L. A., III (1976). *Biophys. J.* **16**(2), 101a.
Schmidt, R. J., Gooch, V. D., Loeblich, A. R., III, and Hastings, J. W. (1978). *J. Phycol.* **14**, 5–9.
Schmitter, R. E., Njus, D., Sulzman, F. M., Gooch, V. D., and Hastings, J. W. (1976). *J. Cell. Physiol.* **87**, 123–134.
Seliger, H. H., Biggley, W. H., and Swift, E. (1970). *Photochem. Photobiol.* **10**, 227–232.
Staples, R. F. (1966). Tech. Rep. TR-184. U.S. Nav. Oceanogr. Off., Washington, D.C.
Sweeney, B. M. (1963). *Biol. Bull. (Woods Hole, Mass.)* **125**, 177–181.
Sweeney, B. M. (1969a). "Rhythmic Phenomena in Plants." Academic Press, New York.
Sweeney, B. M. (1969b). *Can. J. Bot.* **47**, 299–308.
Sweeney, B. M. (1971). *J. Phycol.* **4**, 53–58.
Sweeney, B. M. (1972). *Proc. Int. Symp. Circadian Rhythmicity, Wageningen, 1971* pp. 137–156.
Sweeney, B. M. (1976). *In* "The Molecular Basis of Circadian Rhythms" (J. W. Hastings and H.-G. Schweiger, eds.), pp. 77–83. Dahlem Konf., Berlin.
Sweeney, B. M. (1978). *J. Phycol.* **14**, 116–120.

Sweeney, B. M., and Bouck, G. B. (1966). *In* "Bioluminescence in Progress" (J. H. Johnson and Y. Haneda, eds.), pp. 331–348. Princeton Univ. Press, Princeton, New Jersey.

Sweeney, B. M., and Hastings, J. W. (1957). *J. Cell. Comp. Physiol.* **49,** 115–128.

Sweeney, B. M., and Hastings, J. W. (1958). *J. Protozool.* **5,** 217–224.

Sweeney, B. M., and Herz, J. (1977). *Proc. Int. Conf., Int. Soc. Chronobiol., 12th, Washington, D.C.,* pp. 751–761.

Sweeney, B. M. Haxo, F. T., and Hastings, J. W. (1959). *J. Gen. Physiol.* **43,** 285–299.

Swift, E., and Reynolds, G. T. (1968). *Biol. Bull. (Woods Hole, Mass.)* **135,** 439–440.

Swift, E., and Taylor, W. R. (1967). *J. Phycol.* **3,** 77–80.

Swift, E., Biggley, W. H., and Seliger, H. H. (1973). *J. Phycol.* **9,** 420–426.

Taylor, W. R., Seliger, H. H., Fastie, W. G., and McElroy, W. D. (1966). *J. Mar. Res.* **24,** 28–43.

Tett, P. B. (1969). *J. Mar. Biol. Assoc. U.K.* **49,** 245–258.

Tett, P. B. (1972). *J. Mar. Biol. Assoc. U.K.* **51,** 183–206.

Tett, P. B., and Kelly, M. G. (1973). *Oceanogr. Mar. Biol. Annu. Rev.* **11,** 89–173.

Starr, 1970), and these vitamins can increase the growth rate in some isolates of other genera. Perhaps the variety in vitamin requirement found within the species *G. pectorale* by Stein (1966a) is more widespread in the family.

III. VEGETATIVE REPRODUCTION

A. Meiosis

The only diploid cell in the volvocacean life cycle is the zygote. This heavy-walled, dormant cell, capable of survival in dry soil for many years, is an important agent of species maintenance and dispersal, as well as a means of genetic recombination. Meiosis occurs at zygote germination, yielding, in the genus *Gonium* (Schreiber, 1925; Stein, 1958b), four cells which represent the four products of meiosis. Only one of the meiotic products normally survives in all the examples of the other genera so far examined, with rare exceptions (Starr, 1975). Each surviving cell produces a colony of haploid cells by mitosis.

Volvocacean colonies can be propagated by vegetative reproduction indefinitely. One of the major taxonomic criteria of the family is the proportion and location in a colony of the cells which can propagate further (Figure 1). In the majority of genera all cells of a colony are potentially capable of forming daughters, while in the others only a proportion of the cells (called *gonidia*, and recognizable by their greater size) retain the generative ability. The remainder of the cells normally never again undergo division and are termed *somatic* cells.

B. Daughter Colony Formation

Vegetative reproduction is accomplished by 2^n successive mitotic divisions in each of the cells of the colony capable of reproduction, the maximum value of n being a species character. In their rapidity these divisions are entirely comparable to cleavage of a fertilized egg in animals; only in a few species of *Volvox* is there a long interval between cleavages. Furthermore, the pattern of cleavage planes is very exact. After completion of the cleavages, the resultant saucer or hollow sphere of cells, held together by cytoplasmic bridges, undergoes the process of colony inversion which orients the nucleated apical ends of the cells to the outside of the sphere. Meanwhile, two flagella sprout from each cell, and finally the complex cellular and colony envelopes are secreted, binding all the cells together in the precise orientation laid down by the pattern of planes of division

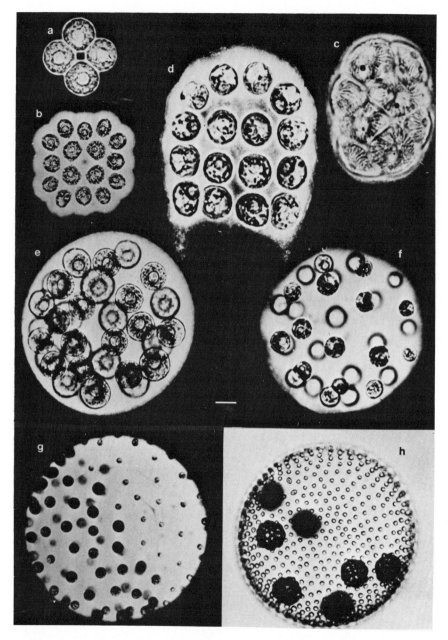

Figure 1. Living colonies, surrounded by india ink except in (h). (a) *Gonium sociale*. (b) *Gonium pectorale*. (c) *Pandorina morum*. (d) *Platydorina caudata*. (e) *Eudorina elegans*. (f) *Eudorina illinoisensis*. (g) *Eudorina (Pleodorina) californica*. (h) *Volvox carteri*. (a–f) Bar = 5 μm; (g) bar = 20 μm; (h) bar = 40 μm.

during the mitoses. The genus *Astrephomene* shows one variation in this pattern; its cell cleavages result in a plate of cells already inverted, i.e., with their apical ends outward, and it does not invert subsequently. After daughter colony formation is complete, the young colonies escape from the parental colony matrix which undergoes gradual dissolution. Colonies increase in size (but not in cell number) until daughter colony formation occurs, 1–3 days later under optimal conditions for the various genera.

IV. STAGES OF THE SEXUAL REACTION: MORPHOLOGY AND PHYSIOLOGY

Sexual reproduction of green flagellate species has been observed in collections from nature on many occasions (e.g., Thompson, 1954). However, identification can be confusing in natural collections (Pocock, 1953), which often contain a mixture of species. Laboratory culture permits the use of clonal material, control of the timing of mating, and genetic analysis of the consequences. These have greatly increased our understanding of the interplay of environment and genome in controlling sexual reproduction in at least some of the many variant patterns in this group.

A. Terminology of Sexual Reproduction

Much confusion can be avoided at the outset by defining the nomenclature of breeding patterns. In genera such as *Eudorina, Platydorina,* and *Volvox,* the sex of a gamete or gamete-producing colony is readily recognizable morphologically. Hence, one speaks of sperm or egg, i.e., male or female gametes, and describes their mating as *heterogamous.* Among the other genera there is no striking morphological difference recognizable at the light microscope level between two fusing gametes; i.e., mating is *isogamous.*

To complicate matters further, in the majority of cases one clone must be mixed with another genetically different partner in order to form zygotes, while with some clones zygotes can be formed in a clonal culture. Therefore, two levels of gametic compatibility must be distinguished, one determined genetically and the second acquired epigenetically. Historically (see Whitehouse, 1949), the following terms have been applied to genetic sex determination among haploid thallophytes:

A *homothallic* organism is genetically capable of forming zygotes within a clone, i.e., it is a "selfer." It may be isogamous or heterogamous. When sexual dimorphism is present, it is easy to distinguish between two subvariants which occur: in one, the two gamete types may form within the same colony (*monoecious*), while in the other they are always in separate colonies (*dioecious*).

A *heterothallic* organism is genetically incapable of forming zygotes within a clone. It must be mixed with a different clone in order to mate, a clone of "opposite" or "complementary" mating type. Among the Volvocaceae, heterothallic clones have been found, upon genetic analysis of mating-type inheritance, to behave as if a single locus with two alleles controls mating-type compatibility. A heterothallic organism may be isogamous or heterogamous. Where sexual dimorphism is lacking, the two compatible clones are often labeled + and −, respectively, corresponding to the arbitrary designation of one *mating-type* allele as + and one as −. Only the combination of + and − produces zygotes.

B. Sexual Reproduction in Isogamous Genera

For purposes of description it is simpler to divide the eight classical genera of colonial green flagellates into two groups, those which reproduce essentially isogamously and those which are heterogamous. Although there is every reason to think that the two groups are part of the same evolutionary series, the aspects of sexual reproduction which have received the most attention in the heterogamous forms are almost entirely different from those investigated in the isogamous forms. A discussion of breeding patterns, which vary enormously, will be delayed for the moment, and only the actual mating reactions will be described. Most have been studied using heterothallic clones.

The isogamous genera, and the maximum number of cells per colony in their species, are *Gonium* (4, 8, 16, 32), *Pandorina* (16, 32), *Volvulina* (16, 32), and *Astrephomene* (128). The stages leading to zygote formation—flagellar agglutination which causes colony and gamete clumping, release of gametic cells from their colonial matrix, pairing, and fusion—are illustrated in Figure 2.

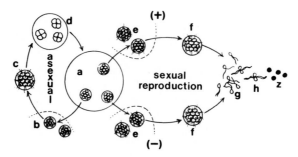

Figure 2.   Reproduction of isogamous genera (*Pandorina*). Division occurs at d. z, Zygotes.

1. Gametogenesis

Cells capable of sexual reproduction arise in the course of growth, whether a compatible mating type is present or not. Colonies capable of mating (e in Figure 2) are not usually distinguishable morphologically from vegetative colonies, but their sexual competence becomes obvious immediately when they are given the opportunity to mate, for the flagellar agglutination reaction occurs (g in Figure 2). Mating activities such as flagellar agglutination and gamete release take place only when sexually competent (+) colonies are mixed with sexually competent (−) colonies of the complementary (sexually compatible) mating type.

Considerable effort has been made to discover factors controlling the initial expression of sexuality. Healthy, rapidly growing cultures in most cases are not sexually competent. In fact, competence seems to arise in the latter part of the growth curve most often, and then only in suitably formulated media (Wilbois-Coleman, 1958; Carefoot, 1966). Some investigators (Carefoot, 1966; Rayburn, 1974) have been successful in eliciting mating competence by resuspending actively growing colonies in various low-nitrogen, high-calcium media. Thus, both methods, the "natural" development of sexual maturity during growth of a culture, and the "induction" of sexual maturity by resuspension in starvation media, can be used.

Although there is considerable danger in overgeneralizing from the various methodologies and organisms investigated, several major factors influencing differentiation in the sexual pathway are apparent. Ammonia, nitrate, or urea can serve equally well for growth prior to mating. Temperature and pH ranges compatible with mating are usually more narrow than those compatible with growth. Nitrogen depletion as a stimulus and adequate calcium and light as necessities are fairly universally agreed upon. However, some strains mate at higher levels of nitrogen than others (Rayburn, 1974), and the heterotroph *Astrephomene* forms zygotes in the dark (Brooks, 1966).

In the phototrophic genera, if *Pandorina unicocca* is typical, there may be two light requirements (Rayburn, 1974), a prolonged one for gametogenesis, and a relatively short one for actual expression of the agglutination reaction after fully differentiated gametes are subjected to an interruptive dark period. Such experiments were done with cultures washed and resuspended in nitrogen-free medium. In undisturbed cultures grown under typical light–dark regimens, where gametogenesis occurs at some point in the growth curve, it has commonly been observed that mating potential undergoes a diurnal periodicity, being highest during the several hours just after the light comes on (Wilbois-Coleman, 1958; Starr, 1962; Brooks, 1966; Carefoot, 1966). This may correspond to the second light requirement discussed by Rayburn (1974). However, in these latter cases, the medium was soil–water or a weakly buffered inorganic medium containing nitrate, and the pH showed a diurnal fluctuation, increasing throughout the light period and decreasing again during the dark (Rayburn and Starr, 1974). Thus, the diurnal periodicity of mating competence might be primarily a consequence of the fluctuating pH pattern. This can be tested either by using high buffer concentrations or by substituting urea as the nitrogen source to avoid diurnal pH fluctuations. *Pandorina morum* in urea medium shows the same diurnal periodicity of mating competence as in nitrate media (A. W. Coleman, unpublished observations).

2. The Mating Reaction—Flagellar Agglutination and Gamete Release

Once sexual competence has appeared, two more activities must occur before colony cells can function as gametes. Each must escape from its colony wall, and each must find a partner. These two activities appear to be sequentially linked in a similar fashion among the isogamous genera, and their control may closely resemble that described for isogamous *Chlamydomonas* species (Goodenough, 1977).

The first overt stage in mating is the mating-type-specific flagellar agglutination reaction. The flagellar tips of sexually competent cells are capable of sticking to each other, and intense reactions between freshly mixed colonies lead to clumping of whole colonies in large aggregates. Simultaneously the colony matrix begins to autolyze, apparently most rapidly at the regions where flagella traverse the wall. It may be that, as suggested for *Chlamydomonas* (Solter and Gibor, 1977; Goodenough and Weiss, 1975; Wiese, 1969), the agglutination reaction itself is necessary for the autolysis of the colonial wall, allowing individual gametes to squeeze out. Flagellar agglutination continues among released gametes, leading to clusters of gametes exactly like those seen in mating *Chlamydomonas* species (Figure 3a). Although these aggregations of first colonies and then free gametes can be very striking, the interaction of mating colonies at high dilution can produce zygotes equally rapidly without noticeable aggregates forming—only transient groups of a few free cells and a colony. Thus the flagellar agglutination reaction seems always to occur, but aggregate size is more a function of colony concentration than a direct measure of reaction intensity.

Divalent cations are clearly required for both flagellar agglutination and colony dissolution, and Rayburn (1974) reported that only the latter exhibited a specificity for calcium. Magnesium sufficed for colony clumping to occur, but no gametes were released. Much remains to be learned about the causes of the making and breaking of flagellar tip associations.

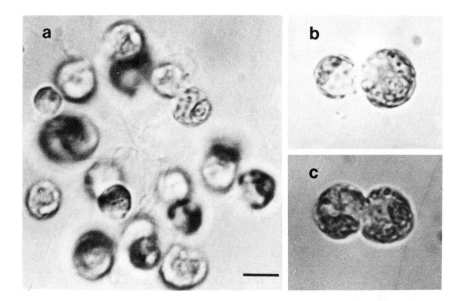

Figure 3. *Pandorina* gametes, living. (a) A clump of gametes, agglutinated by their flagellar tips at a central point. (b) A relatively anisogamous pair of gametes beginning fusion at their apices. (c) An isogamous gamete pair from same mating mixture, beginning fusion. (a–c) Bar = 5 μm.

3. Pairing

The next stage of mating, pairing, may begin within a few minutes after colonies are mixed, or pairs of cells may only gradually emerge from a major aggregate of gametes. The pairs of gametes have their flagellar tips firmly anchored to each other and bob vigorously against this anchoring force. Stein (1958a,b) suggests that in *Gonium, Astrephomene,* and *Volvulina* flagella pair along their entire lengths, but this has not been reported in other species. Instead, the agitations of the gametes seem eventually to lead to contact between the cell apices and instantaneous formation of an intercellular cytoplasmic bridge.

4. Fusion

Formation of a cytoplasmic bridge between the apices of a gamete pair is the first step in cell fusion. The bridge is formed by a special mating structure, a tiny protoplasmic protrusion from the subapical region of one gamete which makes contact and instantaneously fuses with a similar protrusion on the apex of the partner cell. Such mating structures have so far been described among Volvocaceae only in *Volvulina pringsheimii*

(Starr, 1962), *V. steinii* (Carefoot, 1966), and *Astrephomene* (Brooks, 1966), where they are visible with light microscopy on gametes of both mating types. Presumably they are homologous to the mating structures observed with the electron microscope in studies of *Chlamydomonas* (Brown *et al.,* 1968; Friedmann *et al.,* 1968; Goodenough and Weiss, 1975). Two unsolved experimental problems are immediately apparent: (1) Are the elements of mating structures always present in cells expressing the flagellar agglutination reaction? (2) What activates the mating structure to protrude?

5. The Zygote

Once an apical bridge has formed, cell fusion is normally completed within 1–2 min (Figures 3b and 3c). After this, the flagellar agglutination activity disappears, leaving the four flagella of the zygote totally indifferent to any surrounding gametes. The zygote now undergoes a growth period in which carbohydrate and oil accumulate, a heavy, variously sculpted, pigmented wall is laid down within the primary zygote membrane, and the two gamete nuclei fuse. Once the zygote is fully matured, mass germination is difficult to induce without prolonged dormancy. In experimental work investigators generally avoid this problem by removing freshly formed zygotes from the light before zygote maturation is complete. After several days' incubation in the dark, such immature zygotes germinate when dampened and placed in the light.

C. Sexual Reproduction in Heterogamous Genera

The heterogamous genera of colonial green flagellates include *Platydorina, Eudorina* [*sensu latu,* as proposed by Goldstein (1964), encompassing the species of *Pleodorina*], and *Volvox.* While *Platydorina* has a maximum of 32 cells per colony, and *Eudorina* species vary from 32 to 128 cells, *Volvox* species are characterized by from 2^8 to 2^{13} cells per colony. Details of colony organization vary among these organisms, particularly the presence and proportion of two types of cells, gonidial (reproductive) and somatic (nonreproductive), but their manner of sexual reproduction is similar (Figure 4). All form sperm cells, i.e., small biflagellate gametes produced in bundles by multiple mitoses of a mother cell, and these fuse with large unremarkable cells commonly called eggs (though in *Platydorina* and *Eudorina* they still bear functional flagella). Thus sexual dimorphism is obvious in the gametes, and it arises from an "extra" or "precocious" set of rapid mitoses in the male line, which produces the sperm bundle (Figures 4 and 5a and b).

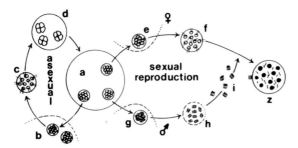

Figure 4. Reproduction of heterogamous genera (*Eudorina*). Division occurs at d and g. i, Sperm bundles; z, Zygotes.

1. Gametogenesis

The expression of sexual differentiation appears to be a more closely regulated process among the heterogamous genera than among the isogamous ones, for rarely is there a spontaneous wholesale transformation of the population into zygotes. In general, freshly transferred colonies

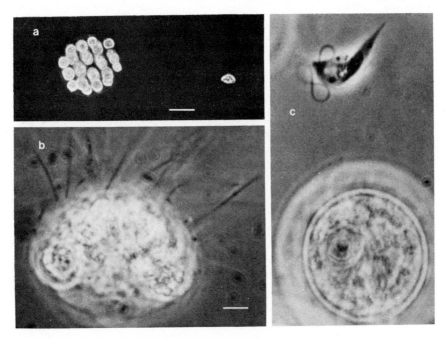

Figure 5. *Eudorina elegans*, living. (a) An asexual colony and sperm bundle. Bar = 50 μm. (b) Sperm bundle. Bar = 5 μm. (c) Biflagellate sperm and recently fertilized egg. Bar = 5 μm.

do not express sexuality at all, and only a small proportion do so as a culture ages (Starr, 1968). Research on hormonal induction of gametogenesis, which has been clearly shown to occur in *Volvox* species and suggested in *Eudorina,* is discussed later after a description of the mating process.

As in the isogamous genera, when sexual reproduction begins, adequate light is important. In fact, the data from Table I in Szostak *et al.* (1973) concerning *Eudorina* male production suggest that levels of light which saturate growth requirements are far below those which might saturate the system controlling whether a reproductive cell makes another asexual colony or a sperm bundle. Furthermore, light intensities greater than 400 footcandles are commonly employed in *Volvox* research on hormones, although they are not usual for vegetative maintenance in this group.

A distinct stimulation of the extent of sexual differentiation in response to *Volvox* hormones has been found with 0.5% CO_2 supplementation. This technique, first discovered in efforts to explain low-response females, has now made it possible to induce *Volvox carteri* f. *weismannia* male colonies with hormone (Starr, 1971a), whereas only the female clone had responded before.

2. Male Differentiation

Under conditions suitable for sexual differentiation, *Eudorina* and *Platydorina* male colonies not visibly different from vegetative colonies form sperm bundles at the next round of mitoses. In *Volvox,* differentiating male colonies are recognizable by their much increased proportion of reproductive cells, each of which gives rise to a bundle of sperm cells (g in Figure 6). The multiple mitoses (2^4, 2^5, or 2^6 usually) which produce sperm bundles are not visibly different in pattern from those responsible for vegetative colony formation, but the resultant plate of cells fails to finish inversion in most species. Instead, the cells elongate after partial inver-

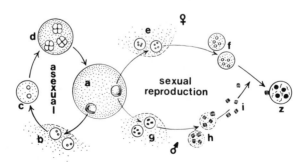

Figure 6.   Reproduction cycles of *Volvox.* Division occurs at d and g. z, Zygotes.

sion and the small cone of tightly packed sperm cells escapes from the parental colony and swims rapidly about (Figure 5a).

3. Female Differentiation

Young female colonies with developing eggs cannot always be differentiated from young asexual colonies bearing developing gonidia, for in *Eudorina, Platydorina,* and a number of species of *Volvox* (e.g., *V. aureus*) the gonidia act as facultative eggs; no morphologically distinctive female colonies are formed. In other *Volvox* species (see Table I), the presence of a hormone secreted by sexually differentiated male colonies causes the female strain to form a morphologically distinct female colony type at the next round of daughter colony formation. The induced female colony has an increased number of large reproductive cells, and only these are capable of acting as eggs.

Unfertilized egg cells eventually undergo daughter colony formation, just like vegetative gonidia, in *Eudorina, Platydorina,* and some *Volvox* species. Furthermore, there are a few examples of clones in which cells expected to act as facultative eggs instead mature into heavy-walled spores (parthenospores) which look like zygotes but develop without fertilization. The significance of these is not clear, since no one has yet reported success in germinating them.

4. Sperm Release and Penetration of the Female Colony

In order for a sperm to fuse with an egg, sperm bundles must locate female colonies, the individual sperm cells must be freed from each other, and the female sheath layers surrounding the egg must be breached. The stimuli for these activities are not understood yet. There is little suggestion of a flagellar agglutination reaction among the heterogamous species comparable to that among the isogamous group and chlamydomonad unicells. Although Hutt (1972) reported an agglutinationlike interaction between male and female flagella in *V. carteri* f. *weismannia* and observed that artificially deflagellated female colonies remained unfertilized, R. T. Huskey (personal communication) has been able to do genetic studies with flagellaless mutants of *V. carteri* f. *nagariensis*. Thus there seems to be no universal requirement for flagellar agglutination.

When the rapidly swimming sperm bundle encounters a female colony, the flagella of the sperm bundle seem to stick to the surface gelatinous materials of the female and slide along the surface for a brief period. Soon a dissolution of the female colony gelatinous matrix appears where the sperm bundle becomes anchored. At this point the sperm bundle generally breaks apart into individual sperm, and these enter the loosening gelatinous material, burrowing through it with a corkscrewlike cell body motion

Table I *Volvox* Developmental Patterns and Hormones

Characteristic	V. carteri f. nagariensis	V. carteri f. weismannia	V. obversus	V. gigas	V. dissipatrix (India)	V. dissipatrix (Australia)	V. rousseletii	V. aureus
Heterothallic, dioecious	x	x	x	x	x		x	
Homothallic								
Dioecious								x
Monoecious								
Precocious division					x	x		x
Gonidia first evident								
At cleavage	x	x	x		x	x		
Before inversion				x				
After inversion							x	x
Female colonies special	x	x	x	x	x	x	x	
Hormone made by	♂	♂	♂	♂	♂, ♀	♀	♂	♂
Hormone(s) induces[a]	♂, ♀	♂, ♀	♂, ♀	♂, ♀	♂, ♀	♀	♂, ♀	♂, par.
Sensitive period[a]	c	Late b + c	c	Last ⅔ of c	?	?	Last ½ of d to a	d to a
Evidence for glycoprotein	x	x	?	x	?	?	x	x
Active concentration (gm/liter)	10^{-11}	10^{-9}					4×10^{-8}	10^{-10}
Reference[b]	1, 2	3, 4	5, 6	7	2, 8	8	2, 9	10, 11, 12

[a] These entries denote stages in the cell cycle (see Figure 6) when the presence of hormone is capable of switching development to the sexual pathway. par, Parthenospores.

[b] Key to references: 1, Starr (1969); 2, Starr (1971a); 3, Kochert (1968); 4, Kochert and Yates (1974); 5, Karn et al. (1974); 6, Starr (1968); 7, Van de Berg and Starr (1971); 8, Starr (1972b); 9, McCracken and Starr (1970); 10, Darden (1966); 11, Darden (1968); 12, Ely and Darden (1972).

quite different from normal flagellar propulsion. The sperm cell is extremely ameboid, undergoing continuous alterations in form as it burrows.

As several authors have suggested, the sperm bundle association with the female colony surface seems to trigger an enzyme release, presumably from the male, since the nearest female cell may lie at some distance, which makes the initial breach in the female colony boundary. Hutt and Kochert (1971) have reported that soybean trypsin inhibitor and other substances can block this activity of the sperm bundles. The extensive dissolution of the female gelatinous matrix found in a mating mixture does not normally occur in the absence of sperm, although sperm bundles break up eventually even in the absence of females.

5. Fertilization

Observations of fertilization itself have rarely been reported, but most accounts agree that the proboscis of the sperm cell contacts and fuses with the apical region of the egg, and then the remainder of the sperm comes to lie along the surface of the egg, gradually melding into it. The outlines of recently fused sperm can often be distinguished lying near the zygote surface.

V. HORMONAL CONTROL OF DIFFERENTIATION

A. *Volvox* Developmental Patterns

While the timing of growth and division is similar in all the other colonial genera and in many *Volvox* species, some species of *Volvox* are different in this respect. In these exceptional species the gonidia do not go through a prolonged growth phase followed by a rapid series of mitoses; instead, the first division in gonidial cells occurs early while they are still relatively small (<40 μm in diameter compared to >80 μm in conventional species), and there follows a regular alternation of growth and division until the final size and cell number are achieved. In a sense, these species begin to divide precociously (Table I).

A number of other variations among *Volvox* species are found. In the *V. carteri* group, where hormones have been the most studied, potential reproductive cells are set aside at particular cleavage stages by unequal divisions of certain cells (Kochert, 1968; Starr, 1969). The division at which this occurs is determined by the development pathway, such that asexual colonies have only 16 or fewer gonidia, induced females have twice as many large cells which act as eggs, and induced male colonies may have as many as 50% of their cells forming sperm bundles.

In other species no inequality of division is detectable, and distinctive large cells, the gonidia, can first be observed at inversion (*Volvox gigas*), after inversion (*V. aureus* and *V. rousseletii*), or even after release from the parent (*V. globator*). The increase in number of reproductive cells in sexual colonies in some of these species appears to result from reclamation of smaller cells which would never have divided under conditions of vegetative reproduction.

Thus, in all variants of the growth cycle, sexual reproduction represents a developmental delay in the final determination of which cells will be permanently somatic. This in turn guarantees that sexual colonies will contain a relatively higher proportion of potentially reproductive cells than vegetative colonies.

B. *Volvox* Hormones

In 1966 Darden reported that the fluid in which sperm bundles had been produced by *V. aureus* colonies was capable of causing the production of male colonies in populations which otherwise would have produced only asexual ones. This was the first of a continuing series of papers on *Volvox* hormones which evoke sexual reproduction. Evidence for hormone action has been presented for seven species, some with several different representatives (Starr, 1968), while it has not yet been possible to demonstrate any hormonal activity in another five species (Starr, 1971a).

The timing of hormone action is different in each of the species studied, but so also is the developmental pattern (Table I). Two extreme examples, which also represent the most studied cases, are those of *V. aureus* and *V. carteri*.

1. *Volvox aureus*

Volvox aureus belongs to the small group of *Volvox* species which exhibit precocious division. When a colony is first released, it has 8 to 12 gonidial cells slightly larger than their somatic sister cells. The gonidia continue to enlarge, meanwhile also cleaving rather than postponing the cleavage period. When the mature size and cell number are reached, the embryonic colonies invert. While they continue to enlarge within the mother colony, the gonidial cells of the next generation become discernible by their increase in size. Prior to inversion they are not easily detectable, although Timpano and Thompson (1977) suggest they derive from unequal embryonic division. Darden (1966) showed clearly that treatment of vegetative colonies with hormone for only 30 min at any point from about the third division of the embryo up to almost the time of inversion (d to a in Figure 6), a period of nearly 40 hr, was sufficient to induce maximal numbers of male colonies. These are already recognizable at inversion by the absence of gonidia, and shortly thereafter by the division of most of the posterior cells in the colony to form sperm bundles.

There is no detectable effect of hormone on female induction, since young, undivided gonidia of vegetative colonies serve as eggs in *V. aureus*. However, in another strain of *V. aureus*, which is capable of forming parthenospores, the same hormone induces parthenospore formation where continued vegetative reproduction would otherwise have occurred (Darden, 1968).

2. Volvox carteri

Asexual colonies of *Volvox carteri* f. *nagariensis* contain 16 gonidia formed by unequal cleavages during embryonic development of the colony. Like most Volvocaceae, these reproductive cells have a prolonged growth stage and only then undergo the series of rapid cleavages which form a new colony. Starr (1969) reported that if a vegetative colony is exposed to hormone from the time of its release from the parent until the early stages of division (b to d in Figure 6), a 30- to 40-hr period, it will form at that set of divisions a new, sexual colony. Females form 30 to 45 eggs, and males delay unequal division until the final mitosis, forming a colony with a 1:1 ratio of somatic and generative cells. The generative cells very rapidly cleave to form sperm bundles. Thus, in *V. carteri* a prolonged and stage-specific exposure to hormone before the onset of cleavage is required to bring about sexual differentiation.

These two examples represent extremes in that, at a time in the generation cycle (middivision, d in Figure 6) when *V. carteri* colonies are already determined and insensitive to hormone treatment, *V. aureus* colonies are fully sensitive to induction. *Volvox aureus* cells already present can be completely converted to the sexual pathway. This led to the suggestion (McCracken and Starr, 1970) that there are two different systems of hormone action.

3. The Assay

The assay of hormone titer, the dilution which induces at least 50% of asexual colonies to produce sexual colonies, is most successful when great care is taken in the preparation of the assay colonies and small numbers are used to preclude gas limitation. The colonies should be in the log phase of growth and just entering the hormone-sensitive stage, e.g., in *V. carteri* f. *nagariensis*, just at release from the parental colony (b and c in Figure 6). A series of 10-fold dilutions of the hormone stock is prepared using fresh culture medium, and five colonies are placed in each 10-ml sample. The inoculum contains an average of 15 young daughters per colony, a total of 75 per assay tube. When these 75 gonidia have in turn formed colonies large enough to classify, the proportion of the ca. 1125 colonies (15 × 75) which are sexual can be determined. Generally, activity disappears over a span of two tubes in the dilution series.

Whether male or female strains are preferentially used for assay is determined by the biological properties of the particular species or strain. In *V. carteri*, females are preferable for the assay because "spontaneous" females are extremely rare. Once obtained, hormone is very useful because the best way to obtain high titer hormone samples is to use hormone on male cultures to synchronize their sexual differentiation.

4. Properties of Hormones

There are many similarities among *Volvox* hormones, as shown in Table I. These hormones are species-specific and often further restricted within a species, e.g., hormone from *V. carteri* f. *weismannia* has no effect on *V. carteri* f. *nagariensis,* and vice versa. Hormones are harvested from males, but induce both male colonies and the special female colonies in species where females differ recognizably from young vegetative colonies. Only in *Volvox dissipatrix* has hormone activity been obtained from a female clone so far, and its action is the same as that of the male hormone. Furthermore, the highest titer of hormone from males is found when sperm bundles have been released and have disintegrated. The medium harvested from populations of such differentiated males can, in *V. carteri*, be diluted as much as 10^7 and still induce 100% females in the standard assay.

This high natural titer plus the remarkable stability of *Volvox* hormones harvested from raw culture fluid has greatly aided in their chemical study, although their tendency to dimerize and polymerize under many conditions has been a drawback. All but *V. gigas* hormone bind to carboxymethylcellulose, a property useful for purification and indicative of their basic net charge. All are sensitive to Pronase, but are not significantly altered by other common proteolytic enzymes or by DNase or RNase. Darden and Sayers (1971) report that *V. aureus* hormone is sensitive to brief exposures of germicidal wavelengths of ultraviolet light. *Volvox aureus* hormone is also the most heat sensitive, being destroyed by autoclaving, though it survives boiling with half its activity. *Volvox carteri* f. *nagariensis* hormone can retain considerable activity after autoclaving, e.g., a reduction in titer from 10^8 to 10^6. For this reason, in assay work baked glassware must be used and great care must be taken to avoid imperfect mixing and surface binding phenomena.

The two hormones purified sufficiently to show a single band on sodium dodecyl sulfate (SDS) polyacrylamide gel electrophoresis are both from *V. carteri*, the two varieties, *V. carteri* f. *weismannia* (Kochert and Yates,

1974) and *V. carteri* f. *nagariensis* (Starr and Jaenicke, 1974). The single band containing the biological activity is a glycoprotein consisting of about 60% protein and 40% carbohydrate. Analysis of the amino acid composition showed that hydroxyproline is low or absent, which is significant because the colony gelatinous sheath materials are known to contain up to 20% hydroxyproline (Miller *et al.*, 1972), and these would be prime candidates as contaminating materials. The carbohydrate material consists of six or seven types of sugar residues, but nothing is known about the linkages or numbers of chains.

Volvox carteri hormones, like those of other species, form dimers and polymers easily, but there seems to be a consensus now that the monomers from both *V. carteri* varieties have apparent molecular weights of about 30,000 (Kochert and Yates, 1974; Starr and Jaenicke, 1974). This means that hormone at $3 \times 10^{-15} M$ is sufficient for full induction in the *V. carteri* assay. As calculated by Starr and Jaenicke (1974), this concentration is equivalent to several thousand molecules per gonidium in the standard assay. By contrast, Pall (1973) found that the induction response increased exponentially with hormone concentration in *V. carteri* f. *nagariensis*, with a slope suggesting that the induction event(s) required two molecules per gonidium. The distance from this theoretical suggestion to our understanding of the hormone action is very great.

Current work is focused on further characterization of the chemical structure and identification of the hormone and its site of action. ^{125}I-Labeled *V. carteri* hormone retains its inducing activity (Starr and Jaenicke, 1974; Kochert, 1975) and shows specificity in which *Volvox* variety it binds (Noland *et al.*, 1977). Studies by Starr (1971a) and by Meredith and Starr (1975) with the variety *V. carteri* f. *nagariensis* suggest that various F_1 and F_2 female strains all have the same threshold of response to a given sample of hormone; yet sibling F_1 and F_2 male strains may vary by as much as 10^4 in the level of hormone they produce under standard conditions. These investigators conclude that the male potency phenotype is probably under multigenic control. A low-response female strain (which responds poorly even to 10^4 times normal hormone levels) is known, and this trait segregates at a single locus (Starr, 1971a).

5. Nature of Hormone Action

The primary action of the sexual hormone system in *Volvox* is obscured by the variety of growth cycles among the species studied and the male-specific and female-specific changes which follow hormone induction of sexual differentiation. Major characteristics of both the hormones and the species so far reported are presented in Table I. A moment's study shows that when gonidial development is sufficiently advanced that more than one cell size can be recognized in the colony, it is too late for added hormone to alter the course of events in that generation. The only exception to this generalization is found in *V. rousseletii,* where hormone can be added to young colonies when size differences are apparent and still induce sexual differentiation. However, when hormone is added to *V. rousseletii* late, just as the first gonidia begin to divide, these cells sometimes abort as a consequence (McCracken and Starr, 1970).

In general, the switch to sexual reproduction represents (1) suppression or slowing of gonidial development; (2) an increase in the proportion of reproductive, as compared to somatic, cells in the colony; and (3) in the female, suppression of division for a period of time and, in the male, division to form a sperm bundle after only a brief period of cell enlargement.

As noted by Darden (1970), the first hormone-induced alteration visible in *V. aureus* forming male colonies is the suppression of gonidial cell enlargement, so that induced colonies after inversion have cells all the same size. In *V. carteri* also the delay of unequal division during cleavage to form sexual colonies could be interpreted to represent a suppression of gonidium formation, but in *V. rousseletii* and *V. gigas* the large cells are formed in hormone-treated colonies just as in asexually reproducing colonies. These large cells plus additional smaller cells are converted into eggs or sperm bundles under the influence of hormone. Hence the hormone effect must lie in some more primary alteration of metabolism to which the remaining effects are entrained.

Some extremely interesting experiments which may have a direct bearing on the control mechanism have been presented by Kochert. His earlier studies (Kochert and Yates, 1970) on vegetative reproduction showed that a 15-min exposure to uv irradiation prior to cleavage could result in colonies lacking one or more of their normal complement of eight gonidia, with little or no detectable effect on colony morphology or on the somatic population. More recently (Kochert, 1975) he has studied sexually induced colonies and reported that briefer doses of uv just prior to cleavage can totally suppress the effects of inducer and result in colonies with gonidia, not eggs, while 100% of the controls shielded from uv produce eggs. Further studies of this sort, combined with the analysis of *Volvox* developmental patterns now possible with the diverse mutant strains available (Sessoms and Huskey, 1973), may help pinpoint the switch which operates between vegetative and sexual reproduction. Results with mutants have already made it clear that an increase in reproductive cell number (Section V,B,2) per se has no obligatory association with sexuality (Starr, 1971a), since mutants are known which make increased numbers of reproductive cells, but these are not automatically sexual.

An operon model, in which hormone acts as a corepressor of a master operon responsible simultaneously for production of a repressor of sexual reproduction and maintenance of asexual reproduction has been outlined by Starr (1972a). Furthermore, what might be interpreted as "constitutive female mutant strains" are known, each apparently caused by mutation at a single locus. They form typical (in terms of egg number and appearance) female colonies whether or not inducer is present, and are often difficult to propagate because the eggs show little regenerative capacity. Supplementation of the medium with acetate and thiamine (Starr, 1972a) encourages daughter colony formation by the eggs, however. One of these mutations is linked to the mating-type locus, but the other segregates and could be found in male progeny of a cross; a constitutive male, however, cannot be propagated.

6. The Problem of the First Male

The final question is, Where does the first male colony come from that releases the hormone, which in turn induces the surrounding male and female colonies to differentiate sexually? The observed answer is that within a few rounds of division after a population is introduced into a fresh culture tube, one or a very few male colonies appear, followed at the next round of division by an increased number, presumably induced by the first. There are two explanations proposed for the initial formation of a male colony in a presumably hormone-free environment: epigenetic accident and mutation (Starr, 1971a; Kochert, 1975). The first presumes a cellular control mechanism, subject to environmental accident or innate fluctuation, such that an occasional colony manages to express the sexual pathway. The second explanation, while more unorthodox, actually has some experimental backing. Starr (1972a) estimates the rate of mutation to the constitutive female condition to be roughly in the range of 1 in 20,000. If the same rate occurred in male colonies, it might account for some or all "spontaneously" formed males. Experiments using small culture volumes and single male colony inocula might give more precise data on the frequency of such events.

C. Hormones in Other Species

Harris and Starr (1969) sought evidence for the existence of a hormone produced by sperm and sperm bundles in *Platydorina,* but obtained only negative results. For the third heterogamous genus, *Eudorina,* there has been only one suggestion of hormone activity. Szostak *et al.* (1973) found that culture medium from sperm-producing colonies of *E. elegans* promoted sperm bundle production in males when they were treated for one generation. The effect was nondialyzable and could be detected when the used medium was diluted as much as 1 : 30. It would be valuable to know if this effect is specific for the *Eudorina* mating types from which it was isolated since its low potency, compared to *Volvox* hormone levels, does not encourage hope for isolation and purification.

Among the isogamous genera there has been less activity in searching for extracellular hormonal influences. This is perhaps for the obvious reason that colonies of opposite mating type can interact immediately when mixed, if they are already at the sexually competent stage. However, the observation that *P. morum* + and − mating types consistently mated 1–3 days earlier if grown together, compared to being grown separately and mixed daily, suggested some kind of premating interaction between mating types. In a preliminary study (Coleman, 1959; Wilbois-Coleman, 1958), cell-free supernatants of *Pandorina* colonies of one mating type were found to cause cell release and even some flagellar agglutination when applied to both the same and to the complementary mating type. The released cells always proved to be gametic, when suitable cells of the complementary mating type were available for testing. Only the combination of supernatant with the colonies from which it was harvested always failed to cause gamete release.

The lack of mating-type specificity in this reaction was curious at the time, but more recent work with other genera suggests several further areas of research. Claes (1971) and Schlösser *et al.* (1976), using *Chlamydomonas,* and Saito (personal communication), using *Gonium,* have described "hatching" enzymes which can be harvested from cell supernatants at the time of vegetative reproduction or gamete release, and which act on both mating types, degrading the wall. Furthermore, both Schlösser and Saito observe these enzymes to be active only on organisms belonging to the same mating group (syngen; see Section VII). The conditions required for optimal activity of these enzymes have not been well defined as yet.

In addition, the original mating-type-specific, agglutination-inducing "isoagglutinin" of *Chlamydomonas* (Wiese, 1969) has been found to consist of small, membranous vesicles presumably budded off from the tips of gamete flagella (McLean *et al.,* 1974), and the source of such material is the supernatant fluid of gametic cells. Such supernatant materials obtained from *Pandorina* gametes also cause mating-type-specific agglutination (Coleman, unpublished observations). Thus far, all results parallel those obtained with isogamous *Chlamydomonas* species.

VI. CELL CYCLE EVENTS AND DIFFERENTIATION

Since gametes are expected for basic genetic reasons to contain only the haploid, 1C, amount of nuclear DNA, it is valuable to review what is known about the timing of major cell cycle events in the colonial flagellates. Once a nucleus has entered the S phase, the cell has to wait until after the next mitosis to satisfy the nuclear DNA criteria for being a gamete (Coleman, 1961). Furthermore, recent work on the control of mating in yeast (Bücking-Throm et al., 1973) and in Tetrahymena (Wolfe, 1974) has revealed that, in the first case, hormones, and in the second case, some cell–cell interactions, act primarily to synchronize the nuclear cycle of the potential gametes.

A. DNA Replication in Nuclei and Chloroplasts

As early as 1960, biochemical studies of isogamous chlamydomonads suggested a cell cycle with nuclear DNA synthesis essentially coincident with the period of successive mitoses. Similar studies with Eudorina and Volvox species generally agreed that a major portion of DNA synthesis was delayed until division, but consistently showed significant amounts of DNA synthesis during earlier stages of growth (Kochert, 1975; Lee and Kemp, 1976; Margolis-Kazan and Blamire, 1976; Tautvydas, 1976; Yates, 1974).

The early DNA synthesis may represent organelle replication. The data of Yates (1974) and Margolis-Kazan and Blamire (1976), using two varieties of V. carteri, support the notion that chloroplast DNA synthesis occurs predominantly, if not exclusively, shortly after inversion of the young colony. A similar periodicity of cytoplasmic DNA increase can be observed by fluorescence microscopy in developing Pandorina and Eudorina colonies (Coleman, 1978), and early G_1 synthesis of chloroplast DNA has been observed in Chlamydomonas (Chiang and Sueoka, 1967).

The relatively massive contribution of organelle DNA synthesis during the G_1 period of the cell cycle in colonial flagellates can be illustrated using Eudorina as an example. Since colonial genera with 32 cells must make enough chloroplast DNA for 32 daughters, and assuming that the basic chloroplast genome amounts to at least 2% of that in the nucleus (which is almost threefold lower than the lowest values reported for Chlamydomonas), then a cell near the end of G_1 would have undergone a 64% (2 × 32) increase in DNA per cell (!), and 39% (ca. 64/164) of its DNA would be chloroplast associated. What is true for chloroplast DNA is also true for mitochondrial DNA, although its quantity would be much smaller. Consideration of these phenomena may help in interpretating data from biochemical analyses.

It now appears possible that the cell cycles of all colonial green flagellates, except those species of Volvox with precocious division, are the same and rather unusual. There are, in fact, two kinds of cell cycles during vegetative reproduction. The first type, during which organelle

DNA synthesis proceeds, lasts until within approximately 1–2 hr before daughter colony formation. At this point nuclear DNA synthesis begins and, as Feulgen microspectrophotometric studies have shown, in *Eudorina* (Lee and Kemp, 1975; Kemp, 1977) and in *Pandorina* and *Volvox* (Coleman, 1979), S periods alternate with mitoses until cleavage ends. During the cleavage period, the second type of cell cycle occurs; it is similar to that in cleaving animal eggs, and there is no detectable G_1 or G_2 (Lee and Kemp, 1975). Only by using microspectrophotometric measurements on single nuclei has it been possible finally to distinguish that polyteny does not occur normally, but rather the regular alternation of S and mitotic periods during cleavage.

B. Fertility versus Cell Cycle Events

Volvocacean cells increase considerably in size before the cleavage period begins, and cells throughout this size range have been observed to act as gametes in the isogamous genera (Coleman, 1959). The prolonged G_1 period helps to explain why there are gametes of such variable size (Figure 3b), but leaves unresolved the question of whether there is a subsection of G_1, or a decision step during its progression, which restricts fertility. In *Chlamydomonas*, evidence has been presented (Schmeisser *et al.*, 1973) that young cells recently released from their cell walls are optimally fertile. However, in the species examined, this point in the life cycle corresponds to the beginning of the daily light cycle. The latter has been observed to be the optimal mating time for all clones of *Pandorina*, whether they characteristically divide in the "afternoon" or after the light goes out (Coleman, 1977). It should be possible with colonial forms to test the importance of light cycle versus cell cycle stage in fertility by using highly synchronized cultures obtained by sizing colonies through various weaves of silk screen nylon cloth.

An additional problem raised by the cell cycle studies concerns cell organelle significance in the zygote, since the nuclear DNA replication cycle is different from the chloroplast DNA cycle. Presumably, in the isogamous genera, a gamete in late G_1 with a 16X content of chloroplast DNA could fuse with an early G_1 gamete containing only a 2X or 4X complement. Or is there a mechanism for synchronizing organelle replication at gamete production? Or is this problem entirely circumvented by another characteristic, uniparental inheritance of chloroplast DNA through the zygote? The occurrence of this phenomenon has been studied extensively in *Chlamydomonas* (Gillham, 1969; Sager and Ramanis, 1973) and shown for *Eudorina elegans* by Mishra and Threlkeld (1968). Despite the uniparental inheritance, the quantity of organelle DNA entering the zygote with each gamete may be significant, for uniparental inheritance is not 100% effective, and the reasons for the exceptions are not yet understood. Cell cycle effects might be important.

The heterogamous genera appear to have avoided such problems. They have achieved nuclear synchronization of their gametes by suppressing progress through G_1 in the female and accelerating it in the male line, resulting in a precocious set of cleavages and a crop of sperm fresh from mitosis and presumably in the early G_1 state. *Volvox* hormones appear to activate the master switch for this process, a switch that we unfortunately do not yet understand.

VII. MATING TYPES AND SPECIES STRUCTURE

A. Syngens

Although the classical taxonomic categories of colonial green flagellates have generally been accepted as reflecting natural relationships, since there is little difficulty in identifying field material to genus and species levels, studies of sexuality have revealed unsuspected complexities. Within single taxonomic species exist one or more sexually isolated gene pools, or *syngens* (*sensu* Sonneborn, 1957), each characterized by a unique pair of + and −, or male and female, mating types. All clones, whatever their origin, which are capable of intercrossing are considered to belong to the same syngen. Organisms belonging to different syngens cannot mate with each other. In the most extensively studied species, *P. morum,* there are 20 known syngens, some worldwide in distribution (Coleman, 1977). Often more than one syngen is found in the same pond (Coleman, 1977; Stein and McCauley, 1976).

Multiple syngens are known for species of all the genera of colonial green flagellates (Brooks, 1966; Carefoot, 1966; Coleman, 1959; Goldstein, 1964; Rayburn and Starr, 1974; Starr, 1971a; Stein, 1958b, 1966b) except *Platydorina* (Harris and Starr, 1969). The latter, a relatively rare alga, was studied only from North American material, although it occurs also in Europe (Korshikov, 1938). Furthermore, although most syngen limits are smaller than those of species, syngen boundaries sometimes cross classical taxonomic species limits. Goldstein (1964) reported several crosses between *Eudorina* species and even a cross between *Eudorina* and *Pleodorina,* which produced viable, though abnormal, progeny. This was a major justification for his taxonomic consolidation of the two genera (Goldstein, 1964).

The primary barrier to cross-breeding, so far as has been observed, lies at the level of the flagellar agglutination reaction or earlier in the isogamous species, since agglutination and gamete release do not occur in mixtures of incompatible mating types. For the heterogamous species, Goldstein (1964) describes halos of sperm around the surface of female colonies in incompatible *Eudorina* pairings, and Starr (1971a) reports that *V. carteri* hormones not only are variety-specific but also show patterns of limited inductive ability even within a variety, i.e., among *V. carteri* f. *nagariensis* clones from Japan, India, and Australia.

The significance of this inhomogeneity of species is not at all clear, and the possibility exists that such a classification is an artifact of collecting practices (Stein, 1965, 1966a; Coleman, 1977). However, members of any one syngen resemble each other in a variety of characters examined, characters which differ among syngens. These include zygote distribution pattern (Carefoot, 1966; Coleman, 1959; Goldstein, 1964), daily division time (Coleman, 1977), isozyme pattern (Fulton, 1977), flagellar antigens (Coleman, 1963; Herbst and Kemp, 1976), and hormone specificity (Starr, 1971a). They also include chromosome number, which may be the most important character of all. Wherever it has been checked, clones which mate successfully also have the same chromosome number. Furthermore, although not all syngens have been examined, zygotes from fully compatible pairings germinate successfully. All these observations suggest that syngens represent groups of organisms potentially capable of interbreeding.

B. The Relationship between Syngens and Karyotypes

In *P. morum,* haploid chromosome numbers can range from 2 to 14 (Coleman and Zollner, 1977), and other colonial species, particularly *A. gubernaculifera* (Cave and Pocock, 1956) and *E. elegans* (Goldstein, 1964), have diverse chromosome numbers also (see Table II). The variation in chromosome numbers within a species may not be primarily a result of polyploidy mechanisms, but may represent instead a circus of breakage and reforming. If so, diverse chromosome numbers might all still represent one basic genome in content—the larger the numbers, the smaller the individual chromosomes. This description exactly fits the karyotypes of *P. morum* and was tested directly. Using microspectrophotometry, it was found that one could not distinguish among *P. morum* nuclei containing 2, 5, or 12 chromosomes, although all three could be distinguished from those of a known diploid strain (Coleman, 1979). Some further suggestive evidence for chromosome rearrangement was found by Stein (1958a). She

Table II Chromosome Numbers Known for Colonial Green Flagellates

Flagellate	Chromosomes	Reference
Volvocaceae		
Gonium sociale	10 ± 1	Rayns (1961)
Gonium pectorale	17	Cave and Pocock (1951)
	17	Stein (1958b)
	17	Rayns (1961)
Pandorina morum	ca. 12	Dangeard (1900)
	12, 12 ± 2	Coleman (1959)
	10 ± 1	Rayns (1961)
	2, 3, 5, 6, 8, 10, 11, 12	Coleman and Zollner (1977)
Pandorina charkowiensis	>12	Coleman and Zollner (1977)
	17	Rayns (1961)
Volvulina pringsheimii	7 ± 1	Starr (1962)
Volvulina steinii	7, 8	Cave and Pocock (1951)
	7	Stein (1958a)
	7	Carefoot (1966)[a]
Platydorina caudata	14	Harris and Starr (1969)[a]
	15 ± 1	Rayns (1961)
Eudorina elegans	10	Hartmann (1921)
	12	Cave and Pocock (1951)
	16	Rayns (1961)
	12	Sarma and Shyam (1973)
	14	Goldstein (1964)
Eudorina spp.	5, 6, 7	Goldstein (1964)
Eudorina unicocca	7	Godward (1966)
	7	Cave and Pocock (1951)
Pleodorina illinoisensis	10	Hovasse (1937)
	12	Merton (1908)
	ca. 14	Goldstein (1964)
	15	Rayns (1961)
Pleodorina californica (*indica*)	10	Doraiswami (1940)
	14	Cave and Pocock (1951)
Volvox tertius	13	Cave and Pocock (1951)
Volvox africanus	13	Cave and Pocock (1951)
Volvox spermatosphaera	13	Cave and Pocock (1951)
Volvox carteri f. *weismannia*	14	Cave and Pocock (1951)
Volvox carteri f. *nagariensis*	14	Starr, Cave, and Pocock (unpublished observations)
Volvox powersii	15	Cave and Pocock (1951)
Volvox gigas	16	Cave and Pocock (1951)
Volvox obversus	16	Starr, Cave, and Pocock (unpublished observations)
Volvox aureus	12	Zimmermann (1921)
	14	Cave and Pocock (1951)
	14	Rayns (1961)

(Continued)

Table II (*Continued*)

Flagellate	Chromosomes	Reference
Volvox pocockiae	13	Starr, Cave, and Pocock (unpublished observations)
Volvox dissipatrix	12	Starr, Cave, and Pocock (unpublished observations)
Volvox globator	5	Cave and Pocock (1951)
	5	Rayns (1961)
Volvox barberi	5	Cave and Pocock (1951)
Volvox merrilli	5	Cave and Pocock (1951)
Volvox rousseletii	5	Cave and Pocock (1951)
Astrephomeneaceae		
Astrephomene gubernaculifera	4, 8 (6, 7)	Cave and Pocock (1956)
	4, 6, 7, 8	Stein (1958a)
	4, 7, 8	Brooks (1966)

[a] Studies in which organisms from at least three different collection sites were examined and all were found to have the same chromosome number.

observed in *Astrephomene* examples of chromosomes lagging at mitosis, and reported clones which varied intraclonally in chromosome number ($n = 6$, 7, or 8), sometimes even within a colony. These exceptional clones came from one small area and seemed to represent a cytogenetically unusual situation, compared to other isolates of the species studied by Brooks (1966).

Only when the clones of a syngen have originated at considerable geographic distance from each other is zygote germination and/or F_1 vigor reduced (*Pandorina*, Coleman, 1977; *Eudorina* × *Pleodorina*, Goldstein, 1964; *V. carteri* f. *nagariensis*, Starr, personal communication). In the *Eudorina* × *Pleodorina* example, diploidy and anomalies of sex distribution were found in the hybrid and in backcrosses (see Section VII,A). These various observations on the correlation of sexual characters with chromosome variation led Coleman (1977) to suggest that chromosomal reorganization might be a cause of mating-type variation.

Whatever the cause, bipolarity of sexual dimorphism has little to do with the barriers to inbreeding in these organisms. It is the mating-type system, expressed either at the cell agglutination level or earlier, which most limits mating affinity. Until its basis is determined, we will have no full explanation for such phenomena as syngens, apparently asexual clones, and homothallic clones, all belonging to the same morphologically uniform species.

VIII. GENETIC ANALYSIS OF SEXUALITY

A. Nutritional, Developmental, and Fertility Mutants

Genetic analysis of the control of the stages in the mating reaction in *Chlamydomonas* has made considerable progress in recent years (Goodenough, 1977), but information concerning colonial forms is still fragmentary (Wiese, 1976). The work of Mishra and Threlkeld (1968) and Toby and Kemp (1977) with *Eudorina* suggests that an array of nutritional mutants can be obtained and used as markers both for nuclei and for chloroplasts. In addition, more than 12 types of developmental mutants of *V. carteri* have been obtained by Sessoms and Huskey (1973). These are in addition to the various *Volvox* mutants in which the development of egg and sperm are affected, as discussed by Starr (1971a) and Van de Berg and Starr (1971) and presented in Section V,B,4. More extensive analyses of the genetic relationship of these many mutant loci in *V. carteri* have not yet been published, but in this alga, where $n = 14$, Huskey (personal communication) has identified 10 linkage groups with at least two markers each and four other unlinked markers. Analysis of these *Volvox* mutants has shown that the syndrome of characteristics associated with either female or male development can be dissected, and further work of this type should help to define the critical steps in sexual development.

B. Genetics of Bipolar Sexuality

A long-standing question involving the green flagellate line of evolution concerns the genetic control of sex determination: What features of the system allow both heterothallic and homothallic clones to exist within the same species? Interestingly, although homothallic clones have been isolated from nature in three of the same species where heterothallic clones have been studied extensively (*G. pectorale*, Stein, 1965; *P. morum*, Coleman, 1959; *E. elegans*, Goldstein, 1964), in none of these have zygotes from homothallic strains been germinated. Yet Darden (1966) reported that the zygotes of homothallic *V. aureus*, a species where heterothallic strains are unknown, germinate readily.

Another species in which both homothallic and heterothallic forms are known is *Volvox africanum* (Starr, 1971b). The variety in sexual expression in various isolates of the species is bewildering, extending from typical heterothallic forms to homothallic forms in which sperm packets and eggs are made in the same colony at the same time. No information has been published yet on zygote germination or on possible intercrosses between various clones, but further study of this species should prove interesting.

The combination of two haploid genomes of complementary mating type into the same cell might reveal how the mating system operates, but the only two such items of experimental information are not entirely helpful. In one case involving *P. morum,* where heterothallic clones from Iowa and Thailand hybridized to give an F_1 with a doubled chromosome number (Coleman and Zollner, 1977), the F_1 behaved as a normal, fully fertile heterothallic clone. The second case, which gave no evidence of one sex being dominant over the other, occurred in the cross between heterothallic *E. elegans* and *E. (Pleodorina) illinoisensis* which Goldstein (1964) obtained. Some weakly homothallic F_1 progeny were found, in addition to males and females, and these could be both backcrossed and selfed. The zygotes the homothallic clones formed could germinate, and yielded females and more homothallics. A cytological examination of these and the other F_1 and F_2 progeny of the hybridization revealed that all contained approximately double the normal number of chromosomes of the original parental clones.

IX. MAINTENANCE OF CLONES AND THEIR FERTILITY

Although rarely mentioned in print (but see Gowans, 1976), there are a disconcerting number of losses of algal clones through accidents at culture facilities. For example, a number of the *G. pectorale* clones studied by Stein (personal communication) were lost in a temperature control failure. To avoid such losses, the most important research strains should be forwarded to the major algal culture collections, where they will be maintained by vegetative propagation. No detrimental aging effects are detectable after more than 20 years of such maintenance in most clones of *Pandorina,* but fertility was apparently lost within a few years after isolation in a few clones of this and other genera (Brooks, 1966; Carefoot, 1966; Coleman, 1975).

The solution to the problem of maintaining fertile clones is to store zygotes, from which new, fertile clones can be obtained at will. Most suitable for storage are those zygotes formed in soil–water tubes (Starr, 1964) and allowed to dry slowly after maturation; such material has proven viable after more than 20 years (Coleman, 1975).

Even when clones appear to have lost their fertility, it may be possible for them to recover. Szostak *et al.* (1973) mention the use of a prolonged series of frequent transfers to rejuvenate an apparently sterile *E. elegans* clone, and exactly the same technique has been used to revive some of the

P. morum clones in the culture collection of algae maintained by Starr (Coleman, 1975). In one case this method only succeeded after nearly 2 years of monthly transfers in soil–water.

X. SUMMARY

Colonial green flagellates, although a small and relatively uniform group of organisms, display a rich variety of approaches to sexual reproduction. Some genera form isogametes, while others produce classic examples of sperm and eggs. Some clones must outbreed, while others are self-fertile; and among the self-fertilizing forms there is variation in the timing of epigenetic sex determination so that some clones are monoecious and some are dioecious.

The factors which influence colonies to express sexual activity are a major subject of research. In several species of *Volvox*, there are hormones which act as signals between the sexes to initiate sexual differentiation. These have been isolated and characterized as glycoproteins. In other species and genera, only the major environmental factors influencing sexuality are known.

These observations have laid the foundation for much more extensive investigations. The life cycles of these organisms can be controlled in the laboratory; the cells can be synchronized to a high degree by combining a suitable light–dark periodicity with simple sieving procedures which select the appropriate size range of colonies; and classical genetic studies are clearly possible. Consequently, these organisms are prime material for studying the basis of sexual differentiation, both genetic and epigenetic. Such knowledge may then reveal the significance of the systems of multiple mating types, syngens, widespread in the group, and their role in the evolution of the volvocine line.

ACKNOWLEDGMENTS

The author acknowledges with great pleasure the hospitality of Dr. Nils Ringertz and the Institute of Medical Cell Research and Genetics of the Karolinska Institute, Sweden, where the microspectrophotometric studies were made. Part of both the research and the manuscript preparation was made possible by NSF grants NSF DEB-76-82919 and NSF PCM-76-80784.

REFERENCES

Brooks, A. E. (1966). *J. Protozool.* **13,** 367–375.
Brooks, A. E. (1972). *J. Protozool.* **19,** 195–199.

Brown, R. M., Johnson, C., and Bold, H. C. (1968). *J. Phycol.* **4**, 100–120.
Bücking-Throm, E., Duntze, W., Hartwell, L. H., and Manney, T. R. (1973). *Exp. Cell Res.* **76**, 99–110.
Carefoot, J. R. (1966). *J. Phycol.* **2**, 150–156.
Carefoot, J. R. (1967). *J. Protozool.* **14**, 15–18.
Cave, M. S., and Pocock, M. A. (1951). *Am. J. Bot.* **38**, 800–811.
Cave, M. S., and Pocock, M. A. (1956). *Am. J. Bot.* **43**, 122–134.
Chiang, K.-S., and Sueoka, N. (1967). *Proc. Natl. Acad. Sci. U.S.A.* **57**, 1506–1513.
Claes, H. (1971). *Arch. Mikrobiol.* **78**, 180–188.
Coleman, A. W. (1959). *J. Protozool.* **6**, 249–264.
Coleman, A. W. (1961). *Q. Rev. Biol.* **36**, 247–253.
Coleman, A. W. (1963). *J. Protozool.* **10**, 141–148.
Coleman, A. W. (1975). *J. Phycol.* **11**, 282–286.
Coleman, A. W. (1977). *Am. J. Bot.* **64**, 361–368.
Coleman, A. W. (1978). *Exp. Cell Res.* **114**, 95–110.
Coleman, A. W. (1979). *J. Phycol.* **15**, 216–220.
Coleman, A. W., and Zollner, J. (1977). *Arch. Protistenk.* **119**, 224–232.
Dangeard, P. A. (1900). *Botaniste* **7**, 193–211.
Darden, W. H. (1966). *J. Protozool.* **13**, 239–255.
Darden, W. H. (1968). *J. Protozool.* **15**, 412–414.
Darden, W. H. (1970). *Ann. N.Y. Acad. Sci.* **175**, 757–763.
Darden, W. H., and Sayers, E. R. (1971). *Microbios* **3**, 209–214.
Doraiswami, S. (1940). *J. Indian Bot. Soc.* **19**, 113–139.
Ely, T. H., and Darden, W. H. (1972). *Microbios* **5**, 51–56.
Friedmann, I., Colwin, A. L., and Colwin, L. H. (1968). *J. Cell Sci.* **3**, 115–128.
Fulton, A. B. (1977). *Biochem. Syst. Ecol.* **5**, 261–264.
Gillham, N. W. (1969). *Am. Nat.* **103**, 355–387.
Godward, M. B. E. (1966). "Chromosomes of the Algae." St. Martins Press, New York.
Goldstein, M. (1964). *J. Protozool.* **11**, 317–344.
Goodenough, U. W. (1977). *In* "Microbial Interactions, Receptors, and Recognition" (J. L. Reissig, ed.), Series B, Vol. 3, pp. 324–350. Chapman & Hall, London.
Goodenough, U. W., and Weiss, R. L. (1975). *J. Cell Biol.* **67**, 623–637.
Gowans, C. S. (1976). *In* "The Genetics of Algae" (R. A. Lewin, ed.), pp. 145–173. Univ. of California Press, Berkeley.
Harris, D. O., and Starr, R. C. (1969). *Arch. Protistenk.* **111**, 138–155.
Hartmann, M. (1921). *Arch. Protistenk.* **43**, 223–286.
Herbst, D. S., and Kemp, C. L. (1976). *J. Phycol.* **12**, 170–172.
Hovasse, R. (1937). *Bull. Biol. Fr. Belg.* **71**, 220–237.
Hutt, W. (1972). Ph.D. Thesis, Univ. of Georgia, Athens.
Hutt, W., and Kochert, G. (1971). *J. Phycol.* **7**, 316–320.
Karn, R. C., Starr, R. C., and Hudock, G. A. (1974). *Arch. Protistenk.* **116**, 142–148.
Kemp, C. L. (1977). *J. Phycol.* **13**, 35A.
Kochert, G. (1968). *J. Protozool.* **15**, 438–452.
Kochert, G. (1975). *In* "The Developmental Biology of Reproduction" (C. L. Markert and J. Papaconstantinou, eds.), Symposium, Society for Developmental Biology, Vol. 33, pp. 55–90. Academic Press, New York.
Kochert, G., and Yates, I. (1970). *Dev. Biol.* **23**, 128–135.
Kochert, G., and Yates, I. (1974). *Proc. Natl. Acad. Sci. U.S.A.* **71**, 1211–1214.
Korshikov, A. (1938). *Viznacnik Prisnovodnich Vocloroskj. V.R.d.R. (Kiev)* **IV**, 170.
Lee, K. A., and Kemp, C. L. (1975). *Phycologia* **14**, 247–252.

Lee, K. A., and Kemp, C. L. (1976). *J. Phycol.* **12**, 85–88.
McCracken, M. D., and Starr, R. C. (1970). *Arch. Protistenk.* **112**, 262–282.
McLean, R. J., Laurendi, C. J., and Brown, R. M., Jr. (1974). *Proc. Natl. Acad. Sci. U.S.A.* **71**, 2610–2613.
Margolis-Kazan, H., and Blamire, J. (1976). *Cytobios* **15**, 201–216.
Meredith, R. F., and Starr, R. C. (1975). *J. Phycol.* **11**, 265–272.
Merton, H. (1908). *Z. Wiss. Zool.* **90**, 445–477.
Miller, D. H., Lamport, D. T. A., and Miller, M. (1972). *Science* **176**, 918–920.
Mishra, N. C., and Threlkeld, S. F. H. (1968). *Genet. Res.* **11**, 21–31.
Noland, T., Yates, I., and Kochert, G. (1977). *J. Phycol.* **13**, 49.
Pall, M. (1973). *J. Cell Biol.* **59**, 238–241.
Palmer, E. G., and Starr, R. C. (1971). *J. Phycol.* **7**, 85–89.
Pocock, M. A. (1953). *Trans. R. Soc. S. Afr.* **34**, 103–127.
Pringsheim, E. G. (1935). *Naturwissenshaften* **23**, 110–114.
Pringsheim, E. G., and Pringsheim, O. (1959). *Biol. Zentralbl.* **78**, 937–971.
Rayburn, W. R. (1974). *J. Phycol.* **10**, 258–265.
Rayburn, W. R., and Starr, R. C. (1974). *J. Phycol.* **10**, 42–49.
Rayns, D. G. (1961). Ph.D. Thesis, Univ. of London, London.
Sager, R., and Ramanis, Z. (1973). *Theor. Appl. Genet.* **43**, 101–108.
Saito, S., and Ichimura, T. (1975). *Bot. Mag.* **88**, 245–247.
Sarma, Y. S. R. K., and Shyam, R. (1973). *Nucleus (Calcutta)* **16**, 93–100.
Schlösser, U. G., Sachs, H., and Robinson, D. G. (1976). *Protoplasma* **88**, 51–64.
Schmeisser, E. T., Baumgartel, D. M., and Howell, S. H. (1973). *Dev. Biol.* **31**, 31–37.
Schreiber, E. (1925). *Z. Bot.* **17**, 337–376.
Sessoms, A. H., and Huskey, R. J. (1973). *Proc. Natl. Acad. Sci. U.S.A.* **70**, 1335–1338.
Solter, K. M., and Gibor, A. (1977). *Nature (London)* **265**, 444–445.
Sonneborn, T. M. (1957). *In* "The Species Problem" (E. Mayr, ed.), pp. 125–324. Am. Assoc. Adv. Sci., Washington, D.C.
Starr, R. C. (1962). *Arch. Mikrobiol.* **42**, 130–137.
Starr, R. C. (1964). *Am. J. Bot.* **51**, 1013–1044.
Starr, R. C. (1968). *Proc. Natl. Acad. Sci. U.S.A.* **59**, 1082–1088.
Starr, R. C. (1969). *Arch. Protistenk.* **111**, 204–222.
Starr, R. C. (1970). *J. Phycol.* **6**, 234–239.
Starr, R. C. (1971a). *Dev. Biol., Suppl.* **4**, 59–100.
Starr, R. C. (1971b). *In* "Contributions in Phycology" (B. C. Parker and R. M. Brown, eds.), pp. 59–66. (Additional copies may be obtained from the editors, Virginia State University, Blacksburg.)
Starr, R. C. (1972a). *Soc. Bot. Fr. Mem.* pp. 175–182.
Starr, R. C. (1972b). *Br. Phycol. J.* **7**, 279–285.
Starr, R. C. (1975). *Arch. Protistenk.* **117**, 187–191.
Starr, R. C., and Jaenicke, L. (1974). *Proc. Natl. Acad. Sci. U.S.A.* **71**, 1050–1054.
Stein, J. R. (1958a). *Am. J. Bot.* **45**, 388–397.
Stein, J. R. (1958b). *Am. J. Bot.* **45**, 664–672.
Stein, J. R. (1965). *Am. J. Bot.* **52**, 379–388.
Stein, J. R. (1966a). *J. Phycol.* **2**, 23–28.
Stein, J. R. (1966b). *Am. J. Bot.* **53**, 941–944.
Stein, J. R., and McCauley, M. J. (1976). *Can. J. Bot.* **54**, 1126–1130.
Szostak, J. W., Sparkuhl, J., and Goldstein, M. E. (1973). *J. Phycol.* **9**, 215–218.
Tautvydas, K. J. (1976). *Differentiation* **5**, 35–42.
Thompson, R. H. (1954). *Am. J. Bot.* **41**, 142–145.

Timpano, P., and Thompson, R. H. (1977). *J. Phycol.* **13,** 68.

Toby, A. L., and Kemp, C. L. (1977). *J. Phycol.* **13,** 368–372.

Van de Berg, W. J., and Starr, R. C. (1971). *Arch. Protistenk.* **113,** 195–219.

Whitehouse, H. L. K. (1949). *Biol. Rev. Cambridge Philos. Soc.* **24,** 411–447.

Wiese, L. (1969). *In* "Fertilization" (C. B. Metz and A. Monroy, eds.), Vol 2, pp. 135–188. Academic Press, New York.

Wiese, L. (1976). *In* "The Genetics of Algae" (R. A. Lewin, ed.), pp. 174–197. Univ. of California Press, Berkeley.

Wilbois-Coleman, A. D. (1958). Ph.D. Thesis, Indiana Univ., Bloomington.

Wolfe, J. (1974). *Exp. Cell Res.* **87,** 39–46.

Yates, I. (1974). Ph.D. Thesis, Univ. of Georgia, Athens.

Zimmerman, W. (1921). *Jahr. Wiss. Bot.* **60,** 256–294.

Learning in Protozoa

11

PHILIP B. APPLEWHITE

I. INTRODUCTION TO LEARNING

I like protozoa, but they do not like me. They have infected me, and they behave badly. It is to this latter point that I wish to address myself with regard to their learning abilities. Learning is defined here as a change in behavior caused by being trained to respond to specific stimuli. In addition, the learning must be due to changes taking place within the organism and not to changes in the environment that then cause a change in behavior. Consider, for example, a protozoan placed into a chamber that is half dark and half light, with electrodes present in the light half. Initially, assume the protozoan shows no preference for either the dark or the light half, but after being shocked in the light half it eventually stays in the dark. Until one can show that avoidance of the light half was not due to avoidance of electrolysis changes in the environment (culture medium) around the electrodes, it is not possible to say that the protozoan learned to associate light with shock. Protozoa may certainly learn during their existence in natural habitats without the aid of an experimenter but this

BIOCHEMISTRY AND PHYSIOLOGY OF PROTOZOA
SECOND EDITION, VOL. 1

chapter considers only laboratory attempts to train them. Of the many types of animal learning (Mackintosh, 1974), habituation (the diminution of a response because of repeated stimulation) can be considered a form of learning under certain circumstances (Thorpe, 1963). Since habituation has clearly been shown to exist in protozoa (Applewhite, 1971; Hamilton *et al.*, 1974; Patterson, 1973; Taylor and Chiszar, 1976; Wood, 1970), it has been omitted from this chapter, which deals with more complex types of learning. The very thorough review of protozoan learning by Corning and Von Burg (1973) serves as a benchmark for this chapter, which covers in less detail the papers we both discuss but updates and reorients them. Among those who have reviewed protozoan learning in one form or another there is disagreement whether these organisms learn (Warden *et al.*, 1940; Hyman, 1940; Thorpe, 1963; McConnell, 1966; McConnell and Jacobson, 1973; Poskocil, 1966) or do not learn (Wichterman, 1953; Jensen, 1964; Jahn and Bovee, 1967; Corning and Von Burg, 1973; Dryl, 1974).

In the discussion to follow, the learning experiments are grouped into categories that best describe the behavior, rather than into the standard learning classification categories. It should be mentioned that some of the early workers in the field such as Verworn (1889), Jennings (1906), Mast (1911), and Loeb (1918) performed a variety of experiments on protozoan behavior, but they were not well enough controlled to have demonstrated learning. If indicated in the original papers, the species used is mentioned here.

II. PATTERN LEARNING

Bramstedt (1935, 1936, 1939) reported that paramecia could learn the shape of a container they were swimming in. When paramecia were placed into a triangle or square-shaped container within their own culture medium, Bramstedt reported that they swam around the edges, following the shape of the container's walls. When the bottomless and topless container was carefully lifted up away from the medium, the protozoa continued to delineate a triangular or square-shaped swimming pattern for a while. Grabowski (1939) offered several objections, ranging from the claim that Bramstedt imagined the patterned swimming to the suggestion that chemical changes occurring in the culture medium were responsible. One could imagine that if the protozoa did follow the perimeter of the compartment, chemical changes could take place in this swimming area (because of material taken up from or discharged into the medium by the concentration of protozoa there) that would remain in place before diffusing away after the compartment was lifted away. Paramecia could be attracted to the chemical changes.

I attempted to confirm one of Bramstedt's (1939) experiments with *Paramecium caudatum*. A topless, bottomless square, 1 cm on a side, was constructed of glass and placed into the culture medium (pond water). To the square compartment was added about 200 protozoa. Even after 3 hr, with 15 separate replications, no patterned swimming within the compartment was observed. As one could say about all failures to replicate someone else's findings, there may have been factors unknown to Bramstedt that influenced his experiments. But since those factors were necessarily unreported, it becomes nearly impossible to replicate the exact conditions for success. In any case, pattern learning was not successfully demonstrated.

III. MAZE LEARNING

A logical extension of running rats through mazes was to test protozoa also. Lepley and Rice (1952) made small T-shaped mazes (only 0.03 mm deep) by etching glass with hydrofluoric acid. They put a few dozen *Paramecium multimicronucleatum* at the base of the T and observed how many protozoa turned to the left or right at the top (arms) of the T. Based on about 600 protozoa, 50% turned left and 50% turned right, as would be expected by chance. When, however, about equal numbers of protozoa were placed into a new maze with a right turn necessary before entering the T maze, 86% turned left at the arms rather than the expected 50%. Furthermore, when a left turn was necessary before entering the T maze, 83% turned right at the arms rather than the expected 50%. This "reactive inhibition," as it is called, has been noted in rats running a maze, but its cause is not understood (Mackintosh, 1974).

Lachman and Havlena (1962) performed a reactive inhibition experiment with *P. caudatum* similar to that of Lepley and Rice but failed to confirm the latter's results. Not only did Lachman and Havlena use another species of protozoa, but also their mazes were of a different size and were Y-shaped rather than T-shaped. They did not carry out an exact replication. However, when the experiments of Lepley and Rice were replicated exactly, their reactive inhibition findings were confirmed (Rabin and Hertzler, 1965).

I also replicated two of Lepley and Rice's experiments, using their procedures and methods. One at a time, 120 *P. multimicronucleatum* were put into a T maze; 48% were observed going to the left arm, and 52% going to the right arm. When a left turn preceded the T maze, 51% then turned right in the T (based on a new group of 120 protozoa). When a right turn preceded the T maze, 54% then turned left in the T (based on another group of 120 protozoa). Neither of these results supports the findings of Lepley and Rice, and Rabin and Hertzler. I made four of each kind of maze (running 30 protozoa on each) and averaged the results of each, since under the microscope the bottom surfaces of the mazes were irregular because of the uneven etching action of the hydrofluoric acid. In some mazes, the protozoa "preferred" going left in the T, and in other mazes they went mostly right. These maze experiments need to be investigated more fully to verify how the bottom texture of the maze contributes to the behavior. Thus far, learning has not been clearly demonstrated here.

IV. TUBE ESCAPE LEARNING

Smith (1908) reported sucking a *Paramecium* up from its culture medium with a small capillary tube and noted that if the tube diameter was smaller than the length of the protozoan, it would after several hours of practice turn more quickly when it reached each end of the tube where the meniscus was. The mechanics of this behavior have recently been reanalyzed (Fukui and Asai, 1976), independently of learning. Day and Bentley (1911) conducted the same type of experiment but in a quantitative way and supported Smith's observations. Buytendijk (1919) also found similar results but suggested that the facilitated turning of paramecia over time had nothing to do with practice (learning) but with changes in the flexibility of the pellicle which enabled them to turn more easily in the tube. Who knows?

French (1940) improved upon the procedures by placing a single *Paramecium* in a drop of its culture medium on a piece of glass. A capillary tube of diameter larger than the length of the protozoan was placed vertically over the *Paramecium,* which was drawn up by capillary action. The protozoan could then swim down the tube, back into the drop of medium. This event was timed, the liquid in the tube was removed, and 15 sec was allowed to elapse until another trial began, for 30 trials. In trial 1 the *Paramecium* took a mean time of about 30 sec to escape the tube, but in trial 30 only about 9 sec. French measured no difference in swimming speed before trial 1 and after trial 30, so a change in activity could not explain the findings.

It was not until the 1970s that these tube escape experiments were replicated (Hanzel and Rucker, 1971, 1972; Huber *et al.,* 1974). Rucker's laboratory confirmed French's results, using *P. caudatum.* Within the ranges tested, Rucker *et al.* found that tube diameter and the time interval between trials had no real effect on the time required to escape from the tube, which was at a minimum by trial 12. In addition, an increase in the speed of swimming over trials was not found to be an explanation of faster and faster escape times over trials. Applewhite and Gardner (1973) replicated one of Rucker's experiments (Hanzel and Rucker, 1972) and confirmed that over trials paramecia do indeed escape faster out of a capillary tube without any increase in swimming speed over trials. Applewhite and Gardner performed an additional experiment to test the influence of the capillary tube on behavior. Ordinarily, a protozoan would be drawn up a tube from a drop of culture medium and timed until it swam out into its culture medium. After an intertrial interval of 30 sec (during which time the liquid in the tube was discarded), the protozoan was drawn up the tube, and so on for a total of 12 trials. The additional experiment set up many groups of three separate protozoans each in its own drop of culture medium. One tube was used throughout the experiment; the first protozoan was drawn up in the tube and timed until it swam out; a 30-sec intertrial interval followed, and the tube liquid was discarded. This was repeated for three more trials for a total of four. Trials 5 through 8 were then carried out in exactly the same way, except with the second protozoan, and trials 9 through 12 with the third protozoan. As the trials progressed, it was found that each protozoan escaped more quickly from the tube than the preceding protozoan, suggesting that some change in the tube was building up over the trials that caused the protozoan to leave the tube faster. Since the protozoan's environment was changing, the criteria for learning were not met.

The tube escape procedure was tried on *Stentor* by Bennett and Francis (1972), and these organisms also escaped more quickly as the trials progressed. These authors attribute the behavior of swimming down the tube, however, not to learning but to a delayed geotaxis, inhibited initially by bumping into the tube during the first few trials. Support for this comes from their failing to find *Stentor* escaping faster over trials from a horizontal tube or from a vertical tube with the exit at the top rather than at the bottom. If *Stentor* is really learning the way out of the tube, it seems reasonable that it could learn it in a variety of directions—but it does not. While tube escape behavior is a real phenomenon, it has not been shown to be learning.

V. FOOD SELECTION LEARNING

Metalnikov (1907) reported that when paramecia were placed into a culture medium containing the inert substances carmine or india ink, they ingested relatively large quantities of them initially, but after 18 days in the same culture very few organisms contained these substances. Metalnikov suggested that they learned not to ingest inert substances, and demonstrated this by adding fresh carmine to paramecia which no longer were ingesting carmine. These paramecia still would not take up the carmine. Schaeffer (1910) replicated these experiments and reported that paramecia did in fact ingest less and less carmine over time, but when new carmine was added they always ingested it—they had not learned to discriminate against it. Schaeffer felt that over time the carmine particles became stuck together into larger clumps because of mucus secreted by the protozoa, and as the carmine particles became larger, it became more difficult for the protozoa to ingest them. Therefore, any time new carmine was added, the protozoa always ingested it. Bozler (1924a,b) was able to show, in fact, that larger grains of carmine were not ingested at all by paramecia compared to smaller ones which were.

Metalnikov (1912, 1913a,b, 1914, 1916, 1917) continued his observations with essentially the same results as those reported in his earlier paper. Wladimirsky (1916) also repeated some of Metalnikov's experiments and found, as did Schaeffer, that paramecia continued to ingest carmine particles when constantly exposed to them. Wladimirsky's interpretation of Metalnikov's findings was that paramecia were harmed after remaining in the same culture for several days and therefore they eventually could not take in carmine particles. Lozina-Losinskij (1929, 1937), in demonstrating that paramecia behave as Metalnikov indicated, showed, however, that the carmine they were exposed to did not injure them, as Wladimirsky suggested, since their rate of division and ability to feed on other substances were not diminished. Furthermore, Bragg (1936, 1939) commented that, if paramecia were really harmed or fatigued, then they would cease ingesting bacteria as well as carmine, but they did not. It is not convincing that paramecia learned to avoid ingesting carmine. Given these criticisms, it becomes difficult to accept another of Metalnikov's findings. He fed paramecia on carmine under red light until they ingested it no longer. When the protozoa were moved to a yeast-containing medium under red light, they ingested less yeast than they would have under normal light. Adequate controls were not run to help interpret the results, which he claimed showed that paramecia learned to associate red light with unpleasant "food."

VI. FOOD–WIRE ASSOCIATION LEARNING

This section deals with the experiments on *Paramecium aurelia* by Gelber (1952, 1954, 1956a,b, 1957, 1958a,b, 1962, 1964; Gelber and Rasch,

1956) and the replications and critiques by Jensen (1957a,b, 1959, 1964) and Katz and Deterline (1958). The exchange between these people is detailed elsewhere (Corning and Von Burg, 1973; Poskocil, 1966), so I will only briefly summarize the claims and counterclaims.

Gelber's (1952) initial experiment was designed to determine whether paramecia could learn to associate food (*Aerobacter aerogenes*) with the wire with which it was introduced into the protozoan culture. About 100 protozoa were placed into a drop of culture medium. Before the protozoa were trained, a small platinum wire was inserted into the drop for 3 min and the number of protozoa clinging to it was counted. The wire was inserted again, this time for 15 sec, and then removed for 25 sec, and so on for a total of 40 trials. In every third trial in this series, the wire was coated with bacteria as a food reward, and then the wire was cleaned after withdrawal. As the trials progressed, Gelber reported that the number of protozoa in a 3-mm zone around the wire increased. A control group receiving a wire with no bacteria on it showed no such increase in the number of protozoa in this zone. After the 40 trials were completed, a clean wire was again inserted into the culture medium for 3 min, and the number of protozoa clinging to the wire was counted. Compared to the number of protozoa clinging in the initial 3-min test, there was a large increase. No such increase was found in the control groups. Gelber found that the activity of the paramecia from the initial 3-min test period to the final 3-min test period did not change, so that the increased clinging to the wire was not due to more protozoa swimming by and sticking to the wire just by chance.

Jensen (1957a) demonstrated in an experiment (similar but not identical to Gelber's) that the insertion of a wire with bacteria on every third trial created a bacteria-rich zone in the area where the wire was inserted. Jensen found that paramecia were attracted to bacteria-rich areas and concluded that they were attracted to the area around the wire in the testing procedure for this reason and not because of any learning. Since more protozoa were in this zone as the trials progressed, more clung to the test wire just by chance. Gelber (1957b) then performed an experiment and found that paramecia were *not* attracted to bacteria-rich zones, although she used fewer bacteria than Jensen. To eliminate the possibility of a bacteria-rich zone where the wire was inserted, Gelber then reported that she stirred the drop of protozoa (and bacteria) by rotating the slide they were on before inserting the wire to test their clinging after training. Jensen reported trying to mix the drop using Gelber's procedure and found that it did not work. The center of the drop was not mixed at all, and the bacteria-rich zone where the wire was inserted was still intact. There was always some doubt, therefore, that Gelber's procedures ruled out alternative explanations to learning.

A crucial test of Gelber's initial report was carried out by Katz and Deterline (1958). They replicated her experiments exactly and confirmed that, after training, significantly more paramecia were attached to the wire and present in the 3-mm counting zone around the wire. The crucial addition to the procedure was stirring the drop after training, and then inserting the wire and counting the protozoa on or near it. When this was done, it was found that no more protozoa approached the wire than in control groups that received just the wire with no bacteria on it. It is not possible to say whether the stirring eliminated a bacteria-rich zone (suggesting that the culture medium and not the protozoa had changed) or whether its intensity disrupted the memory of paramecia (suggesting that the protozoa really had learned). Gelber's results seem to be valid, but many experiments with different and better controls need to be run to determine if learning really occurs.

VII. APPROACH AND AVOIDANCE LEARNING

Smith (1908) observed that when a capillary tube containing paramecia was heated at one end and made cold at the other end, the protozoa moved to the cold end. When the hot and cold ends were alternated, over time the paramecia moved to the cold end more quickly with each trial. Since Smith gave no data and provided no controls, his experiment cannot be evaluated.

Mast and Pusch (1924) reported that when *Amoeba proteus* (kept in the dark in a drop of culture medium) were presented with light passing through a very narrow slit, they avoided advancing into the light. I replicated the experiment as best I could, given that certain experimental procedures were lacking in the paper, such as the intensity of the light used, the size of the narrow light slit, and the length of a trial. A single *A. proteus* was placed in a drop of medium and kept in the dark 10 min. Then a narrow light slit (0.5 mm wide and 4 mm long) was placed at a distance approximately half the length of the ameba away from its approaching pseudopods. A light was turned on to shine through the slit (the built-in substage illuminator of a Bausch & Lomb Galen microscope was set at maximum intensity with a 10X objective and a 10X eyepiece). If an advancing pseudopod came in contact with the light and did not stop, the experiment was terminated. If it stopped, the number of other pseudopods moving into the light was counted as well, for a total of 5 min (an estimate for the mean trial length of Mast and Pusch). Three minutes later, the procedure was repeated for a total of 27 trials. Counting the mean total number of pseudopods moving into the light for each ameba for 3 con-

Table I Mean Total Contacts with Light for Each Amoeba in Groups of Three Consecutive Trials

Trials	Mast and Pusch[a]	Applewhite[b]
1–3	6	5
4–6	4	8
7–9	5	9
10–12	5	6
13–15	4	9
16–18	3	4
19–21	2	8
22–24	3	6
25–27	2	7

[a] $N = 5$.
[b] $N = 30$.

secutive trials (as did Mast and Pusch) over the 27 trials, I found no avoidance of the light (Table I). My experimental conditions may certainly have varied from those of Mast and Pusch, but it is disappointing not to be able to replicate their findings. Could they be confirmed, controls would have to be added to rule out alternatives to learning.

Bramstedt (1935) put a single *P. caudatum* in a culture drop on top of a special chamber that kept one half of the drop at 42°C (in the light) and the other half at 15°C (in the dark). Eventually, a protozoan was found mostly in the dark 15°C side. When the entire drop was kept either at 15°C or at 42°C with one side light and the other dark, protozoa showed no preference for either side. After a 90-min training session with the 15°C-dark and 42°C-light conditions in the drop, Bramstedt found that when the entire drop was 15°C, protozoa still preferred the dark side, suggesting that paramecia learned to associate the 15°C side with the dark. Although another control showed that paramecia preferred 15°C to 42°C if both sides of the drop were dark, they preferred it much more after the 90-min training. About 15 min later, the protozoa "forgot" their trained association. A confirmation of these results was published by Alverdes (1937a,b) and by Grabowski (1939), although Grabowski thought something other than learning was responsible for the results. Grabowski demonstrated that when he subjected a drop of culture medium with no paramecium to the 15°C-dark and 50°C-light conditions (he did not use Bramstedt's 42°C) and then kept the entire drop at 15°C, a newly introduced paramecium behaved as though it had been trained—it stayed in the dark side. Consequently, something must have changed in the medium during the training procedures—perhaps the 50°C side became partly deoxygenated and paramecia avoided it. Dembowski (1950) said that even at Bramstedt's 42°C the same chemical changes in the medium took place. Koehler (1939) also joined in the criticism of Bramstedt's results. Diebschlag (1940) changed the culture medium temperatures again, using 38°C rather than Bramstedt's 42°C, and reported that when the entire drop was changed to 15°C paramecia avoided the lighted end (as Bramstedt had found). Diebschlag added some new controls and concluded that there was no chemical change in the medium and that the protozoa avoided the lighted end because they had learned to. Perhaps using a different culture medium and a lower high temperature eliminated medium effects. Mirsky and Katz (1958) repeated some of the experiments of Bramstedt and Grabowski and found neither learning nor evidence of culture medium effects. However, the results are not directly comparable, since different temperatures and species of paramecia were used. Best (1954) moved the experimental setting from drops on slides to capillary tubes in order to investigate heat and light associations in *P. aurelia,* having failed to confirm Bramstedt's drop findings. Best found that paramecia did not learn to associate heat and light—which must be the conclusion drawn from all these experiments.

Bramstedt (1935) reported training *Stylonychia mytilus* to associate light with a rough bottom surface in its training chamber. He kept these protozoa in a chamber divided into a smooth bottom–dark half and a rough bottom–light half. When placed into a smooth bottom chamber with dark and light halves, *Stylonychia* stayed mostly in the dark. Necessary controls to establish this as learning were not performed. Nevertheless, in a modification of this procedure, Machemer (1966a,b) was unable to obtain any rough bottom–light association in *Stylonchia.* On the other hand, Soest (1937) claimed to have trained this protozoan to avoid a rough bottom area by shocking it whenever it entered—eventually, without electric shocks, it avoided the area.

Soest (1937) claimed to have obtained learning also in *Paramecium, Spirostomum,* and *Stentor* by pairing light with electric shock. In a chamber half dark and half light, a protozoan was shocked whenever it entered the lighted side, and if it learned it then avoided the lighted side even when no shocks were given. At about the same time, Tschakhotine (1938) claimed that *P. caudatum* could learn to avoid a spot in its culture medium where an ultraviolet microbeam had been focused up to 30 min before. Grabowski (1939) attributed this behavior to the avoidance of chemical or physical changes in the irradiated area. Whether there were any changes was never established. Soest's findings became somewhat questionable in view of Sgonina's (1939) report that applying electric shocks to parts of the culture medium (with protozoa absent) produced avoidance reactions to those parts when the protozoa were introduced. Diebschlag (1940) did not find such an effect, however. In any case, the failure of Soest to use controls in testing for shock effects on the medium for all his protozoa weakens his findings considerably. Had he run controls, they may have come out in his (the protozoan's?) favor and he would have presented a learning demonstration worth replicating. Wawrzyńczyk (1938) moved the light–shock experiments to a capillary tube for *P. caudatum*. Electrodes were placed at either end, and when a protozoan swam into a central area between two lights of certain colors, it was shocked. He reported that paramecia avoided this area for up to 50 min after the shocks were withdrawn. Dembowski (1950) replicated these experiments and confirmed the results, but did not attribute them to learning. Control experiments showed that paramecia avoided the central area after being shocked in the absence of light and their location had nothing to do with light—hence no association was formed. Again, culture medium effects seem to have been involved, as confirmed by Jensen (1955) and Dabrowska (1956), who found no evidence of association learning in *Paramecium, Stentor,* or *Spirostomum.*

Voss (1975) reported he had trained paramecia to avoid the lighted half of a chamber. He paired light with shock in one chamber, and then moved the paramecia to another chamber to measure their light avoidance. He showed that the amount of avoidance depended upon the intensity of the light paired with the shock and that the acquired response could be extinguished by flashes of light (without shock reinforcement). Controls subjected to light flashes only, shock only, or no shock and light apparently produced no light avoidance behavior, but graphs for these protozoa were not given. In view of the problems with the other light–shock experiments, replication of these results is necessary before they can be accepted as evidence of learning.

Bergström (1968a,b, 1969a,b) reported that *Tetrahymena* could be trained to avoid light if previously shocked in it. In a small dark chamber containing over 1000 protozoa, an electric shock was given right after the onset of light. Fifteen pairings were given in all over a period of 15 min. For testing, a screen with 15 holes was placed between the protozoa and a light source, and the number of *Tetrahymena* in these lighted areas was measured over a period of 15 min. Bergström found that fewer and fewer protozoa stayed in the lighted area. Appropriate controls showed no such avoidance of the light. This was a well-designed experiment, like Voss's (except that Bergström's sample size was too small), but a replication of it by Applewhite *et al.* (1971) failed to find any evidence of a light–shock association.

VIII. TOUCH–LIGHT ASSOCIATION LEARNING

Plavilstchikov (1928) was able to train *Carchesium lachmanni* to con-
tract to a red or blue light. The protozoa were kept in the dark and then
presented with a red (or blue) light followed by touch, which caused
contraction. Eventually presentation of the red (or blue) light alone elicited contraction. In addition, Plavilstchikov took a trained stalk of *Carchesium* and grafted it onto an untrained organism, which then began to
respond sooner to training (only five experiments of many attempts succeeded). I replicated one of Plavilstchikov's experiments. I used the same
size of training chamber and put the *Carchesium* in it in the dark for 10 min
before starting the experiment in the dark. For the red light, I used one
layer of red cellophane (650-nm peak) placed over a Bausch & Lomb
31-33-53 illuminator (set at 3) placed 10 cm from the substage mirror of a
stereoscopic microscope. Plavilstchikov did not give a light intensity or
wavelength. The red light was turned on, and 2 sec later an individual
Carchesium was touched with a fine platinum wire. The red light was left
on for a total of 1 min. Fifteen minutes later, the procedure was repeated
so as to obtain 20 trials per day per organism for a total of 180 trials.
Plavilstchikov ran from 150 to 200 total trials this way, and trained 62
protozoa with the red light. He found that all the protozoa contracted to
the red light alone, ranging from the 79th to the 184th trial at first contraction. I ran 20 protozoa, all of which also contracted to the light alone,
ranging from the 55th to the 123rd trial at first contraction. However,
these 20 protozoa never contracted to the light any more times than the
controls used ($N = 20$), which received red light only and no touch. The
mean total number of contractions to red light of those receiving light and
touch was 7, and for those receiving only light was 9. Plavilstchikov never
ran a red light-only control, so his results must stand in question.

IX. CONCLUSIONS

Learning (beyond habituation) has not been adequately demonstrated in
protozoa. Claims of learning have either not been confirmed in replications, or controls are lacking to rule out alternative explanations to learning. In spite of this, I find no a priori reason why protozoa cannot learn.
Most of the protozoan experiments reported on here are relatively simple
to perform and I would encourage their replication by others. As for me, I
give up.

ACKNOWLEDGMENTS

The author thanks his students Mark Ganisin and Elena Cirellio for their help in replicating some of the experiments reported here.

REFERENCES

Alverdes, F. (1937a). *Z. Tierpsychol.* **1**, 35–38.
Alverdes, F. (1937b). *Zool. Anz.* **120**, 90–95.
Applewhite, P. B. (1971). *Nature (London), New Biol.* **230**, 285–287.
Applewhite, P. B., and Gardner, F. T. (1973). *Behav. Biol.* **9**, 245–250.
Applewhite, P. B., Gardner, F., Foley, D., and Clendenin, M. (1971). *Scand. J. Psychol.* **12**, 65–67.
Bennett, D. A., and Francis, D. (1972). *J. Protozool.* **19**, 484–487.
Bergström, S. R. (1968a). *Scand. J. Psychol.* **9**, 215–219.
Bergström, S. R. (1968b). *Scand. J. Psychol.* **9**, 220–224.
Bergström, S. R. (1969a). *Scand. J. Psychol.* **10**, 16–20.
Bergström, S. R. (1969b). *Scand. J. Psychol.* **10**, 81–88.
Best, J. B. (1954). *J. Exp. Zool.* **126**, 87–99.
Bozler, E. (1924a). *Z. Vgl. Physiol.* **2**, 82–90.
Bozler, E. (1924b). *Arch. Protistenk.* **49**, 163–215.
Bragg, A. N. (1936). *Physiol. Zool.* **9**, 433–442.
Bragg, A. N. (1939). *Turtox News* **17**, 41–44.
Bramstedt, F. (1935). *Z. Vgl. Physiol.* **22**, 490–516.
Bramstedt, F. (1936). *Forsch. Fortschr.* **12**, 176.
Bramstedt, F. (1939). *Zool. Anz., Suppl. (Verh. Dtsch. Zool. Ges.* No. 41) **12**, 111–132.
Buytendijk, F. J. (1919). *Arch. Neerl. Physiol.* **3**, 455–468.
Corning, W. C., and Von Burg, R. (1973). "Invertebrate Learning," Vol. 1, pp. 49–122. Plenum, New York.
Dabrowska, J. (1956). *Folia Biol. (Warsaw)* **4**, 77–81.
Day, L. M., and Bentley, M. (1911). *J. Anim. Behav.* **1**, 67–73.
Dembowski, J. (1950). *Acta Biol. Exp. (Warsaw)* **15**, 5–17.
Diebschlag, E. (1940). *Zool. Anz.* **130**, 257–271.
Dryl, S. (1974). *In* "Paramecium—A Current Survey" (W. J. Van Wagtendonk, ed.), pp. 165–218. Elsevier, New York.
French, J. W. (1940). *J. Exp. Psychol.* **26**, 609–613.
Fukui, K., and Asai, H. (1976). *J. Protozool.* **23**, 559–563.
Gelber, B. (1952). *J. Comp. Physiol. Psychol.* **45**, 58–65.
Gelber, B. (1954). *Am. Psychol.* **9**, 374.
Gelber, B. (1956a). *J. Genet. Psychol.* **88**, 31–36.
Gelber, B. (1956b). *J. Comp. Physiol. Psychol.* **49**, 590–599.
Gelber, B. (1957). *Science* **126**, 1340–1341.
Gelber, B. (1958a). *J. Comp. Physiol. Psychol.* **51**, 110–115.
Gelber, B. (1958b). *Am. Psychol.* **13**, 405.
Gelber, B. (1962). *Psychol. Rec.* **12**, 165–192.
Gelber, B. (1964). *Anim. Behav.* **1**, Suppl., 21–29.
Gelber, B., and Rasch, E. (1956). *J. Comp. Physiol. Psychol.* **49**, 594–599.

Grabowski, U. (1939). *Z. Tierpsychol.* **2**, 265–282.
Hamilton, T. C., Thompson, J. M., and Eisenstein, E. M. (1974). *Behav. Biol.* **12**, 393–407.
Hanzel, T. E., and Rucker, W. B. (1971). *J. Biol. Psychol.* **13**(2), 24–28.
Hanzel, T. E., and Rucker, W. B. (1972). *Behav. Biol.* **7**, 873–880.
Huber, J. C., Rucker, W. B., and McDiarmid, C. G. (1974). *J. Comp. Physiol. Psychol.* **86**, 258–266.
Hyman, L. H. (1940). "The Invertebrates," Vol. 1. McGraw-Hill, New York.
Jahn, T. L., and Bovee, E. C. (1967). *In* "Research in Protozoology" (T. Chen, ed.), Vol. 1, pp. 41–200. Pergamon, Oxford.
Jennings, H. S. (1906). "Behavior of the Lower Organisms." Indiana Univ. Press, Bloomington. (Reprint, 1962.)
Jensen, D. D. (1955). M.A. Thesis, Univ. of Nebraska, Lincoln.
Jensen, D. D. (1957a). *Science* **125**, 191–192.
Jensen, D. D. (1957b). *Science* **126**, 1341–1342.
Jensen, D. D. (1959). *Behaviour* **15**, 82–122.
Jensen, D. D. (1964). *Anim. Behav.* **1**, Suppl. 9–20.
Katz, M., and Deterline, W. A. (1958). *J. Comp. Physiol. Psychol.* **51**, 243–247.
Koehler, O. (1939). *Zool. Anz., Suppl. (Verh. Dtsch. Zool. Ges.* No. 41) **12**, 132–142.
Lachman, S. J., and Havlena, J. M. (1962). *J. Comp. Physiol. Psychol.* **55**, 972–973.
Lepley, W. M., and Rice, G. E. (1952). *J. Comp. Physiol. Psychol.* **45**, 283–286.
Loeb, L. (1918). "Forced Movements, Tropisms, and Animal Conduct." Lippincott, Philadelphia, Pennsylvania.
Lozina-Losinskij, L. K. (1929). *Rep. Acad. Sci. USSR, Ser. A* **17**, 403–408.
Lozina-Losinskij, L. K. (1937). *Arch. Protistenk.* **74**, 18–120.
McConnell, J. V. (1966). *Annu. Rev. Physiol.* **28**, 107–136.
McConnell, J. V., and Jacobson, A. L. (1973). *In* "Comparative Psychology" (D. A. Dewsbury and D. A. Rethlingshafer, eds.), pp. 429–470. McGraw-Hill, New York.
Machemer, H. (1966a). *Z. Tierpsychol.* **6**, 641–654.
Machemer, H. (1966b). *Arch. Protistenk.* **109**, 245–256.
Mackintosh, N. J. (1974). "The Psychology of Animal Learning." Academic Press, New York.
Mast, S. O. (1911). "Light and the Behavior of Organisms." Wiley, New York.
Mast, S. O., and Pusch, L. C. (1924). *Biol. Bull. (Woods Hole, Mass.)* **46**, 55–59.
Metalnikov, S. (1907). *Trans. Soc. Imp. Nat. St. Petersburg* **38**, 181–187.
Metalnikov, S. (1912). *Arch. Zool. Exp. Gen.* **49**, 373–498.
Metalnikov, S. (1913a). *C. R. Soc. Biol.* **74**, 701–703.
Metalnikov, S. (1913b). *C. R. Soc. Biol.* **74**, 704–705.
Metalnikov, S. (1914). *Arch. Protistenk.* **34**, 60–78.
Metalnikov, S. (1916). *C. R. Soc. Biol.* **79**, 80–82.
Metalnikov, S. (1917). *Russ. J. Zool.* **2**, 397.
Mirsky, A. F., and Katz, M. S. (1958). *Science* **127**, 1498–1499.
Patterson, D. J. (1973). *Behaviour* **45**, 304–311.
Plavilstchikov, N. N. (1928). *Arch. Russ. Protistol.* **7**, 1–14.
Poskocil, A. (1966). *Worm Runner's Dig.* **8**(1), 31–42.
Rabin, B. M., and Hertzler, D. R. (1965). *Worm Runner's Dig.* **7**(2), 46–50.
Schaeffer, A. A. (1910). *J. Exp. Zool.* **8**, 75–132.
Sgonina, K. (1939). *Z. Tierpsychol.* **3**, 224–247.
Smith, S. (1908). *J. Comp. Neurol.* **18**, 499–510.
Soest, H. (1937). *Z. Vgl. Physiol.* **24**, 720–748.
Taylor, S. V., and Chiszar, D. (1976). *Behav. Biol.* **16**, 105–111.

Thorpe, W. H. (1963). "Learning and Instinct in Animals." Methuen, London.
Tschakhotine, S. (1938). *Arch. Inst. Prophylac., Paris* **10**, 119–131.
Verworn, M. (1889). "Psycho-physiologische Protistenstudien." Jena.
Voss, H. J. (1975). *Mikrokosmos* Nov., No. 11, 348–351.
Warden, C. J., Jenkins, T. N., and Warner, L. H. (1940). "Comparative Psychology," Vol. 2. Ronald Press, New York.
Wawrzyńczyk, S. (1938). *Trav. Soc. Sci., Wilno* **12**, 1–28.
Wichterman, R. (1953). "The Biology of Paramecium." Blakiston, New York.
Wladimirsky, A. P. (1916). *Russ. J. Zool.* **44**, 4.
Wood, D. C. (1970). *J. Neurobiol.* **1**, 363–377.

Physiological Ecology of Red Tide Flagellates

12

HIDEO IWASAKI

I. INTRODUCTION

Red tide is the popular name given to the discoloration of water caused by microscopic organisms. Species of dinoflagellates are most commonly responsible, though other flagellates, and even ciliates, can also cause red tides. The color of the water mostly depends on the causative planktonic organisms, being sometimes brown, green, or yellow. In general, a red tide first appears as colored patches of water, varying in size. The affected water masses become soupy or cloudy as the numbers of organisms increase. The water changes color progressively, sometimes developing a distinct reddish tinge, and becomes more or less viscous or slimy; the

BIOCHEMISTRY AND PHYSIOLOGY OF PROTOZOA
SECOND EDITION, VOL. 1

highest mortality of marine life usually occurs at this stage. The red tide in the Seto Inland Sea in 1972 killed great numbers of farmed fish, and the damage was about $35 million. Harmful effects of red tides are not confined to fish. Kills of barnacles, oysters, coquinas, shrimps, crabs, porpoises, and turtles have also been reported (Gunter *et al.,* 1947).

Some red tides cause illness resulting from the consumption of poisonous clams, mussels, and fish. The commonest type is the periodic acute paralytic shellfish poisoning. Shellfish become toxic when they ingest large amounts of *Gonyaulax,* generally during the summer months.

The occurrence of red tides seems to be restricted to certain areas, generally neritic, such as the coasts of India and Arabia along the Indian Ocean, the southwestern and northwestern coasts of Africa, the Gulf of Mexico, the coast of Brazil along the Atlantic Ocean, the coast of California, the coast of Japan, and the central Pacific Ocean. Most of these areas coincide with regions of upwelling (Brongersma-Sanders, 1957; Eppley *et al.,* 1968; Eppley and Harrison, 1975; Blasco, 1975, 1977) or with highly polluted regions (Braarud, 1945; Iwasaki, 1973). Thus, red tides have attracted more and more interest as a consequence of increased knowledge of planktology and of their economic and social effects.

II. BEHAVIOR RELATIVE TO PHYSICAL FACTORS

A. Light

Since light is the energy source of photosynthesis, light intensity, quality, and the light–dark cycle play a dominant role in photosynthetic organisms.

Most dinoflagellates prefer light intensities higher than those preferred by other groups of algae, though they have species-specific preferences. The saturation illuminance for photosynthesis by several dinoflagellates has been reported (Ryther, 1956; Moshkina, 1961; Thomas, 1966a). Thomas (1966a) found the compensation illuminance for photosynthesis for two tropical *Gymnodinium* species to be 350 lx. The basic photosynthetic zone of dinoflagellates in the sea seems to be between 0 and 25 m. It is not certain, however, that the same light intensity is optimal for both photosynthesis and growth. Loeblich (1969) showed that the saturating light intensity for growth of *Cachonina niei* was 10,000 lx, and for photosynthesis slightly less than 20,000 lx. Kain and Fogg (1960) found that the lag phase in *Prorocentrum micans* was reduced by increasing the light intensity above the saturation value for exponential growth.

Cell division in *Gonyaulax polyedra* (Sweeney and Hastings, 1958) and

Peridinium triquetrum (Braarud and Pappas, 1951) grown under alternating light–dark conditions is restricted to certain hours during a 24-hr light–dark cycle. In these organisms, maximal division occurs at about the beginning of the light period. Maclean (1977) observed that the cell division rate of *Pyrodinium bahamense*, collected from Port Moresby Harbor, New Guinea, reached a maximum during midmorning (0800–1000 hr) and a minimum at 0200 hr. In contrast, *Ceratium* species divide in their natural environment only at night (Gough, 1905; Jorgensen, 1911; Hasle, 1954).

Most dinoflagellates respond behaviorally to light. Detailed studies of the phototactic movements of several dinoflagellates were reported by Nordli (1957) and Halldal (1958). Phototactic vertical migrations of dinoflagellates in response to changing light intensity may be quite common in marine environments. In general, cells migrate toward the surface during the day, and away from the surface at night. *Gonyaulax polyedra, P. micans, Ceratium furca, P. triquetrum* (Hasle, 1950, 1954), *Exuviaella baltica* (Wheeler, 1964), and *Gymnodinium* spp. (Hirano, 1967) all migrated toward the surface during the day, and away at night; only *C. fusus* and *C. tripos* migrated downward during the day (Hasle, 1950). All migrations took place within the upper 5 m of water. *Pyrodinium bahamense* showed two diurnal vertical migrations per day; maximum concentrations at the surface occurred at about 2200 and 1400 hr, with a minimum at 0900 hr (Soli, 1966). But, in Port Moresby Harbor, Maclean (1977) observed that *P. bahamense* concentrations had a minimum in the upper 5 m between 1800 and 2400 hr, followed by a gradual rise until dawn, 0600 hr. The proportion and quantity of *Pyrodinium* then declined, with an anomalous peak in number at 5 m and 1400 hr. The discrepancy between the two observations on the behavior of *P. bahamense* suggests that the response of the organism to light intensity differs even in the same species, although the difference in light and hydrographic conditions between the two waters is unknown. According to Seliger *et al.* (1975), *Gymnodinium breve* remains at the surface at night as well as by day.

As mentioned above, the response of dinoflagellates to light differs with the species. The phototactic behavior suggests, however, that most dinoflagellates migrate easily to their preferred level of illumination, i.e., where light is optimal for reproduction. It is probable, therefore, that the phototactic vertical migration of dinoflagellates plays a role in the accumulation of red tide organisms in the water column.

B. Temperature

In nature, dinoflagellates occur over a wide range of temperatures. *Ceratium tripos, C. fusus,* and *C. furca* are found in seawater at temperatures varying from $-1.0°$ to $29.5°C$, and *C. lineatum* occurs from $3°$ to $21°C$ (Nordli, 1957). The temperature which supports optimal growth for marine species is generally close to the upper range of tolerated temperatures. For instance, the colorless dinoflagellate *Crypthecodinium* (*Gyrodinium*) *cohnii* was sensitive to temperatures below $20°C$; growth was inhibited between $15°$ and $20°C$, and no growth occurred at $10°C$. The optimum was $20°–28°C$. This organism grew moderately well at $32°C$, and at $35°C$ it barely survived (Provasoli and Gold, 1962; Gold and Baren, 1966). The temperatures at which blooms occurred and the optimal temperatures for the growth of flagellates are shown in Table I.

The optimum temperature for the dinoflagellates studied is about $20°C$, except for *C. fusus* and *Gonyaulax catenella*, which have an optimum at $15°C$. Even in the same species, differences in the optimum temperature have been observed. *Peridinium trochoideum* from Golfo di Napoli (Braarud, 1961) and *Prorocentrum micans* from the coast of California (Barker, 1935) have an optimum temperature $(25°C)$ higher than that of strains from Oslofjorden.

Clear relationships to temperature have also been found in regard to the distribution of red tides. In the Seto Inland Sea, in Ohmura Bay, and along the coast of Shima Peninsular, Japan, most red tides have occurred at temperatures from $20°$ to $27°C$—most frequently at $24°–26°C$. The blooms of *Pyrodinium* in Jamaica, Florida, and New Guinea waters have been found at temperatures higher than $22.2°C$ (Buchanan, 1971; Steidinger and Williams, 1970; Maclean, 1977). However, a red tide of toxic *G. catenella* occurred at $12°–14°C$ in Owase Bay, in January.

C. Salinity

Dinoflagellates are widely distributed in both fresh water and marine water. Some species are euryhaline and can withstand wide salinity changes, whereas others are stenohaline and are sensitive to salinity variations.

Salinity tolerances of marine flagellates are shown in Table II. Most red tide flagellates are euryhaline, except subtropical and oceanic species. *Pyrodinium bahamense, Gonyaulax monilata, G. tamarensis,* and *Gymnodinium breve* prefer high salinities, and red tide blooms occur in salinities higher than $30‰$ (Buchanan, 1971; Maclean, 1977; Steidinger and Williams, 1970; Wardle *et al.,* 1975; Yentsch *et al.,* 1975; Aldrich and Wilson, 1960). However, salinity, temperature, and light intensity interact in their influence on growth (Norris and Chew, 1975). At optimum salinity, *G. catenella* could tolerate a temperature variation of at least $12°C$, but at the extreme salinity of $40‰$, the tolerable temperature variation was

Table I Temperature and the Growth of Red Tide Dinoflagellates

Species	Area of occurrence	Bloom temperature (°C)	Temperature range for growth (°C)	Optimum temperature (°C)	Reference
Amphidinium carterae	Near Woods Hole, Massachusetts	—	18–33	24	Jitts et al. (1964)
Amphidinium conradi	Matoya Bay, Japan	24.5–25.5	—	—	Sato et al. (1965)
Cachonina niei	Salton Sea, California	—	—	19–23	Loeblich (1969)
Ceratium furca	Oslofjorden	—	12–26[a]	20	Nordli (1957)
Ceratium fusus	Oslofjorden	—	9–24[a]	15	Nordli (1957)
Ceratium lineatum	Oslofjorden	—	10–26[a]	20	Nordli (1957)
Ceratium tripos	Oslofjorden	—	<5–24<[a]	20	Nordli (1957)
Dissodinium lunula	?	—	17–28	25	Swift and White (1973)
Exuviaella baltica	Oslofjorden	—	14–30[a]	20–26	Braarud (1961)
Gonyaulax catenella	Sequim Bay, Washington	—	—	13–17	Norris and Chew (1975)
	Owase Bay, Japan	12–14	—	—	Adachi (unpublished)
Gonyaulax monilata	Western Florida, offshore	28–32 (29)	—	—	Wardle et al. (1975)
Gonyaulax polyedra	Oslofjorden	—	13–29[a]	20	Braarud (1961)
	Scripps Pier, California	—	13–28[a]	25	Hastings and Sweeney (1964)
Gonyaulax tamarensis	Cape Ann, Massachusetts	17.5	—	—	Yentsch et al. (1975)
	Boothbay Harbor, Maine	15.5	—	15–20	Yentsch et al. (1975)
Gonyaulax tamarensis	Gulf of Maine	9.5–15	—	15–19	Prakash (1967)
	Florida	27–28	—	—	Steidinger and Williams (1970)
Gymnodinium ochraceium	Ohmura Bay, Japan	26–28	23–31	—	Irie (1965)
Gymnodinium breve	Florida	16–27	15–30[b]	26–28[c]	Rounsefell and Nelson (1966)
					Aldrich (1959)
					Finucane (1960)

(Continued)

Table I (*Continued*)

Species	Area of occurrence	Bloom temperature (°C)	Temperature range for growth (°C)	Optimum temperature (°C)	Reference
Gymnodinium spp.	Tropical	—	—	23–29	Thomas (1966a)
Peridinium trochoideum	Oslofjorden	—	10–26[a]	18–20	Braarud (1961)
	Golfo di Napoli	—	14–30<[a]	25	Braarud (1961)
Prorocentrum gracile	Coast of California	—	13–21[a]	18	Barker (1935)
Prorocentrum micans	Coast of California	—	15–28[a]	25	Barker (1935)
	Oslofjorden	—	12–27[a]	20	Braarud (1961)
Pyrodinium bahamense	Florida	22.2–29.2	—	—	Steidinger and Williams (1970)
	Jamaica	27–35	—	—	Buchanan (1971)
	New Guinea	26.2–30.7	—	—	Maclean (1977)

[a] Temperature permitting growth beyond half-maximum.
[b] Aldrich (1959).
[c] Finucane (1960).

Table II Salinity Tolerances of Marine Flagellates

Species (clone)	Origin	Salinity permitting growth (%)			Reference
		Beyond half-maximum	Max-imum	Negli-gible	
Dinoflagellates					
Amphidinium sp.	Oslofjorden	6–43	16–20	≤10	Braarud (1951)
Amphidinium carterae	Near Woods Hole, Massachusetts	13–35	25	≤10	McLachlan (1961)
Ceratium furca	Oslofjorden	13–40	20–27	≤10	Nordli (1957)
Ceratium fusus	Oslofjorden	15–33	20	≤10	Nordli (1957)
Ceratium lineatum	Oslofjorden	13–28	20	<10, >40	Nordli (1957)
Ceratium tripos	Oslofjorden	13–32	20		Nordli (1957)
Dissodinium lunula	?	—	35	<20, >45	Swift and White (1973)
Exuviaella baltica	Oslofjorden	3–42	10	>44	Braarud (1951)
Exuviaella cassubica	Near Woods Hole, Massachusetts	10–20	—	<0.8, >35	Provasoli et al. (1954)
Glenodinium foliaceum	Long Island, New York	22–26	24	<6, >56	Prager (1963)
Glenodinium sp.	Nagasaki Bay, Japan	28–37	—	<20	Uchida (1975)
Gonyaulax catenella	Sequim Bay, Washington	20–37	—	<15, >40	Norris and Chew (1975)
	Owase Bay, Japan	17–35	24–29	≤12	Iwasaki (unpublished)
Gonyaulax catenella 1B	Ise Bay, Japan	16–30	24	≤12	Iwasaki (unpublished)
Gonyaulax monilata	Galveston, Texas, offshore	(30–40)[a]	—	—	Wardle et al. (1975)
Gonyaulax tamarensis	Cape Ann, Massachusetts; Boothbay Harbor, Maine; Gulf of Maine	(29.5–31.0)[a]	—	—	Yentsch et al. (1975)
Gonyaulax tamarensis	Off Florida	(35.5–35.9)[a]	15–23		Prakash (1967)
Gymnodinium breve	Coast of Florida	27–37	—	<22.5, >46.0	Steidinger and Williams (1970)
Gymnodinium nelsoni	Seto Inland Sea	11–35	19–22	≤6	Aldrich and Wilson (1960)
Gymnodinium nelsoni SF	Seto Inland Sea	15–31	18–25	≤15	Iwasaki (unpublished)
Gymnodinium sp. SA1	Seto Inland Sea	15–30	18–24	≤10	Ueno et al. (1977)
Gymnodinium sp. SF4	Seto Inland Sea	12–33	16–22	≤8	Iwasaki (unpublished)

(Continued)

Table II (*Continued*)

Species (clone)	Origin	Salinity permitting growth (%)			Reference
		Beyond half-maximum	Maximum	Negligible	
Gymnodinium type 65	Nagasaki Bay, Japan	20–33	—	—	Hirayama and Numaguchi (1972)
Peridinium balticum	Near Woods Hole, Massachusetts	8–20	—	<3, >35	Provasoli et al. (1954)
Peridinium hangoei	Coast of Shima Peninsular, Japan	17–29	21–24	≦8	Iwasaki (1969a)
Peridinium triquetrum	Oslofjorden	6–40	12–22	—	Braarud and Pappas (1951)
Peridinium trochoideum	Oslofjorden	9–35	18–24	<6, >40	Braarud (1951)
Polykrikos schwartzi (Gyrodinium ?)	Seto Inland Sea	15–30	17–19	≦11	Iwasaki (1971a)
Prorocentrum micans	Oslofjorden	12–37	20	—	Braarud and Rossavik (1951)
Prorocentrum minimum var. triangulata	Ise Bay, Japan	13.5–29	22–25	5	Iwasaki (unpublished)
Pyrodinium bahamense	Jamaica	(30–36)[a]	—	—	Buchanan (1971)
	Florida	(30.6–36.5)[a]	—	—	Steidinger and Williams (1970)
	Coast of New Guinea	(28.5–36.8)[a]	—	—	Maclean (1977)
Other flagellates					
Chattonella (Heterosigma) akashiwo	Seto Inland Sea	21–30	27	10	Iwasaki et al. (1968)
Chattonella (Heterosigma) inlandica	Coast of Shima Peninsular, Japan	7–35	10–14	4	Iwasaki and Sasada (1969)
Chattonella japonica (Exuviaella sp.)	Seto Inland Sea	20–35	24–30	15	Iwasaki (1971b)
Hornellia sp. J1 (Eutreptiella sp. I)	Seto Inland Sea	14–30	16–21	—	Iwasaki (1971b)

[a] Salinity at bloom.

reduced to less than 9°C. Similarly, the wide range of salinities tolerable at optimal temperatures was reduced to a much narrower range at suboptimal temperatures. Variation in light intensity from 1000 to 3500 lx had no significant effect on the tolerance ranges for temperature or salinity (Norris and Chew, 1975).

Ryther (1955) reported that salinity seemed not to be a limiting factor for dinoflagellate blooms. It is likely that salinity does not restrict the distribution, except in marine waters of extremely low salinity, of euryhaline species. However, salinity seems to be an important factor for stenohaline species. Nordli (1957) found that the distribution and mass occurrence in the North Sea and near the coast of Norway of three species of *Ceratium* were controlled by salinity and temperature; the laboratory data on salinity and temperature parallel the ecological data.

Slobodkin (1953) assembled impressive evidence that outbreaks of red tide followed exceptionally heavy rainfalls on land, which implies that red tide organisms may prefer low salinity and possibly nutrients washed into the sea from the land. Steven (1966) reported that a bloom of *Exuviaella* sp. developed when the area in question had been diluted with fresh water. In fact, in limited areas, high correlations between heavy rainfalls and the occurrence of red tides have been found. Present knowledge of growth responses of red tide dinoflagellates to salinity may partly support Slobodkin's hypothesis. It is unquestionable that the dilution of seawater by heavy rainfalls prepares a suitable environment for the growth of most neritic red tide flagellates.

It is worth mentioning that salinity has a great influence on the production of toxins in the chrysomonad *Prymnesium parvum* (Schilo and Rosenberger, 1960).

D. Hydrogen Ion Concentration (pH)

Hydrogen ion concentration or pH is generally maintained within narrow limits by the buffer effect of dissolved salts; hence its range of variation may have only a moderate influence as a limiting factor for planktonic algae. However, during periods of high diatom production a marked rise in pH occurs as a result of the rapid utilization of carbonic acid by the plants during photosynthesis. On such occasions an excessive rise might tend to act as a natural check on further proliferation of some species, although this has not been shown for dinoflagellates. However, the pH of coastal waters adjacent to industrial zones or large cities may be expected to fluctuate widely. In these areas, the pH of seawater has ecological significance.

The pH tolerances of red tide flagellates are shown in Table III. Most

Table III pH Tolerances of Red Tide Flagellates

| Species (clone) | Origin | pH permitting growth | | | Reference |
		Beyond half-maximum	Maximum	Negligible	
Dinoflagellates					
Exuviaella baltica	Coast of Angola, Africa	7.5–8.3	7.8	>8.5	Paredes (1967–1968)
Gonyaulax catenella	Owase Bay, Japan	7.7–9.0	8.3–8.5	<7.0, >9.2	Iwasaki (unpublished)
Gonyaulax catenella IB	Ise Bay, Japan	7.7–8.7	8.2	<7.2, >8.9	Iwasaki (unpublished)
Gymnodinium nelsoni	Seto Inland Sea	7.5–8.8	8.4–8.5	<7.0, >9.0	Iwasaki (unpublished)
Gymnodinium nelsoni SF	Seto Inland Sea	7.8–8.4	8.2	<7.5, >8.6	Ueno *et al.* (1977)
Gymnodinium sp. SA1	Seto Inland Sea	7.8–8.4	8.0	<7.5, >7.9	Iwasaki (unpublished)
Gymnodinium sp. SF4	Seto Inland Sea	7.2–8.8	8.5–8.6	<6.2, >9.0	Iwasaki (unpublished)
Peridinium hangoei	Gokasho Bay, Japan	7.5–8.3	7.9–8.1	<6.6, >8.4	Iwasaki (1969a)
Prorocentrum micans	Coast of California	7.7–9.1	8.5	<7.3, >9.3	Barker (1935)
	Coast of England	7.5–8.8	—	—	Kain and Fogg (1960)
Polykrikos schwartzi (*Gyrodinium* ?)	Seto Inland Sea	7.9–8.9	8.5	<7.3, >9.2	Iwasaki (1971a)
Other flagellates					
Chattonella akashiwo	Seto Inland Sea	6.9–8.4	7.4	<6.5, >9.0	Iwasaki *et al.* (1968)
Chattonella inlandica	Gokasho Bay, Japan	7.7–9.5	8.6–9.0	<6.9	Iwasaki and Sasada (1969)
Chattonella japonica	Seto Inland Sea	7.5–9.5	8.5	<7.0	Iwasaki (1971b)

red tide flagellates grow moderately well at pH 7.8–8.4, which is the normal pH of seawater.

III. NUTRITIONAL REQUIREMENTS

A. Nitrogen

All red tide flagellates, with a few exceptions, can utilize nitrate, nitrite, ammonia, urea, and uric acid as nitrogen sources when these compounds are supplied at suitable concentrations. At times ammonia is preferentially or more rapidly absorbed. *Gymnodinium nelsoni* cannot use inorganic nitrogen at all (Iwasaki, 1973) and needs yeast extract for growth. Since nitrate reduction requires energy, if the energy supply is limited, more growth will occur with ammonium salts. Consequently, at a low light intensity, cell synthesis is faster with ammonium salts than with nitrate (Bongers, 1956).

In culture, the growth rate of *G. catenella* was approximately equal at lower concentrations of nitrate, ammonium ions, and urea, although the generation time was slightly shorter with ammonium ions than with the others. The latter two were toxic at high concentrations (Norris and Chew, 1975). Many investigations indicate that nitrate is the best nitrogen source for red tide flagellates under laboratory conditions. Some amino acids are also utilized by dinoflagellates. However, growth is no better with amino acids than with nitrate (McLaughlin and Provasoli, 1957; Provasoli and McLaughlin, 1963; Thomas, 1966b; Iwasaki and Sasada, 1969; Norris and Chew, 1975; Mahoney and McLaughlin, 1977; Ueno *et al.*, 1977).

Several workers suggested that dinoflagellates could be grown in media with low concentrations of nitrogen and phosphorus. Barker (1935) concluded that nitrogen was probably limiting only at 1–10 μg/liter. In culture experiments with *G. catenella*, Norris and Chew (1975) showed that the growth rate was higher at low nitrogen concentrations than at higher concentrations. The experiment also indicated that total growth of *G. catenella* in nitrate, ammonium salts, and urea was proportional to the available nitrogen at concentrations up to 0.84–1.4 mg nitrogen/liter. Ammonium salts and urea are toxic for the various species at different levels, ranging from 0.5 to 20 mg nitrogen/liter for the former, and in concentrations higher than 3 mg nitrogen/liter for the latter (Iwasaki, 1973; Norris and Chew, 1975).

B. Phosphorus

Phosphorus and nitrogen are generally present in very low concentrations in nature; they are the most evident limiting factors. Most flagellates can utilize not only inorganic phosphate but also organic phosphate (Provasoli and McLaughlin, 1963; Loeblich, 1966; Iwasaki, 1969a,b; 1971a,b; Mahoney and McLaughlin, 1977). Glycerophosphate is more effective for several species than inorganic phosphate. Mahoney and McLaughlin (1977) found that glycerophosphate in all concentrations tested, fructose 1',6'-diphosphate at 0.5 mg phosphorus/liter, and 3'-cytidylic acid, 3'-adenylic acid, and 5'-adenylic acid at 0.1 mg phosphorus/liter supported far higher growth levels of *P. micans* than potassium phosphate.

Barker (1935) showed that the division rate was not affected between concentrations of 0.01 and 10 mg phosphorus/liter in *P. micans* and suggested that phosphate, like nitrate, limited division only at concentrations below about 10 μg phosphorus/liter. On the contrary, using the ^{14}C method to determine the effect of various phosphate concentrations on *Glenodinium hallii,* Gold (1962) showed that division was enhanced at 0.3 μg–3 mg phosphorus/liter. Unfortunately, similar observations are lacking for other dinoflagellate species. "Luxury uptake" has also been found in dinoflagellates. According to Iwasaki (1971a), *Polykrikos schwartzi* cells grown in phosphorus-rich medium increased about 100-fold when transferred to phosphorus-depleted media.

In general, increasing the concentration of nutrient salts increases the yield of organisms when other ecological and nutritional conditions are suitable.

C. Vitamin Requirements

Droop (1957) and Provasoli (1958a,b, 1963) reviewed from an ecological point of view the needs of marine flagellates and algae. The known vitamin requirements of flagellates are shown in Table IV. The requirements of red tide flagellates are limited to three vitamins, B_{12}, thiamine, and biotin, alone or in combination, as in other photosynthetic algal groups; only some of them require thiamine and biotin. The requirement for thiamine is not widespread among dinoflagellates, but *G. nelsoni* SF needs only thiamine (Ueno *et al.,* 1977). These results suggest that the three vitamins, especially B_{12}, may play a significant role in red tide outbreaks.

There are a number of naturally occurring analogs of B_{12} which can be utilized by some organisms but not by others (Kon, 1955). The requirements of 16 bacteria-free flagellates, including seven species of red tide organisms, have been determined in chemically defined culture media (Table V). The majority of red tide flagellates investigated show the wide specificity typical of *Escherichia coli.* These results indicate that vitamin B_{12} analogs may have an importance in outbreaks equal to that of the vitamin itself.

Table IV Vitamin Requirements of Flagellates[a]

Species	B_{12}	Thia-mine	Bio-tin	Reference
Dinoflagellates				
Amphidinium carterae	+	+	+	Droop *et al*. (1959)
Amphidinium rhychocephalum	+	+	+	Droop *et al*. (1959)
Cachonina niei	+	+	−	Loeblich (1969)
Crypthecodinium cohnii	−	s	+	Provasoli (1963)
Exuviaella cassubica	+	−	−	Provasoli and McLaughlin (1955)
Exuviaella sp.	−	−	−	McLaughlin and Zahl (1959)
Glenodinium foliaceum[b]	+	−	−	Droop (1958)
	+	(s)	(s)	Prager (1963)
Glenodinium hallii	+	s	−	Gold (1962)
Gonyaulax catenella[b]	+	−	−	Iwasaki (unpublished)
Gonyaulax catenella IB[b]	±	±	s	Iwasaki (unpublished)
Gonyaulax polyedra[b]	+	−	−	Provasoli (1963)
Gymnodinium breve[b]	+	+	+	Wilson and Collier (1955)
Gymnodinium nelsoni[b]	+	−	−	Iwasaki (1973)
Gymnodinium nelsoni SF	s	+	−	Ueno *et al*. (1977)
Gymnodinium splendens[b]	+	−	−	Provasoli (1963)
Gymnodinium sp. SF4[b]	+	(s)	(s)	Iwasaki (unpublished)
Gyrodinium californicum	+	−	−	Provasoli and Pintner (1953)
Gyrodinium resplendens	+	−	−	Droop *et al*. (1959)
Gyrodinium uncatenum	+	−	−	Droop *et al*. (1959)
Oxyrrhis marina	+	+	+	Droop (1958)
Peridinium baltica	+	−	−	Provasoli (1963)
Peridinium chattonii	+	−	−	Provasoli (1963)
Peridinium hangoei[b]	+	−	−	Iwasaki (1969a)
Peridinium trochoideum[b]	+	−	−	Droop (1958)
Polykrikos schwartzi	+	+	+	Iwasaki (1971a)
Prorocentrum micans[b]	+	?	?	Droop (1957)
Symbiodinium microadriaticum	−	−	−	McLaughlin and Zahl (1959)
Woloszynskia limnetica	+	+	?	Provasoli and Pintner (1953)
Other flagellates				
Chattonella akashiwo[b]	+	−	−	Iwasaki *et al*. (1968)
Chattonella japonica[b]	+	−	−	Iwasaki (1971b)
Chattonella inlandica[b]	+	−	−	Iwasaki and Sasada (1969)

[a] +, Required; s, stimulated growth; −, not needed for growth; and (s), stimulated growth in the presence of B_{12}.

[b] Species appeared in red water.

Table V Specificity of Red Tide Flagellates to B_{12}-like Compounds[a]

Species	B_{12}	DCA	BA	FI	FA	FH	PS	FB	FZ_1	FZ_2	FZ_3	Reference
Dinoflagellates												
Gonyaulax catenella[b]	+	+	+	+	+		+	+	+	+	+	Iwasaki and Kojima (unpublished)
Gymnodinium nelsoni[c]	+	+	+	+	?		+	+	+	+	+	Iwasaki (1973)
Peridinium hangoei[d]	+	?	+	+	+		+	+	?	+	+	Iwasaki (unpublished)
Gonyaulax catenella IB[b]	+	+	+	+	+		+	+	0	0	0	Iwasaki and Kojima (unpublished)
Polykrikos schwartzi[b]	+	?	?	0	+	+	0	+	0	0	0	Iwasaki (1971a)
Amphidinium carterae	+	+	+	+	+	+	0	0				Provasoli and McLaughlin (1963)
Amphidinium klebsii	+	+	+	+	+	+	0	0				Droop et al. (1959)
Amphidinium rhynchocephalum	+	+	+	+	+	+	0	0				Droop et al. (1959)
Peridinium balticum	+	+	+	0	+	0	0	0				Droop (1957)
Gyrodinium californicum	+	+	+	+	0	0	0	0				Provasoli (1958a)
Gyrodinium resplendens	+	+	+	+	0	0	0	0				Provasoli (1958a)
Gyrodinium uncatenum	+	+	+	+	0	0	0	0				Provasoli (1958a)
Woloszynskia limnetica	+	+	+	+	0	0	0	0				Droop et al. (1959)
Other red tide flagellates												
Chattonella akashiwo[b]	+	+	+	+	+		+	+	0	0	0	Iwasaki et al. (1968)
Chattonella inlandica[b]	+	+	+	+	0		0	0				Iwasaki and Sasada (1969)

[a] B_{12}, Cyanocobalamin containing 5,6-dimethylbenzimidazole in the nucleotide; DCA, 5,6-dichlorobenzimidazole analog; BA, benzimidazole analog; FI, factor I (B_{12} III); FA, factor A (with 2-methyladenine); FH, factor H (with 2-methylhypoxanthine); PS, pseudovitamin B_{12} (with adenine); FB, factor B (with no nucleotide); FZ_1, FZ_2, and FZ_3, factors Z_1, Z_2, and Z_3 (Neujahr, 1956). +, Active; 0, inactive.
[b] Species appeared in red tides.
[c] In the presence of 50 μg% xanthine in the medium.
[d] In cobalt-free medium.

IV. ECOLOGY OF RED TIDES

A red tide is a complicated phenomenon, affected not only by biological and chemical factors but also by hydrographic and meteorological conditions. Frequently, observations reported in the literature note that the outbreaks are preceded by a heavy rainfall, an influx of estuarine water, unusually hot, calm sunny weather, smooth seas, or a meeting of dissimilar water masses. These observations led to several hypotheses.

The development of red tide blooms seems to comprise at least two main components: (1) biological growth, and (2) physical continuation and concentration. Of course, the division is not strict; the formation of stratification in water masses may precede growth (Wyatt, 1973; Mulligan, 1975; Hartwell, 1975; Zubkoff and Warinner, 1975), and the positive phototactic behavior of many dinoflagellates may contribute to the concentration mechanisms (Seliger et al., 1970; Eppley and Harrison, 1975; Seliger et al., 1975; Kamykowski and Zentara, 1977). Studies on red tides have mainly sought a limiting factor, rather than studying interactions of multiple factors or seeking a correlation between blooms and environmental factors. With these approaches, however, it is difficult to find correlations, because there are too many variable factors. Steidinger and Ingle (1972) suggested that G. breve existed as a resident cyst (spore) population in coastal sediments and that initial increases were due to excystment. Steidinger (1973, 1975) emphasized the possibility of benthic seed populations and suggested that factors influencing their development be primary research objectives.

A. Growth Mechanisms

In nature, organisms are restricted by many environmental factors: temperature, salinity, essential nutrients, growth factors, and interspecific competition or predatory pressures. For the initiation of growth, release from these ecological pressures is needed, as well as the presence of favorable conditions. Although favorable conditions for G. breve growth have been reported by several workers (Wilson and Collier, 1955; Aldrich and Wilson, 1960; Wilson, 1966; Rounsefell and Nelson, 1966; Steidinger and Ingle, 1972), more information about other red tide species is needed. Knowledge of the physiological characteristics of the causative organisms is indispensable in understanding the growth mechanisms of red tides, and helps to make fieldwork more meaningful and to verify observations made in the field. From this viewpoint, physiological studies on red tide organisms have been carried out by the author; the results are summarized in Table VI.

Table VI Physiological Requirements of Red Tide Flagellates

Requirement	Type I[c]				Type II	
	Chattonella akashiwo	Rhodomonas ovalis	Gymnodinium nelsoni SF	Gymnodinium sp. SF4	Hornellia sp. J1	Chattonella japonica
Salinity, optimal range (‰)[a]	21–30	>5.5	15–31	12–33	14–30	20–35
pH, optimal range[a]	6.9–8.5	7.8–9.0	7.8–8.4	7.1–8.9	6.5–8.5	7.7
B_{12}	+	+	s	+	+	+
Optimal concentration	10 ng/liter	10 ng/liter		3 ng/liter	10 ng/liter	10 ng/liter
Thiamine	–	+	+	s	+	–?
Biotin	–	+	–	s	+	–?
B_{12} specificity[b]	Ochromonas			Escherichia	Escherichia	Euglena
Growth-promoting substances	NH_4 nitrogen, PO_4 phosphorus, B_{12}				Iron, manganese	Iron, manganese, purines, pyrimidines, plant hormones

Type III

Requirement	Chattonella inlandica	Gonyaulax catenella	Gonyaulax catenella (var.)	Gymnodinium nelsoni	Peridinium hangoei	Polykrikos schwartzi	Prorocentrum micans
Salinity, optimal range (%o)[a]	7–35	17–35	16–30	11–35	17–29	15–30	17–34
pH, optimal range[a]	7.6	8.2–8.7	7.8–8.5	7.4–8.8	7.5–8.4	7.7–8.8	6.6–9.0
B_{12}	+	+	±	+	+	+	+
Optimal concentration	10 ng/liter	5 ng/liter	5 ng/liter	20 ng/liter	10 ng/liter		
Thiamine	–	–	±	–	–	+	
Biotin	–	–	s	–	–	+	
B_{12} specificity[b]	Ochromonas	Escherichia	Euglena	Escherichia	Escherichia	Escherichia	
Growth-promoting substances	Purines, pyrimidines	Purines, pyrimidines, kinetin, yeast extract, Yeastolate, trypticase, Hy-Case, Thiotone, DNA	Guanine, hypoxanthine, kinetin, sodium glutamate, Hy-Case, Thiotone, DNA	Yeast extract +, Thiotone, or guanine or DNA	Yeastolate, trypticase, sodium glutamate	Yeastolate, yeast extract, Thiotone, Hy-Case	Adenine, glycine, Hy-Case

[a] Range permitting growth beyond half-maximum.
[b] Patterns of growth response to vitamin B_{12} analogs.
[c] s, Stimulatory.

Though about half of the organisms studied were chloromonads and cryptomonads, their physiological characteristics, especially their nutrient requirements, were very similar to those of dinoflagellates. All the organisms investigated grew well at temperatures higher than 20°C, except two strains of *Gonyaulax*. The optimal temperature for growth was 21°–26°C. Most red tide flagellates are euryhaline and preferred comparatively low salinities (one-third to four-fifths oceanic seawater). Half of them grew well at pH 8.2, and four species preferred a pH lower than that of natural seawater. In general, organisms that preferred higher salinities tended to have lower optimal pH values. Nitrate was the best nitrogen source, although ammonium and urea were utilized. High yields of several species were obtained at low concentrations of ammonium salts (<300 μg nitrogen/liter). All organisms utilized both inorganic and organic phosphorus, and some of them stored excess phosphorus. All needed one or more vitamins for growth; the majority required B_{12}, but two species of *Gymnodinium* needed only thiamine. Red tide flagellates were divided into three groups based on their different patterns of specificity toward B_{12} analogs; the majority, however, utilized all known cobalamins in the environment. The growth of all red tide flagellates was enhanced in the presence of soil extract. The important facts are that (1) all red tide organisms are auxotrophic, and some species need organic substances besides vitamins; and (2) each species has specific tolerances and preferences for environmental factors. From the standpoint of nutritional characteristics, these organisms have been classified into three types:

1. Organisms whose multiplication is induced by supplying only inorganic nitrogen, phosphorus, and vitamin B_{12} under optimal conditions for the other ecological factors. Growth is also enhanced by lowering the salinity.
2. Organisms whose growth is accelerated by supplying high concentrations of dissolved iron and/or manganese.
3. Organisms whose growth is stimulated by enrichment with organic biologically active substances such as purines, pyrimidines, and decomposed organic matter.

1. Type 1

The chloromonad *Chattonella akashiwo* Loeblich and the cryptomonad *Rhodomonas ovalis,* which occur frequently in the coastal waters of Fukuyama, Seto Inland Sea, belong to this group. The red tide bloom caused by these organisms may be expected to result from water washed from the land after a heavy rainfall, or from polluted water.

The growth responses of *R. ovalis* to enrichment with selected inorganic

Table VII Growth of *Rhodomonas ovalis* in Seawater Enriched with Various Nutrients[a]

Nutrient	Growth (number of cells $\times$ 10^4/ml)
Filtered seawater	5.4
Additions	
Metal mixture PII (10 ml/liter)	5.7
NTA (50 mg/liter)	6.2
Marine mud extract	6.8
Land soil extract	22.5
Vitamin B_{12} (0.2 μg/liter)	5.4
KH_2PO_4 (0.5 mg/liter)	5.3
NH_4Cl (5 mg/liter)	12.2
NH_4Cl + KH_2PO_4	46.8
NH_4Cl + KH_2PO_4 + B_{12}	122.1

[a] After 7 days.

nutrients, nitrilotriacetic acid (NTA), B_{12}, and soil extracts, are shown in Table VII. All the elements were added aseptically individually and in combinations to sterile, filtered seawater (using a glass-fiber filter) collected from red tide waters. Though seawater alone allowed high growth, individual enrichment, except for ammonium salts and land soil extract, did not enhance growth. Enrichment with a combination of ammonia and phosphate supported higher growth (a 9-fold increase) and this combination plus B_{12} stimulated growth remarkably (a 23-fold increase). The results seem to be important in showing an interactive effect of multiple nutrient factors.

Field observations showed that the horizontal distribution of population densities in the bloom of *R. ovalis* which occurred along the Fukuyama coast in August 1968 corresponded with the isopleth of ammonium nitrogen in the waters (Iwasaki, 1972). A similar feature was observed in the red tide of *C. akashiwo* in the same waters (Iwasaki *et al.*, 1968). Nitrogen was apparently derived from the land. Inoue *et al.* (1973) found with *Lactobacillus leichmannii* that the coastal waters of Fukuyama, Seto Inland Sea, contained 0.56–6.4 ng B_{12}/liter throughout the year. Though the seasonal variation was not conspicuous, the B_{12} concentration was generally high during months of higher water temperature (ca. 18°C or above); the maximum occurred in June or July when salinity dropped as a result of the rainy season. It is possible, therefore, that red tide blooms caused by organisms of this type require an enhanced nutrient supply, resulting for instance, from an increase in land runoff or a discharge of domestic waste or sewage at temperatures of 20°–26°C and low salinities ($S < 30\%_{oo}$). This may also apply to the *Exuviaella* sp. blooms in Kingston Harbor, Jamaica. Steven (1966) had also shown a close relationship between nitrate concentrations and red tide blooms.

Many investigations of dinoflagellates, however, suggest that these organisms can grow at low concentrations of nitrogen and phosphorus, and that growth rates are not affected by these concentrations (Gran, 1929; Barker, 1935; Murphy *et al.*, 1972; Norris and Chew, 1975; Yentsch *et al.*, 1975). According to McAllister *et al.*, (1960), nutrient depletion did not limit the growth of mixed algal population until a nitrate concentration of less than 14 µg nitrogen/liter was reached. Similar results were obtained with coastal phytoplankters (McAllister *et al.*, 1961; Antia *et al.*, 1963). Thomas (1966b) showed that concentrations of nitrate below approximately 70 µg nitrogen/liter probably limited the growth of *Gymnodinium simplex*. Doig and Martin (1974) found that, at concentrations of orthophosphate typical of Florida coastal waters, the growth-promoting potential of the medium was a linear function of the ammonium ion concentration (10–110 µg nitrogen/liter). However, it is impossible to generalize about rate-limiting concentrations of nutrients and their relationship to ecology, because each species and even local strains may have special preferences (Guillard and Ryther, 1962).

Workers studying the importance of inorganic macronutrients to red tides in the Gulf of Mexico placed initial emphasis on phosphorus and *G. breve* growth but were unable to find a relationship between total phosphorus levels and red tide outbreaks (Bein, 1957; Donnelly *et al.*, 1966). Wilson (1966), in laboratory experiments with *G. breve,* determined that a minimum of about 9 µg nitrogen/liter was necessary for significant growth. Rounsefell and Dragovich (1966) found no statistical relationship between the abundance of *G. breve* and levels of phosphate, nitrate, nitrite, or ammonium ion in data for southwestern Florida waters for 1–6 years. The finding of a maximum growth rate for *G. catenella* even at low concentrations of nitrate (Norris and Chew, 1975) is compatible with the field data which indicate a generally low level (0–140 µg nitrogen/liter) of nitrate (Sribhibhadh, 1963) during months in which larger *G. catenella* populations are observed. It is suggested that these facts may help to explain the occurrence of *G. catenella* after the decline of diatom pulses which may have reduced the available nitrogen to low levels. It is difficult, however, to determine whether the occurrence of *G. catenella* is due only to the lowering of available nitrogen or to the production of essential vitamins (Carlucci and Bowes, 1970) and/or stimulating substances in the secretions (Iwasaki, 1977).

These discrepancies are probably due to a species-specific preference for nutrient concentrations, and to rate-limiting concentrations combined with the effects of other parameters. Many laboratory experiments on red tide flagellates (Kain and Fogg, 1960; Iwasaki, 1973; Norris and Chew, 1975) show that total growth is proportional to nitrogen concentration up to certain levels.

It is known that the coastal surface waters off southern California are enriched by the admixture of nutrient-rich water from a lower depth. This enrichment promotes phytoplankton growth, and either diatom or dinoflagellate blooms may result (Eppley and Harrison, 1975). Holmes et al. (1967) found nitrate concentrations (per liter) in a red water patch as follows: surface, 1 μg nitrogen; 7 m, 8 μg nitrogen; 12 m, 118 μg nitrogen. Armstrong and LaFond (1966) found similar high nutrient levels associated with the thermocline in southern California coastal waters. However, recent evidence suggests that, even though inorganic macronutrients may be available in excess, other factors are critical to primary production and growth (Provasoli, 1963; Barber and Ryther, 1969; Raymont, 1971; Small and Ramberg, 1971). These factors include chelators, trace metals, vitamins, and other biologically active substances.

There are still problems concerning the role of vitamin B_{12} and its analogs in red tide outbreaks. Though data on the geographical and vertical distribution of vitamins in the sea are still limited, clearly coastal and bay waters, especially surface waters, are richer in B_{12} than open-sea waters (Lewin, 1954; Droop, 1955; Cowey, 1956; Kashiwada et al., 1957; Menzel and Speath, 1962; Carlucci and Silbernagel, 1966; Ohwada and Taga, 1969, 1972; Inoue et al., 1973). It was suggested that these results excluded B_{12} as a limiting factor in neritic water (Droop, 1962). However, Carlucci and Bowes (1970) showed that in the waters off La Jolla, California, during September 1967, fluctuation in the concentration of dissolved B_{12} corresponded fairly well with that of the dominant organism, G. polyedra. More investigations of the relationship between fluctuations in vitamin concentration and fluctuations in red tide organisms are desirable for evaluating the role of vitamins in red tide blooms. It is worth mentioning that a supply of B_{12} in the presence of favorable concentrations of other nutrient factors enhanced the growth of red tide flagellates even when it was not the principal limiting factor (Table VII).

2. Type 2

The chloromonads Chattonella japonica and Hornellia sp. are examples of this group. Gymnodinium breve and Gonyaulax tamarensis also resemble this type in their responses.

A large-scale red tide bloom of Hornellia sp. and C. japonica occurred after a violent vertical mixing in shallow waters by a typhoon in the Seto Inland Sea in August 1970. Data on conditions before the outbreak are few; however, 3 months afterward the seawater still contained dissolved iron, 0.14–0.45 mg/liter (Inoue, 1977), which coincided with concentrations permitting maximal growth *in vitro* (Iwasaki, 1971b). Since the marine sediments in the area abounded in metals, e.g., 16–48 mg iron/gm and 0.13–1.0 mg manganese/gm (Hiroshima-Kenchō, 1974), and all other environmental factors were within optimal ranges, it is highly probable that the abundant supply of chelated metals from the bottom induced exponential growth of the organisms. From the results, Iwasaki suggested that a supply of these metals at high concentrations, from the sea bottom or land, induced a bloom of these species, and pointed out the important role of soluble organic compounds in seawater as chelators.

Martin and Martin (1973) also showed suggestive evidence for involvement of organometallic compounds in red tide outbreaks. Analysis of the onset of *G. breve* blooms (1844–1960) indicated that blooms generally occurred following diatom blooms during periods of nutrient depletion, and that outbreaks tended to occur after or during a heavy seasonal rainfall. Certain rivers in Florida, e.g., the Peace River, have increased iron concentrations following periods of heavy rainfall. Prakash and Rashid (1968) noted that substantial amounts of humic substances entered the Bay of Fundy, where blooms of *G. tamarensis* were prominent, just before the occurrence of maximum dinoflagellate populations. Ingle (1965) pointed out the association of red tide outbreaks along the Florida coast with Peace River constituents (iron plus tannic acid, humic acid, and related chelating agents). These observations led to the proposal by Ingle and Martin (1971) of a new means of predicting the occurrence of Florida red tide outbreaks: the iron index. Their conclusions were based on the statistical analysis of data collected over 25 years. They did not imply that the trace metal iron was the "triggering" factor, but only suggested it as an index.

Several trace metals are known to be essential for the growth of red tide organisms. All defined media that support the growth of dinoflagellates include a trace metal mixture, but investigations on the requirements for trace elements are limited. Kramer and Ryther (1960) found the optimum concentration of iron for *Amphidinium carterae* to be 65 μg/liter. Wilson (1966) showed that chelated iron (Fe · EDTA) enhanced *G. breve* growth. The chloromonads *Hornellia* sp. and *C. japonica* needed soluble iron and manganese for growth, and growth was stimulated remarkably at high concentrations of chelated iron (Fe · EDTA) and manganese (Mn · EDTA), i.e., concentrations at least five times greater than those in oceanic seawater (Iwasaki, 1971b). Levandowsky and Hutner (1975), who added iron to a defined marine medium with various concentrations of several chelators, found that a heavy growth of *C. cohnii* at pH 7.5–7.7

occurred with salicylhydroxamic acid (SHAM), aurintricarboxylic acid (ATA), EDTA, NTA, and humic acid, and at pH 7.9–8.1 with SHAM and ATA. In the culture of *P. micans,* Kain and Fogg (1960) reported that providing a suitable chelator was very important for the organisms; EDTA was fairly satisfactory, but glycylglycine was better. Yentsch *et al.* (1975), using cultures of *G. tamarensis,* ascertained that increased growth was achieved with the addition of low levels of NTA—10 mg/liter—while levels of 50 mg/liter and above were inhibitory to growth when added to filtered seawater. No biostimulation, however, was observed for *G. breve,* the red tide organism of the Florida coast, by the addition of NTA (Doig and Martin, 1974b). It is not clear, however, whether the growth-promoting actions of chelators are due to a reduction in toxicity or enhancement of the availability of trace elements. Collier (1958) pointed out the significant role of chelators in the detoxification of copper ions and suggested that sulfides substituted for organic chelators. He suggested that the interplay of copper ions and hydrogen sulfide, which is produced in mangrove-covered marshes along the western coast of Florida, was possibly active in establishing the level of toxicity of untreated seawater to *G. breve* and other organisms. Anderson and Morel (1978) found extreme copper sensitivity in *G. tamarensis,* and concluded that it would not survive at the calculated copper activity of seawater, if only inorganic complexation occurred.

3. Type 3

The majority of neritic red tide flagellates and dinoflagellates are included in this group. The growth responses of red tide flagellates to organic substances are shown in Tables VIII and IX. Although each species has specific preferences, growth is accelerated in the presence of organic substances. Apparently these substances are derived from other organisms or metabolites. In the Seto Inland Sea, frequent red tide blooms have always been observed in organically polluted waters having chemical oxygen demand (COD) values higher than 1 ppm. Though the biostimulative constituents are not precisely clear, because of the difficulty of analysis, several types of evidence have been reported. Okaichi and Yagyu (1969) found that pulp wastes diluted 250 to 500 times were effective in promoting the growth of *Prorocentrum triestinum* and the euglenid *Eutreptiella* spp. which are the organisms causing red tides in waters containing pulp wastes. The optimum concentration of wastes lay in a narrow COD range from 2 to 5 ppm, and concentrations above a COD value of 10 ppm were unfavorable for growth. Aerobically decomposed pearl oyster feces also promoted the growth of *Peridinium hangoei* and *Chattonella inlandica,* which occurred as a red tide bloom in pearl oyster culture grounds

Table VIII Growth Response of Red Tide Flagellates to Purines, Pyrimidines, and Plant Hormones in ASP$_2$ NTA

		Growth (number of cells $\times$ 10^3/ml)[a]				
		Dinoflagellates			Other flagellates	
Substance	Concentration (mg/liter)	*Gymnodinium nelsoni,* 20 days[b]	*Gonyaulax catenella,* 15 days	*Gonyaulax catenella* 1B, 17 days	*Chattonella japonica,* 18 days	*Chattonella inlandica,* 15 days
None	—	1.9	2.5	1.0	4.7	32.0
Adenine	0.5	—	—	—	10.8	—
Guanine	0.5	—	9.8	3.9	16.0	111.0
Guanine	1.0	3.5	6.3	4.1	22.8	62.0
Xanthine	0.3	—	—	—	22.5	135.0
Hypoxanthine	1.0	—	7.5	2.8	11.8	110.0
Cytosine	1.0	2.1	—	—	10.4	143.0
Methylcytosine	0.3	2.8	3.4	—	11.9	42.7
	1.0	—	—	—	14.2	—
Thymine	1.0	—	3.6	—	18.3	117.0
Uracil	0.3	—	3.6	—	—	160.0
	1.0	2.1	3.4	—	7.5	117.0
Indoleacetic acid	0.05	—	—	—	19.5	—
Kinetin	0.2	—	9.5	3.8	15.7	—
	2.0	—	11.0	3.4	—	—
Gibberellic acid	0.4	—	—	—	22.1	—

[a] Days after inoculation are indicated.
[b] *Gymnodinium nelsoni* needs yeast extract for growth; culture medium contained 10 mg/liter of yeast extract.

380

Table IX Growth Response of Red Tide Dinoflagellates to Organic Substances in ASP_2NTA

Substance added	Concentration (mg/liter)	Growth (number of cells $\times$ 10^3/ml)[a]				
		Peridinium hangoei, 20 days	*Polykrikos schwartzi*, 12 days	*Gymnodinium nelsoni*, 20 days[b]	*Gonyaulax catenella*, 15 days	*Gonyaulax catenella* IB, 17 days
None	—	0.9	0.2	—	2.4	0.6
Yeast extract	10	—	2.7	1.9	3.9	—
	30	—	1.8	—	3.7	—
Yeastolate	10	—	3.2	2.8	4.1	—
	30	35.7	2.3	—	4.0	—
Thiotone	10	—	1.9	3.5	4.4	1.3
	30	—	2.6	2.5	4.3	1.7
Hy-Case	10	—	1.4	—	3.5	4.3
	30	—	1.3	—	4.4	—
Trypticase	100	33.4	—	—	3.6	—
DNA	0.3	—	—	3.1	5.0	1.3
	1.0	—	—	2.1	—	—
Sucrose	1000	—	—	—	3.4	1.2
Sodium glutamate	500	41.0	—	—	3.5	1.2

[a] Days after inoculation are indicated.
[b] Culture medium contained 10 mg/liter of yeast extract.

(Iwasaki, 1969b). Recently, more attention has been paid to the role of bottom sediments. Iizuka (1972) found that the Japanese dinoflagellate *Gymnodinium* (type 65) tolerated anoxic or near anoxic conditions in Ohmura Bay and utilized sulfide from the sediments. Hirayama and Numaguchi (1972), in assaying the organism, suggested that dissolution of anaerobically decomposed products of bottom mud into seawater might be one of the causes of red tide outbreaks in Ohmura Bay. Uyeno and Nagai (1973) found that mud extracts promoted the growth of *C. inlandica*, but only during the time of year when natural blooms occurred. In these semiclosed bays it is possible, as Honjo (1974) noted, that biologically active substances are derived from bottom sediments leaking through the discontinuity layer or added directly by wind-induced mixing. There are still problems involving the role of soil extracts in phytoplankton growth. Prakash and Rashid (1968) have stated that the positive effect of humic substances on dinoflagellate growth is, for the most part, independent of nutrient concentration and cannot be attributed entirely to chelation processes; growth enhancement in the presence of humic substances is apparently linked with the stimulation of algal cell metabolism. Humic substances are formed by biochemical degradation and transformation of plant and animal matter in soil or sediments (Martin and Martin, 1973). Plankton decomposition products (*Gelbstoff*) appear to have characteristics similar to those of humic substances (Kalle, 1966). Recently, Morrill and Loeblich (1977) also suggested that the stimulatory effect of soil extracts on certain algae was due to organic compounds and not to chelators or inorganic compounds.

It has been shown that the utilization by phytoplanktonic organisms of organic nutrients in the environment can be significant (Andrews and Williams, 1971) and is related to the development of blooms in certain locales (Ryther, 1954; Provasoli, 1960; North *et al.*, 1972; Prakash, 1975; Mahoney and McLaughlin, 1977). Many organic carbon, nitrogen, and phosphorus compounds are utilized by red tide flagellates, and some of them stimulate growth remarkably (Mahoney and McLaughlin, 1977). Doig and Martin (1974) reported that the enrichment of natural waters with municipal wastes increased populations of *G. breve* significantly. Domestic sewage contains all the elements essential for algal growth, almost in the same proportion as in optimal culture media (Skulberg, 1970). It is clear, therefore, that various organic compounds, especially in substances derived from organisms or their metabolites and contained in domestic and in certain kinds of industrial waste, play an important role not only as potential influences but also as growth promotors in the development of red tide blooms. Also, inorganic nitrogen and phosphorus, through phytoplankton production, contribute indirectly to the organic load in estuary or inshore coastal waters.

B. Community Interactions

Many workers have observed that bloom organisms succeed other organisms. Dinoflagellate blooms often occur in waters depleted of nutrients by preceding organisms. Hornell and Nayudu (1923) found that red tides due to peridinians occurred annually along the Malabar Coast of India after the diatom blooms, heavy rainfall, and southwestern monsoons were over. Menon (1945) reported that annual blooms of *Gymnodinium* sp. occurred along the Trivandrum Coast of India following periods of heavy rain and diatom blooms. *Oscillatoria* (=*Trichodesmium*) *ertharae* occurs prior to most, if not all, *G. breve* blooms, and possibly this alga or associated organisms such as sulfur bacteria precondition neritic waters by adding nutrients or removing undesirable nutrients and/or metabolites (Steidinger and Ingle, 1972). These observations suggest "biological conditioning" by diatoms or bacteria: the production of essential vitamins (Carlucci and Bowes, 1970), biological active secretions (Iwasaki, 1977), and/or biostimulative compounds (Oguri *et al.*, 1975).

Although much progress has been made in determining the basic growth requirements for individual red tide dinoflagellate species in culture, little is known concerning community interactions. In a study on competition between *Skeletonema costatum* and *Chattonella luteus*, Pratt (1966; referred to as *Olisthodiscus*) found that (1) each species was, at least to some degree, autoinhibitory, (2) *Skeletonema* did not inhibit *Chattonella*, and (3) *Skeletonema* was inhibited by high concentrations of *Chattonella*-conditioned medium but was stimulated by low concentrations. From these results, he suggested that *Chattonella* achieved dominance by producing large amounts of an allelochemical (of suggested tannoid nature) that inhibited *Skeletonema*, and that *Skeletonema* achieved dominance primarily by virtue of its superior reproductive rate, but that, in addition, in small competing populations, its growth was accelerated by the *Chattonella* ectocrine—a stimulant at low concentrations. In a similar experiment, Uchida (1977) reported that *P. micans* excreted a diatom-inhibiting substance, which may play an important role in favoring *Prorocentrum* blooms, and showed that the substance had a high molecular weight because it was nondialyzable and that its inhibitory effect disappeared on autoclaving. Iwasaki (1977) extended these observations to several red tide flagellates. In his experiments, 10 ml of the filtrates of media in which

each species grew axenically was added to 100 ml of fresh medium autoclaved before inoculation. The results are summarized in Table X.

The results are suggestive of species interaction, even though they are
qualitative. *Chattonella japonica* is autostimulatory; *Hornellia* sp., *G.
catenella*, *C. inlandica*, and *P. micans* are autoinhibitory to a different
degree; but *C. akashiwo* and *P. minimum* are not affected by their own
secretions. Luxuriant growth of *S. costatum* stimulates the growth of most
red tide flagellates, especially *P. minimum*. Since the control media were not
depleted of nutrients (nitrogen, phosphorus, and vitamins), it is probable
that *Skeletonema* produces some stimulating substances, and this may
help to explain the occurrence of dinoflagellates, especially of *G. catenella*, after the decline of diatom pulses. In contrast with *Skeletonema*,
G. catenella inhibits the growth of many red tide organisms. Some species stimulate or inhibit specific organisms: *C. inlandica* enhances the
growth of *P. micans* and *G. catenella*; *G. catenella* and *Pheodactylum
tricornatum* inhibit completely the growth of *C. japonica*; the growth of
C. inlandica is inhibited considerably by *G. catenella* IB, *P. micans*,
and *Rhodomonas ovalis*. Species interactions among several organisms
comprise three groups (Table XI): (1) those which are mutually advantageous, (2) those which are mutually exclusive, and (3) those which
have no effects (Iwasaki, 1977).

These results, together with differences in the physiological requirements of each species, may help to explain why monospecific flagellate
blooms exclude other phytoplankton species.

C. Concentration Mechanisms

Physical concentration mechanisms of plankton were reported by many
workers (Langmuir, 1938; Woodcock, 1944; Riley, 1952; Sverdrup, 1953;
Kierstead and Slobodkin, 1953; Chew, 1955) and referred to in explanations of red tide phenomena (Slobodkin, 1953; Ryther, 1955; Tsujita, 1955,
1968; Wyatt, 1973, 1975). The vertical stability of water may be a necessary precondition for oceanic red tides, in relation to dissipation by sink-
ing and vertical transport of plankton. A discrete water mass may also be
necessary in the case of inland sea-type red tides. Convergence at the
surface layer gives rise to the aggregation of red tide organisms and causes
red tides. Such surface oceanographic conditions are often seen in the
case of open-sea red tides. Internal waves cause red bands and streaks in
waters along the continental shelf. This concept is based on studies of the
thermocline off California and is introduced in the explanation of red tides
around Japan (Tsujita, 1968). Wyatt (1975) summarized the limitations of
models which produced lines, slicks, or patches of red tide; Eppley and

Table X Effect of Extracellular Products of Phytoplankton on the Growth of Red Tide Flagellates

Source of extracellular product	Growth (relative units)						
	Dinoflagellates			Other flagellates			
	Prorocentrum micans	Prorocentrum minimum	Gonyaulax catenella	Chattonella akashiwo	Chattonella inlandica	Chattonella japonica	Hornellia sp. J2
None	100	100	100	100	100	100	100
Skeletonema costatum	177	343	146	131	112	159	123
Phaeodactylum tricornatum	158	80	195	129	53	0	160
Rhodomonas ovalis	45	122	41	143	10	94	39
Hornellia sp. J1	93	144	84	147	56	120	2
Chattonella akashiwo	45	53	68	106	111	161	4
Chattonella inlandica	330	69	246	130	67	?	24
Chattonella japonica	100	130	100	134	21	171	21
Prorocentrum micans	72	40	90	137	8	176	40
Prorocentrum minimum	94	104	98	141	61	16	51
Gonyaulax catenella	109	95	30	76	66	0	9
Gonyaulax catenella IB	53	117	86	160	2	161	16
Polykrikos schwartzi	89	113	98	166	111	114	11

Table XI Specific Interactions among Organisms[a]

Group I (mutually advantageous relationship)	
Rhodomonas ovalis	$\overset{143}{\underset{183}{\rightleftharpoons}}$ *Chattonella akashiwo*
Chattonella akashiwo	$\overset{106}{\underset{143}{\rightleftharpoons}}$ *Chattonella inlandica*
Group II (mutually exclusive relationship)	
Hornellia sp.	$\overset{56}{\underset{24}{\rightleftharpoons}}$ *Chattonella inlandica*
Hornellia sp.	$\overset{84}{\underset{9}{\rightleftharpoons}}$ *Gonyaulax catenella*
Chattonella akashiwo	$\overset{68}{\underset{66}{\rightleftharpoons}}$ *Gonyaulax catenella*
Group III (no relationship with each other)	
Prorocentrum minimum	*Gonyaulax catenella*

[a] Numbers show growth expressed as a percentage of the amount occurring in basal medium without additions.

Harrison (1975), citing Kamykowski's model, noted that semidiurnal internal tides above a discontinuity layer would produce lines of red tide parallel to shore. These physical factors must be important concentrating agents. But concentrating mechanisms alone cannot explain bloom phenomena; such theories should be applied when causative organisms are dominant and at certain population densities.

It has also been suggested that diurnal vertical migration of dinoflagellates contributes to concentrating mechanisms (Seliger *et al.*, 1970; Seliger *et al.*, 1975; Eppley and Harrison, 1975) and the buildup of near-unialgal populations (Mulligan, 1975). Seliger *et al.* (1970) attributed the daily movement and accumulation of bioluminescent *P. bahamense* in Oyster Bay, Jamaica, to (1) the phototactic nature of the organism, (2) wind moving surface populations, and (3) poor flushing rates of upper bay reaches. Some workers have doubted the capacity of nutrient-poor waters to support bloom concentrations of dinoflagellates or the increase in population in a short time (Rounsefell and Nelson, 1966; Steidinger and Ingle, 1972). Steidinger and Ingle (1972), in observations of the 1971 summer red tide in Tampa Bay, Florida, reported that heavy cell concentrations were aided by physical factors rather than increased cell division. Many workers believe that daily phototactic migration provides a selective advantage for dinoflagellates, in that they can seek out nutrient-rich water at greater depths during the hours not favorable for photosynthesis. Eppley and

Harrison (1975) hypothesize that vertical migration, along with certain idiosyncracies in their nitrogen metabolism, provide an advantage for dinoflagellates over coastal diatoms when favorable physical conditions [steep, shallow thermoclines and a nutrient-depleted, shallow mixed layer (<10 m)] prevail.

Ryther (1955) reviewed the ecological conditions associated with red tide blooms caused by dinoflagellates. He suggested that, for the occurrence of some red tides, it was necessary only to have moderate growth of a given species and the proper hydrographic and meteorological conditions to allow an accumulation of organisms at the surface. Such an explanation may be applicable to limited bays or some offshore waters, however, it does not seem applicable to most Lower New York Bay (Mahoney and McLaughlin, 1977) and Seto Inland Sea flagellate blooms. As pointed out by Mahoney and McLaughlin (1977), with hydrodynamics acting to disperse rather than to concentrate the bloom, persistence in a bay or along a coast must depend largely on the ability of organisms to multiply rapidly.

The red tide phenomenon is not a simple one; there are more than a dozen parameters which participate in the growth of organisms; furthermore, each species has specific tolerances and preferences. However, when growth is considered a function of these variables, if the variables are within the optimal range for growth, growth will be maximal. Thus, the best combination of environmental and nutritional factors leads to an increase in a given species at first, while some biologically active substances accelerate their growth. In understanding the development of red tide blooms, it is essential to realize that different combinations of parameters are responsible for the growth explosion of each specific organism.

A number of laboratory studies indicate that marine flagellates and dinoflagellates generally thrive best under conditions of low salinity and high organic enrichment; therefore, a heavy rainfall and a discharge of domestic waste and/or some kinds of industrial waste, though the contributive rate may differ locally as a result of human activity, are related closely to outbreaks of neritic red tide blooms.

Explanation of the causative mechanisms of red tides requires a combination of biological, chemical, physical, and geological research. Essentially, a red tide is a special event in natural phytoplankton succession. We should pay more attention to the mechanisms controlling the succession. The most common question is, What factors lead to unialgal patches of red tide organisms? The problem may be explained partly by differences in the physiological requirements of each species; however, there are still many questions. For instance, what selects the causative organism— excretion of inhibitory or growth-enhancing compounds, competitive exclusion, or lack of predators or selective grazing? Though evidence is accumulating, further knowledge is required. The following research areas now seem important.

First, detailed life cycles of causative organisms in question should be examined. If cysts or resistant stages form and accumulate in sediments, our outlook on endemism and the mode of initiation of blooms must be revised by fieldwork and may offer new means of control.

Second, community interaction should be investigated. The literature shows that blooms of diatoms or blue-green algae occur prior to most, if not all, red tide blooms. We must clarify the biological conditioning mechanisms. Are they the supply of growth factors, trace metals, organometallic compounds, growth-enhancing compounds, or growth inhibitors for certain species, or the detoxification of seawater? In most cases these mechanisms remain to be determined.

Finally, growth-promoting substances in land drainage should be studied. Most dinoflagellate blooms occur in coastal waters, which suggests that some components of the drainage play an important role in the development of blooms. Therefore identification of such substances is essential in explaining the growth mechanism. For the present, annual changes in known growth factors, dissolved metals, and biostimulating substances need to be established.

ACKNOWLEDGMENTS

Thanks are expressed to Dr. Luigi Provasoli, who read the manuscript and made valuable suggestions, and to Dr. Karen Steidinger, who provided many useful references.

REFERENCES

Aldrich, D. V. (1959). *U.S. Fish Wildl. Serv. Circ.* No. 62, 69.
Aldrich, D. V., and Wilson, W. B. (1960). *Biol. Bull. Woods Hole, Mass.* **119**, 57–64.
Anderson, D. M., and Morel, F. (1978). *Limnol. Oceanogr.* **23**, 283–295.
Andrews, P., and Williams, P. J. LeB. (1971). *J. Mar. Biol. Assoc. U.K.* **51**, 111–125.
Antia, N. J., McAllister, C. D., Parsons, T. R., Stephens, K., and Strickland, J. D. H. (1963). *Limnol. Oceanogr.* **8**, 166–183.
Armstrong, F. A., and LaFond, E. C. (1966). *Limnol. Oceanogr.* **11**, 538–547.
Barber, R. T., and Ryther, J. H. (1969). *J. Exp. Mar. Biol. Ecol.* **3**, 191–199.
Barker, H. A. (1935). *Arch. Mikrobiol.* **6**, 157–181.
Bein, S. J. (1957). *Bull. Mar. Sci. Gulf Caribb.* **7**, 316–329.
Blasco, D. (1975). *In* "Toxic Dinoflagellate Blooms" (V. R. LoCicero, ed.), pp. 113–119. Mass. Sci. Technol. Found., Wakefield, Massachusetts.
Blasco, D. (1977). *Limnol. Oceanogr.* **22**, 255–263.

Bongers, L. H. N. (1956). *Meded. Landbouwhogesch. Wageningen* **56**, 1–56.
Braarud, T. (1945). *Avh. Nor. Vidensk.-Akad. Oslo, Mat. Naturvidensk. Kl.* **1** (11).
Braarud, T. (1951). *Physiol. Plant.* **4**, 28–34.
Braarud, T. (1961). *In* "Oceanography" (M. Sears, ed.), Publ. No. 67, pp. 271–298. Am. Assoc. Adv. Sci., Washington, D.C.
Braarud, T., and Pappas, I. (1951). *Avh. Nor. Vidensk.-Akad. Oslo, Mat. Naturvidensk. Kl.* **1951** (2), 1–23.
Braarud, T., and Rossavik, E. (1951). *Avh. Nor. Vedensk.-Akad. Oslo, Mat. Naturvidensk. Kl.* **1951** (1), 1–18.
Brongersma-Sanders, M. (1957). *In* "Treatise on Marine Ecology and Paleocology" (J. W. Hedgpeth, ed.), Geological Society of America, Memoir No. 67, Vol. 1, pp. 941–1010. Waverly Press, Baltimore, Maryland.
Buchanan, R. J. (1971). *Bull. Mar. Sci. Gulf Caribb.* **21**, 914–937.
Carlucci, A. F., and Bowes, P. M. (1970). *J. Phycol.* **6**, 393–400.
Carlucci, A. F., and Silbernagel, S. B. (1966). *Limnol. Oceanogr.* **11**, 642–646.
Chew, F. (1955). *Trans., Am. Geophys. Union* **36**, 963–974.
Collier, A. (1958). *Limnol. Oceanogr.* **3**, 33–43.
Cowey, C. B. (1956). *J. Mar. Biol. Assoc. U.K.* **35**, 609–620.
Doig, M. R. T., III, and Martin, D. F. (1974). *Mar. Biol.* **24**, 223–228.
Donnelly, P. V., *et al.* (1966). "Observations of an Unusual Red Tide," Mar. Lab. Prof. Pap. Ser., No. 8. Fla. Board Conserv., Tallahassee, Florida.
Droop, M. R. (1955). *J. Mar. Biol. Assoc. U.K.* **34**, 435–440.
Droop, M. R. (1957). *J. Gen. Microbiol.* **16**, 286–293.
Droop, M. R. (1958). *J. Mar. Biol. Assoc. U.K.* **37**, 323–329.
Droop, M. R. (1961). *J. Mar. Biol. Assoc. U.K.* **41**, 69–76.
Droop, M. R. (1962). *In* "Physiology and Biochemistry of Algae" (R. A. Lewin, ed.), pp. 141–159. Academic Press, New York.
Droop, M. R., McLaughlin, J. J. A., Pintner, I. J., and Provasoli, L. (1959). *Prepr., 1st. Int. Oceanogr. Congr. New York* (M. Sears ed.), pp. 916–918. Am. Assoc. Adv. Sci., Washington, D.C.
Eppley, R. W., and Harrison, W. G. (1975). *In* "Toxic Dinoflagellate Blooms" (V. R. LoCicero, ed.), pp. 11–22. Mass. Sci. Technol. Found., Wakefield, Massachusetts.
Eppley, R. W., Holm-Hansen, O., and Strickland, J. D. (1968). *J. Phycol.* **4**, 333–340.
Finucane, J. H. (1960). *U.S. Fish. Wildl. Serv., Circ.* No. 92, 52–54.
Gold, K. (1962). *J. Protozool.* **9**, Suppl., p. 10.
Gold, K., and Baren, C. F. (1966). *J. Protozool.* **13**, 255–257.
Gough, L. H. (1905). *Rep. North Sea Fish, Invest. Comm. (1902–1903)* **1**, 325–377.
Gran, H. H. (1929). *Cons. Int. Exp. Mer. Rap. Proc.-Verb.* **56**, 1–15.
Guillard, R. R. L., and Ryther, J. H. (1962). *Can. J. Microbiol.* **8**, 229–239.
Gunter, G., Williams, R. H., Davis, C. C., and Smith, G. G. W. (1947). *Ecol. Monogr.* **18**, 309–324.
Halldal, P. (1958). *Physiol. Plant.* **11**, 118–153.
Hartwell, A. D. (1975). *In* "Toxic Dinoflagellate Blooms" (V. R. LoCicero, ed.), pp. 47–68. Mass. Sci. Technol. Found., Wakefield, Massachusetts.
Hasle, G. R. (1950). *Oikos* **2**, 162–175.
Hasle, G. R. (1954). *Nytt Mag. Bot.* **2**, 139–146.
Hastings, J. W., and Sweeney, B. M. (1964). *In* "Synchrony in Cell Division and Growth" (E. Zeuthen, ed.), pp. 307–321. Wiley, New York.
Hirano, R. (1967). *Nippon Plankton Kenkyū Renrakukaihō, Commen. Number Dr. Y. Matsui*, pp. 25–29.

Hirayama, K., and Numaguchi, K. (1972). *Nippon Plankton Kenkyūkaihō* (Bull. Plankt. Soc. Jpn.) **19**, 13–21.

Hiroshima-Kenchō (1974). "Kōgai Hakusho" p. 105. Hiroshima-Ken, Hiroshima. (in Japanese).

Holmes, R. W., Williams, P. M., and Eppley, R. W. (1967). *Limnol. Oceanogr.* **12**, 503–512.

Honjo, T. (1974). *Tōkaiku Suisan Kenkyūsho Kenkyū Hōkoku* **79**, 77–121.

Hornell, J., and Nayudu, M. R. (1923). *Bull. Madras Fish.* **17**, 129–197.

Iizuka, S. (1972). *Nippon Plankton Gakkaihō* **19**, 22–33.

Ingle, R. M. (1965). *U.S. Fish Wildl. Serv., Spec. Sci. Rep.-Fish.* No. 521, p. 3.

Ingle, R. M., and Martin, D. F. (1971). *Environ. Lett.* **1**, 69–74.

Inoue, A. (1977). *Kagoshima Daigaku Suisangakubu Kiyō* **26**, 1–6.

Inoue, A, Koyama, H., and Asakawa, S. (1973). *Hiroshima Daigaku Sui-Chikusan gakubu Kiyō* **12**, 13–20.

Irie, H., and Shiokawa, T. (1966). "Akashio-ni-kansuru-Kenkyūkyōgikai," pp. 4–19., Nippon Suisanshigen-hogo-Kyōkai, Tokyo (in Japanese).

Iwasaki, H. (1969a). *Nippon Plankton Gakkaihō* **16**, 132–139.

Iwasaki, H. (1969b). *Nippon Plankton Gakkaihō* **16**, 140–144.

Iwasaki, H. (1971a). *Nippon Suisan Gakkaishi (Bull. Jpn. Soc. Sci. Fish.)* **37**, 606–609.

Iwasaki, H. (1971b). *Nippon Kaiyō Gakkaishi (J. Oceanogr. Soc. Jpn.)* **27**, 152–157.

Iwasaki, H. (1972). *In* "The Cause of Red Tide in Neritic Waters" (T. Hanaoka, ed.), pp. 77–98. Nippon Suisanshigen-hogo-Kyōkai, Tokyo (in Japanese).

Iwasaki, H. (1973). *Nippon Plankton Gakkaihō* **19**, 104–114.

Iwasaki, H. (1977). *Prepr. Nippon Suisangakkai Shunki Taikai, Tokyo,* p. 126. Nippon Suisan Gakkai, Tokyo (in Japanese).

Iwasaki, H., and Sasada, K. (1969). *Nippon Suisan Gakkaishi* **39**, 943–947.

Iwasaki, H., Fujiyama, T., and Yamashita, E. (1968). *Hiroshima Daigaku Sui-Chikusan gakubu Kiyō* **7**, 259–267.

Jitts, H. R., McAllister, C. D., Stephens, K., and Strickland, J. D. H. (1964). *J. Fish. Res. Board Can.* **21**, 139–157.

Jorgensen, E. (1911). *Int. Rev. Gesamten Hydrobiol. Hydrogr.* 4 Biol. Suppl. No. 1, 1–124.

Kain, J. M., and Fogg, G. E. (1960). *J. Mar. Biol. Assoc. U.K.* **39**, 33–50.

Kalle, K. (1966). *In* "Oceanography and Marine Biology, Annual Review 4" (H. Barnes, ed.), pp. 91–104. Allen & Unwin, London.

Kamykowski, D., and Zentara, S. (1977). *Limnol. Oceanogr.* **22**, 148–151.

Kashiwada, K., Kakimoto, D., and Kawagoe, K. (1957). *Nippon Suisan Gakkaishi* **23**, 450–453.

Kierstead, H., and Slobodkin, L. B. (1953). *J. Mar. Res.* **12**, 141–147.

Kon, S. K. (1955). *Biochem. Soc. Symp.* **13**, 17–35.

Kramer, D. D., and Ryther, J. H. (1960). *Biol. Bull. Woods Hole, Mass.* **119**, 324.

Langmuir, I. (1938). *Science* **87**, 119.

Levandowsky, M., and Hutner, S. H. (1975). *Ann. N.Y. Acad. Sci.* **245**, 16–25.

Lewin, R. A. (1954). *J. Gen. Microbiol.* **10**, 93–96.

Loeblich, A. R., III (1966). *Phykos* (Prof. Iyengar Mem. Vol.) **5**, 216–255.

Loeblich, A. R., III (1969). *J. Protozool.* **16**, Suppl., 20–21.

McAllister, C. D., Parsons, T. R., and Strickland, J. D. H. (1960). *J. Cons. Cons. Perm. Int. Explor. Mer.* **15**, 240–259.

McAllister, C. D., Parsons, T. R., Stephens, K., and Strickland, J. D. H. (1961). *Limnol. Oceanogr.* **6**, 237–258.

McLachlan, J. (1961). *Can. J. Microbiol.* **7**, 399–406.

McLaughlin, J. J. A., and Provasoli, L. (1957). *J. Protozool.* **4**, Suppl., p. 7.
McLaughlin, J. J. A., and Zahl, P. A. (1959). *Prepr., 1st. Int. Oceanogr. Congr. New York* (M. Sears, ed.), pp. 930–931. Am. Assoc. Adv. Sci., Washington, D.C.
Maclean, J. L. (1977). *Limnol. Oceanogr.* **22**, 234–254.
Mahoney, J. B., and McLaughlin, J. J. A. (1977). *J. Exp. Mar. Biol. Ecol.* **28**, 53–65.
Martin, D. F., and Martin, B. B. (1973). *In* "Trace Metals and Metal-Organic Interactions in Natural Waters" (P. C. Singer, ed.), pp. 339–362. Ann Arbor Sci. Publ., Ann Arbor, Michigan.
Martin, D. F., Doig, M. T., III, and Pierce, R. H., Jr. (1971). *Fla. Dep. Nat. Resour. Prof. Pap. Ser.* No. 12.
Menon, M. A. S. (1945). *Proc. Indian Acad. Sci., Sect.* B **22**, 31.
Menzel, D. W., and Speath, J. P. (1962). *Limnol. Oceanogr.* **7**, 51–54.
Morrill, L. C., and Loeblich, A. R., III (1977). *J. Phycol.* **13**, Suppl., p. 46.
Moshkina, L. V. (1961). *Fizio. Rast.* **8**, 172–177.
Mulligan, H. F. (1975). *In* "Toxic Dinoflagellate Blooms" (V. R. LoCicero, ed.), pp. 23–40. Mass. Sci. Technol. Found., Wakefield, Massachusetts.
Murphy, E. B., Steidinger, K. A., Roberts, B. S., Williams, J., and Jolley, J. W., Jr. (1972). *Limnol. Oceanogr.* **20**, 481–486.
Neujahr, H. (1956). *Acta Chem. Scand.* **10**, 917.
Nordli, E. (1957). *Oikos* **8**, 200–252.
Norris, L., and Chew, K. K. (1975). *In* "Toxic Dinoflagellate Blooms" (V. R. LoCicero, ed.), pp. 143–152. Mass. Sci. Technol. Found., Wakefield, Massachusetts.
North, W. J., Stephens, G. C., and North, B. B. (1972). *In* "Marine Pollution and Sea Life" (M. Ruvio, ed.), pp. 330–340. Fishing News (Books), Surrey, England.
Oguri, M., Soule, D., Juge, D. M., and Abbott, B. C. (1975). *In* "Toxic Dinoflagellate Blooms" (V. R. LoCicero, ed.), pp. 41–46. Mass. Sci. Technol. Found., Wakefield, Massachusetts.
Ohwada, K., and Taga, N. (1969). *Nippon Kaiȳo Gakkaishi* **25**, 123–136.
Ohwada, K., and Taga, N. (1972). *Mar. Chem.* **1**, 61–73.
Okaichi, T., and Yagyu, A. (1969). *Nippon Plankton Gakkaihō* **16**, 123–132.
Paredes, J. F. (1967–1968). *Mem. Inst. Invest. Cienc. Moçmbique, Ser.* A **9**, (*Cienc. Biol.*) 185–247.
Prager, J. C. (1963). *J. Protozool.* **10**, 195–204.
Prakash, A. (1967). *J. Fish. Res. Board Can.* **24**, 1589–1606.
Prakash, A. (1975). *In* "Toxic Dinoflagellate Blooms" (V. R. LoCicero, ed.), pp. 1–16., Mass. Sci. Technol. Found., Wakefield, Massachusetts.
Prakash, A., and Rashid, M. A. (1968). *Limnol. Oceanogr.* **13**, 598–605.
Prat, D. M. (1966). *Limnol. Oceanogr.* **11**, 447–455.
Provasoli, L. (1958a). *In* "Perspectives in Marine Biology" (A. A. Buzzati-Traverso, ed.), pp. 385–403. Univ. of California Press, Berkeley.
Provasoli, L. (1958b). *Annu. Rev. Microbiol.* **12**, 279–308.
Provasoli, L. (1960). *Trans. Semin. Algae Metropolitan Waters, U.S. Publ. Health Serv., Washington, D.C.*
Provasoli, L. (1963). *In* "The Sea" (M. N. Hill, ed.), Vol. 2, pp. 165–219. Wiley (Interscience), New York.
Provasoli, L., and Gold, K. (1962). *Arch. Mikrobiol.* **42**, 162–203.
Provasoli, L., and McLaughlin, J. J. A. (1955). *J. Protozool.* **2**, Suppl., p. 10.
Provasoli, L., and McLaughlin, J. J. A. (1963). *In* "Symposium on Marine Microbiology" (C. H. Oppenheimer, ed.), pp. 105–113. Thomas, Springfield, Illinois.

Provasoli, L., and Pintner, I. J. (1953). *Ann. N.Y. Acad. Sci.* **56**, 839–851.
Provasoli, L., McLaughlin, J. J. A., and Pintner, I. J. (1954). *Trans. N.Y. Acad. Sci.* **16**, 412–417.
Raymount, J. E. G. (1971). *In* "Fertility of the Sea" (J. D. Costlow, ed.), Vol. 1, pp. 1–16. Gordon & Beach, New York.
Riley, G. A. (1952). *Bull. Bingham Oceanogr. Collect.* **13**, 40–64.
Rounsefell, G. A., and Dragovich, A. (1960). *Bull. Mar. Sci.* **16**, 402–422.
Rounsefell, G. A., and Nelson, W. R. (1966). *U.S. Fish Wildl. Serv., Spec. Sci. Rep.-Fish.* No. 535.
Ryther, J. H. (1954). *Biol. Bull. Woods Hole, Mass.* **106**, 198–209.
Ryther, J. H. (1955). *In* "The Luminescence of Biological Systems" (F. H. Johnson, ed.), pp. 387–414. Am. Assoc. Adv. Sci., Washington, D.C.
Ryther, J. H. (1956). *Limnol. Oceanogr.* **1**, 61–70.
Sato, T., Takeichi, Y., and Adachi, R. (1966). "Akashio-ni-kansuru-Kenkyūkyōgikai" pp. 84–94. Nippon Suisanshigen-hogo-Kyōkai, Tokyo.
Schilo, M., and Rosenberger, F. (1960). *Ann. N.Y. Acad. Sci.* **90**, 866–876.
Seliger, H. H., Carpenter, M., Loftus, M., and McElroy, W. D. (1970). *Limnol. Oceanogr.* **15**, 234–245.
Seliger, H. H., Loftus, E. E., and Subba Rao, D. V. (1975). *In* "Toxic Dinoflagellate Blooms" (V. R. LoCicero, ed.), pp. 181–205. Mass. Sci. Technol. Found., Wakefield, Massachusetts.
Skulberg, O. (1970). *Helgol. Wiss. Meeresunters.* **20**, 111–125.
Slobodkin, L. B. (1953). *J. Mar. Res.* **12**, 148–155.
Small, L. F., and Ramberg, D. A. (1971). *In* "Fertility of the Sea" (J. D. Costlow, ed.), Vol. 1, pp. 475–492. Gordon & Beach, New York.
Soli, G. (1966). *Limnol. Oceanogr.* **11**, 353–363.
Sribhibhadh, A. (1963). Ph.D. Thesis, Univ. of Washington, Seattle.
Steidinger, K. A. (1973). *Crit. Rev. Microbiol.* **3**, 49–68.
Steidinger, K. A. (1975). *In* "Toxic Dinoflagellate Blooms" (V. R. LoCicero, ed.), pp. 153–162. Mass. Sci. Technol. Found., Wakefield, Massachusetts.
Steidinger, K. A., and Ingle, R. M. (1972). *Environ. Lett.* **3**, 271–278.
Steidinger, K. A., and Williams, J. (1970). *Mem. Hourglass Cruises* **2**, 1–251.
Steven, D. M. (1966). *J. Mar. Res.* **24**, 113–123.
Sverdrup, H. U. (1953). *J. Cons., Cons. Perm. Int. Explor. Mer.* **18**, 287–295.
Sweeney, B. M., and Hastings, J. W. (1958). *J. Protozool.* **5**, 217–224.
Swift, E., and White, H. (1973). *J. Phycol.* **9**, Suppl. Abstr., p. 4.
Thomas, W. H. (1966a). *J. Phycol.* **2**, 17–22.
Thomas, W. H. (1966b). *Limnol. Oceanogr.* **11**, 393–400.
Tsujita, T. (1955). *Seikaiku Suisan Kenkyūsho Kenkyū Hōkoku* No. 6, 12–58.
Tsujita, T. (1968). *Nippon Plankton Gakkaihō* **15**, 1–10.
Uchida, T. (1975). *Nippon Plankton Gakkaihō* **22**, 11–16.
Uchida, T. (1977). *Jpn. J. Ecol. (Nippon Seitai Gakkaishi)* **27**, 1–4.
Ueno, S., Iwasaki, H., and Fujiyama, T. (1977). *Nippon Plankton Gakkaihō* **24**, 94–98.
Uyeno, F., and Nagai, K. (1973). *Nippon Plankton Gakkaihō* **19**, 39–45.
Wardle, W. J., Ray, S. M., and Aldrich, A. S. (1975). *In* "Toxic Dinoflagellate Blooms" (V. R. LoCicero, ed.), pp. 257–264. Mass. Sci. Technol. Found., Wakefield, Massachusetts.
Wheeler, B. (1964). *Bot. Mar.* **9**, 15–17.
Wilson, W. B. (1966). *Fla. Board. Conserv., Prof. Pap. Ser.* No. 7.
Wilson, W. B., and Collier, A. (1955). *Science* **121**, 394–395.

Woodcock, A. H. (1944). *J. Mar. Res.* **5**, 196–205.

Wyatt, T. (1973). *Int. Counc. Explor. Sea,* C.M. 1973/L:12 (mimeo).

Wyatt, T. (1975). *In* "Toxic Dinoflagellate Blooms" (V. R. LoCicero, ed.), pp. 81–94. Mass. Sci. Technol. Found., Wakefield, Massachusetts.

Yentsch, C. M., Cole, E. J., and Salvaggio, M. G. (1975). *In* "Toxic Dinoflagellate Blooms" (V. R. LoCicero, ed.), pp. 163–180. Mass. Sci. Technol. Found., Wakefield, Massachusetts.

Zubkoff, P. L., and Warinner, J. E., III (1975). *In* "Toxic Dinoflagellate Blooms" (V. R. LoCicero, ed.), pp. 105–111. Mass. Sci. Technol. Found., Wakefield, Massachusetts.

APPENDIX TO CHAPTER 12

On a Class of Mathematical Models for
Gymnodinium breve Red Tides

M. LEVANDOWSKY

In this appendix we shall examine briefly the properties of a certain restricted class of mathematical models that has been proposed for certain red tides. Kierstead and Slobodkin (1953) proposed a simple initial condition for the appearance of *Gymnodinium breve* blooms off the gulf coast of Florida. They suggested that this species required "conditioned" seawater for growth; such conditioning, which might be due to the presence of vitamins or perhaps to organic metal-binding compounds, was produced in shallow bays and mangrove swamps, and the water was swept out to sea in patches, particularly during heavy runoff after rain.

It was then assumed that *G. breve* could grow only in the patches of suitably conditioned water, and would also tend to diffuse away from these by random swimming. These conditions led to a linear partial differential equation for the population density, $N(\mathbf{p}, t)$, at a point in space $\mathbf{p}$ and time t:

$$\frac{\partial N}{\partial t} = D\nabla^2 N + rN \tag{1}$$

Here $\nabla = (\partial/\partial x) + (\partial/\partial y)$ is the gradient operator, D is a diffusion coefficient, and r is a constant; the first term on the right is a simple diffusion, which can be derived from the assumption that the cells swim in a random walk. Similar models are used elsewhere in physics (and physiology) to describe molecular diffusion and the flow of heat.

Note that this first term does *not* refer to "turbulent diffusion." Kierstead and Slobodkin were quite explicit that this is a *biological* diffusion. The model assumes there is no motion of the medium (and in fact *G. breve* red tides seem to appear mainly in dead calm weather); this important point is emphasized here because subsequent authors have not always appreciated it.

The second term on the right models simple Malthusian, or exponential growth, with a constant growth rate r. Since Kierstead and Slobodkin

BIOCHEMISTRY AND PHYSIOLOGY OF PROTOZOA
SECOND EDITION, VOL. 1

were only interested in initial conditions, this appeared adequate for their purposes (i.e., at low values of N, one would expect the linear term in a power series expansion of a more general growth function to predominate).

The problem is essentially two dimensional (we assume the dinoflagellates remain at the surface as a result of behavioral responses to light and/or gravity), but for purposes of the present analysis the essential features of a radially symmetric two-dimensional patch do not differ qualitatively from those of the corresponding one-dimensional problem,

$$\frac{\partial N}{\partial t} = D \frac{\partial^2 N}{\partial x^2} + bN \tag{1'}$$

(in the actual analysis, it is a question of using Bessel functions for the radially symmetric two-dimensional case and Fourier series for the one-dimensional one; Kierstead and Slobodkin present details of both cases); we will therefore consider the one-dimensional problem in what follows.

Boundary conditions for the problem given by (1), or (1'), are that the density N go to zero at the boundary of the patch. In the one-dimensional version, (1'), if the one-dimensional patch starts at $x = 0$ and stops at $x = L$, we require that

$$N(0,t) = N(L,t) = 0 \tag{2}$$

The problem is now mathematically defined and of a routine kind in applied mathematics, which can be solved easily by standard methods. Kierstead and Slobodkin did this, showing that there is a critical patch size, L_0, such that, for $L < L_0$ diffusion dominates growth and the density N goes to zero everywhere. For $L > L_0$ on the other hand, growth dominates diffusion and N increases without bound within the patch. The patch size threshold L_0 was therefore viewed as an initial condition for the onset of a red tide.

Subsequent writers have considered modifications of this model. Wroblewski et al. (1975) suggested that the effects of predation be considered. They used the predation function of Ivlev (1945):

$$p(N) = b(1 - e^{-c(N-N_0)+}) \tag{3}$$

where N_0 is a threshold below which predation does not occur. This function has the graph shown in Figure 1, and can be derived heuristically from certain assumptions regarding the mechanisms of predation. Biologically, it resembles the mathematically smoother type III predation function of Holling (1965). Adding (3) to (1') then gives the new model:

$$\frac{\partial N}{\partial t} = D \frac{\partial^2 N}{\partial x^2} + rN - b(1 - e^{-c(N-N_0)+}) \tag{4}$$

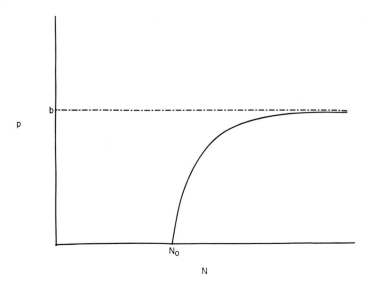

Figure 1. Ivlevian predation, with prey threshold N_0 and predation saturation level b.

Another way to make the original model more realistic is by replacing the Malthusian growth term by self-limiting logistic growth, giving the model

$$\frac{\partial N}{\partial t} = D \frac{\partial^2 N}{\partial x^2} + rN\left(1 - \frac{N}{K}\right) \tag{5}$$

This equation arises also in an entirely different context in genetics, where it is called Fisher's equation; its properties were studied by Kolmogorov *et al.* (1937).

Combining the effects in (4) and (5) gives the composite model

$$\frac{\partial N}{\partial t} = D \frac{\partial^2 N}{\partial x^2} + rN\left(1 - \frac{N}{K}\right) - b(1 - e^{-c(N-N_0)+}) \tag{6}$$

We shall now examine some pertinent mathematical properties of (4), (5), and (6). First, they all have threshold patch sizes for growth of the population.

Next, we inquire as to nonzero steady state solutions. Instead of the density, N, at a point, it is now convenient to follow an extensive variable such as the total patch population, or the average density in the patch; for heuristic reasons we choose instead the maximum density $N_{\max}$ in the patch, which in a steady state ($\partial N/\partial t = 0$) will be found in the center of the patch, where the effects of diffusion are least and those of growth are greatest. $N_{\max}$ is then a function of the patch size L.

In the predation model (4), we look first at the case where $N_0 = 0$ (no predation threshold). In this case, as Wroblewski *et al.* (1975) observed, the critical patch size L_1 is greater than L_0, the critical patch size for (1'). For $L < L_1$, zero is a steady state solution, as before, but now at L_1 the steady state solution *bifurcates;* zero is still a steady state solution (a zero population will not grow), but there is also another branch to the steady state solution. As seen in Figure 2, this branch is unstable: for points above it the predators are effectively "saturated," and the population grows without bound as in (1'); below it the effects of predation and diffusion combine to eliminate the population ($N_{max} = 0$ at steady state).

For Fisher's equation, (5), we have the situation in Figure 3. A stable branch of the steady state solution appears at the threshold L_0, the bifurcation point. This solution is asymptotic to the horizontal line $N_{max} = K$, corresponding to the fact that, for very large patches, diffusion will hardly be noticed in the center of the patch, and N_{max} approaches the logistic carrying capacity K.

For the combined model (6), still with $N = 0$, we have the situation in Figure 4. The upper branch of the bifurcating steady state solution can now have both stable and unstable regions, depending on the relative importance of predation and self-damping.

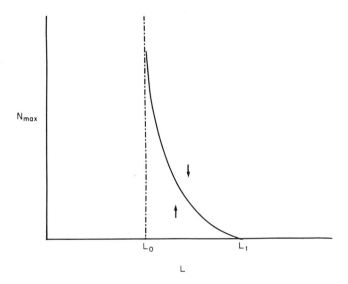

Figure 2. Bifurcation diagram for a model with spatial diffusion, predation, and exponential growth. L_0 is the spatial threshold in the absence of predation (Kierstead–Slobodkin model). The solid curve emanating from L_1 shows the spatial threshold with Ivlevian predation for various values of N_{max} and no predation threshold ($N_0 = 0$).

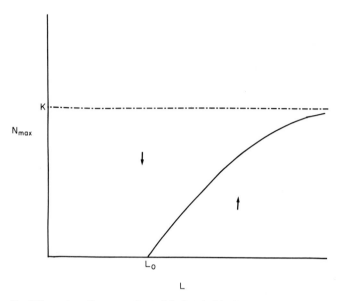

Figure 3. Bifurcation diagram of spatial thresholds for density-dependent (logistic) growth in the absence of predation (Fisher's equation).

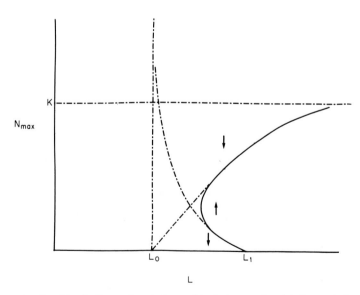

Figure 4. Combined effects of predation and logistic growth, with the predation threshold, N_0, equal to zero.

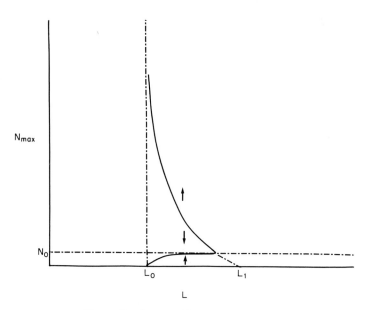

Figure 5. Same as Figure 2, with $N_0 \neq 0$.

Now we consider the somewhat more realistic case where $N_0 > 0$ (a nonzero threshold for predation). The effects of this for Eqs. (4) and (6) are shown in Figures 5 and 6, respectively. The predation threshold, $N_0 > 0$, now permits a small nonzero steady state population in patches below L_1 but above L_0. The nonzero branches of the steady state solutions have both stable and unstable regions. In particular, in Figure 6, we see that for some values of L there are two possible steady state solutions; the lower one is predator limited, and the upper one is resource limited. Between them is an interesting unstable steady state, corresponding to an inoculum size threshold above which the predators are saturated.

Now consider the effects of another little bit of realism; we add a term $\epsilon g(x)$, where $g(x)$ is unknown, but belongs to a large class of positive functions, and ϵ is a suitably small constant. This term could correspond to the effects of recruitment from a residual "seed" population of dormant stages such as spores. The effects of adding such a term are shown in Figures 7 and 8, for Fisher's equation and Ivlev predation with zero threshold, respectively. In Figure 7, we see that the original sharp threshold is now blurred. In Figure 8, the observable threshold is not only blurred, but also can be substantially decreased. Thus, by adding small amounts of realism to our model, step by step, we have caused great changes in the qualitative, topological properties of its steady state solu-

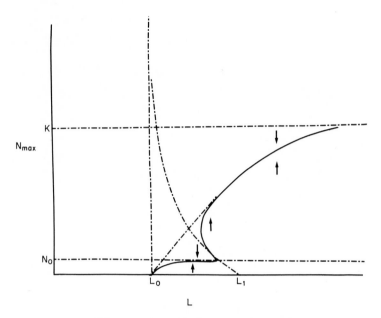

Figure 6. Same as Figure 4, with $N_0 \neq 0$.

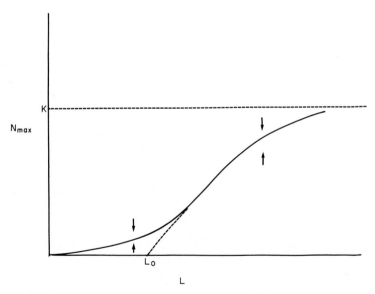

Figure 7. Addition of a small term to the model shown in Figure 3, to model effects of spore recruitment.

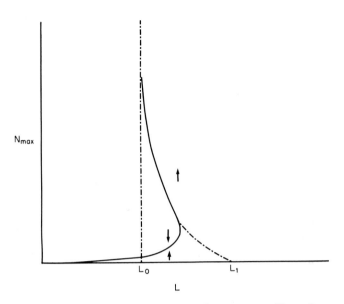

Figure 8. Addition of spore recruitment term to Figure 2.

tions. Since the model is still quite idealized, this seems to suggest that great caution is required in interpreting such solutions.

It should be noted, for example, that we have only considered modifications in the growth term of the model, and have assumed simple diffusion due to random swimming in the first term on the right. Actually, we know from experimental studies that dinoflagellates have a rich behavioral repertoire, and respond to chemosensory cues; their behavior at the boundary of the patch may therefore be far from random.

The original model of Kierstead and Slobodkin only addressed the question of initial conditions for red tides. Later events in such population irruptions may involve other factors, as suggested many years ago by Collier (1958). The upper branches of steady state solutions may therefore be actually somewhat transient conceptions, and a different sort of model may perhaps be required for later stages of a red tide.

It should be stressed also that the class of models described here arose originally in response to the problem of *G. breve* blooms. These appear to have peculiar properties not found with red tides of other species, such as the initial condition for quiet water. It is therefore both unfair and unrealistic to demand that these models apply to other species as well. Other species will no doubt require other sorts of models, which among other things may need to incorporate turbulence and advective mixing.

ACKNOWLEDGMENT

The results presented here arise from a collaboration with Professor E. Reiss.

REFERENCES

Collier, A. (1958). *Limnol. Oceanogr.* **3**, 33–39.
Holling, C. S. (1965). *Mem. Entomol. Soc. Can.* **45**, 5–60.
Ivlev, V. S. (1945). *Usp. Sovrem. Biol.* **19**, 88–120.
Kierstead, H., and Slobodkin, L. B. (1953). *J. Mar. Res.* **12**, 141–146.
Kolmogorov, A., Petrovskii, I., and Piscunov, N. (1937). *Bull. Univ. Moscow, Ser. Int., Sect. Al* **1**, 1–25.
Wroblewski, J., O'Brien, J. J., and Platt, T. (1975). *Mem. Soc. R. Sci. Liege, Collect.* **7**, 43–57.

Poisonous Dinoflagellates   13

EDWARD J. SCHANTZ

The dinoflagellates constitute an important group of one-celled microorganisms that inhabit both fresh and marine waters. Over 1000 different species are known, but only a few of them, mainly marine species, produce substances that are extremely poisonous to humans and most animals. That these tiny organisms produce such deadly poisons became known through reports of food poisoning resulting from the eating of certain shellfish such as mussels and clams that had fed on these organisms. The dinoflagellates that produce poisons and the properties of these poisons are discussed in this chapter.

I. DISTRIBUTION OF KNOWN SPECIES OF POISONOUS DINOFLAGELLATES

Poisonous dinoflagellates are found throughout the world. Those that produce the deadly paralytic poison are the armored type and are usually

BIOCHEMISTRY AND PHYSIOLOGY OF PROTOZOA
SECOND EDITION, VOL. 1

found to bloom in cool waters along seacoasts at latitudes greater than about 30° north and south. Others that produce different poisons bloom in tropical or semitropical waters. The distribution of any one species is by no means general but is limited to areas that provide the various environmental conditions and nutrients required by the particular species. In the northern hemisphere *Gonyaulax catenella,* the predominant poisonous dinoflagellate along the Pacific coast of North America, occurs sporadically and causes shellfish to become poisonous in certain coastal areas of central California, northward along the coasts of Oregon, Washington, British Columbia, and southeastern Alaska, and westward along the Aleutian Islands to Japan. In the southern hemisphere this organism causes shellfish to become poisonous along the southern coasts of Chile and South Africa.

The relationship between a particular dinoflagellate and poisonous shellfish was first observed by Sommer and associates at the University of California Medical Center, San Francisco, during an outbreak of poisoning in humans caused by eating sea mussels collected near San Francisco in 1927 (Sommer and Meyer, 1937). These investigators discovered that a particular dinoflagellate was present in the water surrounding the mussel beds at the time the mussels were poisonous and found that acidified water extracts of these organisms killed mice after producing symptoms similar to those caused by extracts of poisonous mussels. Sommer *et al.* (1937) identified this organism as *G. catenella* Whedon & Kofoid. Sommer placed nonpoisonous mussels in laboratory cultures of *G. catenella* and found that they soon acquired poisonous properties from feeding on the organisms. After the mussels were removed from this culture and placed in a culture of nonpoisonous organisms the poison disappeared within a week or 10 days. These experiments duplicated the natural occurrence of poisonous mussels and explained why they appeared sporadically, remained poisonous for a period of 1–3 weeks, and then became nonpoisonous and again safe to eat.

The discovery of the relationship of *G. catenella* to poisonous mussels in California led to similar discoveries of other dinoflagellates that caused shellfish to become poisonous. Koch (1939) found *Pyrodinium phoneus* Woloszynska & Conrad to be responsible for the extreme toxicity of mussels in Belgium. Needler (1949) and Prakash (1963, 1967) established that the poison in scallops in the Bay of Fundy and in clams and mussels along the North Atlantic coast of the United States and in the St. Lawrence estuary was caused by the dinoflagellate *Gonyaulax tamarensis* Lebour. This organism caused outbreaks of shellfish poisoning along the northeastern coast of England in the early summer of 1968 and has bloomed sporadically in this area since that time. It also caused shellfish to become poisonous along the coast of New England in 1972 and has persisted in this area since then. Prakash and Taylor (1966) found poison in another species, *Gonyaulax acatenella* Whedon & Kofoid, which occurs along the coast of British Columbia and has caused shellfish in this area to become poisonous. This organism has many properties in common with *G. tamarensis,* including size, shape, and arrangement of its armor plates, and also shares some common features with *G. catenella.* Halstead (1965) has presented an excellent description of the toxic marine organisms throughout the world.

The above organisms are the only dinoflagellates known that produce the paralytic type of poison and cause shellfish to become poisonous. There are other dinoflagellates that produce other types of poisons. Schradie and Bliss (1962) reported that *Gonyaulax polyedra* Stein, which occurs along the southern coast of California, was poisonous and that the poison was similar in certain respects to that produced by *G. catenella.* When this organism is grown in culture, it does not produce poison, which indicates that poison is produced only under certain natural conditions if at all. It has never been implicated in any type of shellfish poisoning. Connell and Cross (1950) and Gates and Wilson (1960) reported that *Gonyaulax monilata,* a dinoflagellate common in the Gulf of Mexico, produced a poison that was toxic to fish. Ray and Aldrich (1967) found that this organism was not toxic to chicks and mice. Oysters in the Gulf of Mexico do not filter water when *G. monilata* is present, which may account for their survival. Abbott and Balantine (1957) found that *Gymnodinium veneficum* Ballantine, isolated from the English Channel, produced a poison that was toxic to both fish and mice but had not been known to cause shellfish poisoning. Starr (1958), Ray and Aldrich (1965), Spikes *et al.* (1968), and Martin and Chatterjee (1969) found that one of the naked dinoflagellates, *Gymnodinium breve,* produced a poison toxic to fish, chicks, and mice and often bloomed in the Gulf of Mexico, particularly along the western coast of Florida. *Gonyaulax monilata* and *Gymnodinium breve* are responsible for many of the poisonous red tides and the enormous fish kills in the Gulf of Mexico. Nakazima (1968) has reported another dinoflagellate, *Exuviaella mariae-lebouriae,* occurring in certain areas around Japan, that has caused oysters to become poisonous. There may be other dinoflagellates throughout the world that produce poisons, but we lack information about them. Some evidence indicates that the cause of ciguatera poisoning in humans, which results from eating certain tropical or subtropical coral reef fishes, is due to a poison produced by a marine dinoflagellate (Yasumoto *et al.,* 1978; Schurer 1977) which reaches larger fish through the consumption of small fish that feed on plankton.

It should be mentioned at this point that some algae produce poisons that appear to be much like some of the poisons produced by dinoflagellates. *Aphanizomenon flos-aquae*, a blue-green alga, produces a poison very similar to the poison produced by *G. catenella* (Sawyer *et al.*, 1968; Jackim and Gentile, 1968). Konosu *et al.* (1968) and Noguchi *et al.* (1969) have described a poison in certain crabs from Japan that is identical to the poison produced by *G. catenella*. The latter case could be due to the crabs consuming debris from other toxic organisms that have fed on *G. catenella*. Table I lists the known poisonous dinoflagellates, where they have been found, and some of their important properties. *Gonyaulax catenella* has been grown in axenic culture, demonstrating that the poison is a metabolic product of the organism not requiring the symbiotic effects of other organisms (Burke *et al.*, 1960).

When shellfish such as mussels and clams consume poisonous dinoflagellates, the poison is bound in the dark gland or hepatopancreas and evidently causes no visible change in appearance and no apparent harm to the physiological functions of the shellfish. Along the California coast mussels became too poisonous for human consumption when 200 or more *G. catenella* cells per milliliter of water were present. As counts of this organism arose into the thousands, the mussels became extremely poisonous, and a small mussel weighing about 100 gm remaining in the water for 3 or 4 days was found to contain as much as 25,000 mouse units (MU) or 4–5 mg of poison. As the number of poisonous organisms in the water decreased to a low level, the poison in the mussels decreased, and within 1–2 weeks they were practically free of poison and again safe for human consumption. The amount of poison in a mussel is the result of a dynamic equilibrium between the rate at which it is consumed and bound in the dark gland and the rate at which it is destroyed or excreted. The binding is not particularly strong, because the poison is readily released in a weak acid solution at pH 2–3. The mechanism by which the poison is bound is not known. Usually 95% or more of the poison in a mussel is found in the dark gland. The Alaska butter clam and probably a few other species are exceptions, in that most of the poison, 60–80%, is bound in the siphon. It is believed that the poison is taken up by the dark gland and then moves to the siphon, because poison was found in the dark gland before it was found in the siphon when nonpoisonous clams were transplanted to areas where clams become poisonous (Schantz and Magnusson, 1964). In the siphon the poison is destroyed or detoxified at a very slow rate and, even with a slow rate of uptake of poisonous dinoflagellates, butter clams may remain poisonous for a year or more. If conditions become favorable for the poisonous dinoflagellate to bloom each year in a certain area, clams could remain poisonous for many years at a time. The origin of the poison in the Alaska butter clam has not been definitely established, because of

Table I Distribution of the Known Poisonous Dinoflagellates and Some Important Properties of the Poisons[a]

Dinoflagellate	Usual distribution	Properties of the poisons
Gonyaulax catenella from central California (produces saxitoxin)	Coasts of countries along the North Pacific Ocean—central California to Japan; coasts of Chile and South Africa	Cause of paralytic shellfish poisoning (Sommer and Meyer, 1937); poison is a nitrogenous base of MW 372 as dihydrochloride; structure characterized (Schantz et al., 1975); very water-soluble: heat-stable; among most poisonous substances; acts by blocking sodium channel in nerve and muscle cell membranes
Gonyaulax tamarensis from New England coast (produces 11-hydroxy-saxitoxin sulfate, some saxitoxin some unidentified poisons)	Coasts along New England and Canadian maritime provinces; coasts of countries bordering on the North Sea	Cause of paralytic shellfish poisoning (Needler, 1949; Prakash, 1963); structure characterized (Shimizu et al., 1976; Boyer et al., 1978); physiological action identical to that of saxitoxin
Gonyaulax acatenella (produces poisons similar to *G. tamarensis*)	Coast of British Columbia	Cause of paralytic shellfish poisoning; properties of species similar to those of *G. catenella* and *G. tamarensis* (Prakash and Taylor, 1966); poison not characterized but similar to saxitoxin
Pyronidinium phoeneus	North Sea	Cause of paralytic shellfish poisoning (Koch, 1939); poison not isolated or characterized
Gonyaulax monilata	Gulf of Mexico	Poisonous to fish but not to warm-blooded animals (Ray and Aldrich, 1967); poison not characterized
Gonyaulax polyedra	Coast of southern California	Toxicity reported by Schradie and Bliss (1962). Never known to cause any type of shellfish or fish poisoning or to produce poison in culture
Gymnidinium breve	Gulf of Mexico—western coast of Florida	Cause of shellfish poisoning that appears similar to ciguatera poisoning (McFarren et al., 1965; poison partly characterized: Ray and Aldrich, 1965); action involves depolarization of nerve cells
Gymnidinium veneficum	English Channel	Toxic to fish and mice: poison is water-soluble and of high MW; poison partly characterized: action involves depolarization of nerve cells (Abbott and Ballantine 1957)
Exuviaella mariae-lebouriae	Japan	Causes shellfish poisoning; poison partly characterized: causes fatty degeneration of liver and kidney tissue. (Nakazima, 1968)

[a] Recent reports in the literature indicate considerable confusion about the proper identification of the poisonous dinoflagellates. *Gonyaulax tamarensis* has been confused with *G. excavata* and *G. acatenella* (Loeblich, and Loeblich, 1978). *Exuviaella mariae-lebouriae* has been confused with *Prorocentrum* var. *mariae-lebouriae* (Okaichi and Imatomi, 1978) An unidentified species of a dinoflagellate has caused paralytic shellfish poisoning in the region of Ubatuba, Brazil (Kutner and Sassi, 1978), and in Northeastern Venezula (Reyes-Vasquz and Ferraz-Reyes, 1978). New criteria for the classification of the poisonous dinoflagellates are being considered.

the difficulty in locating a sufficient number of poisonous dinoflagellate cells in the masses of other plankton growing in waters around clam beds to account for the amount of poison in the clams. Some evidence indicates that the source is from the consumption of *G. catenella* (Schantz and Magnusson, 1964; Prakash and Taylor, 1966).

No logical explanation is known for the production of poisons that are lethal to humans and animals, such as the paralytic ones produced by species of *Gonyaulax*. Their predators have evolved through the ages a mechanism to nullify the effects of the poisons. One might speculate that these poisons regulate the passage of certain ions through cell membranes of the dinoflagellate. If this is a necessary function in the life of a dinoflagellate, it is observed that most species carry out the function with a substance that is nonpoisonous to humans and animals.

II. DISEASES OF HUMANS AND ANIMALS CAUSED BY DINOFLAGELLATE POISONS

A. Nature of the Disease

A variety of diseases results from the poisons produced by dinoflagellates. The most lethal of these diseases is paralytic shellfish poisoning which results from the consumption of shellfish that have fed on *G. catenella, G. tamarensis,* or *G. acatenella.* This type of poisoning is often called mussel or clam poisoning, and in some areas the disease is termed mylitointoxication.

The first symptoms of paralytic shellfish poisoning include a tingling sensation and numbness in the lips, tongue, and finger tips and may be apparent within a few minutes after eating poisonous shellfish. This sensation is followed by a feeling of numbness in the legs, arms, and neck, with general muscular incoordination. A feeling of lightness, as though floating in air, is often described by the afflicted person. Other associated symptoms are dizziness, weakness, drowsiness, incoherence, and headache. As the illness progresses, respiratory distress and muscular paralysis become more and more severe, and death, apparently as a result of respiratory paralysis, occurs within 2–12 hr, depending upon the size of the dose. If one survives 24 hr, the prognosis is good, and normal functions are regained within a few days. There is no effective antidote for shellfish poisoning. In cases where humans have had the misfortune to collect and eat poisonous shellfish, emesis should be induced immediately after the first symptoms appear. If respiratory distress becomes apparent, artificial respiration should be applied and continued until recovery or death.

Meyer (1953) believed that artificial respiration saved persons who had consumed marginal doses of the poison, but that larger doses caused death regardless of the treatment given.

The amount of paralytic poison that causes death in humans is not known exactly. On the occasion of some accidental poisonings along the California coast, Meyer (1953) estimated the dose causing death by counting empty shells left by deceased persons to determine the number of mussels eaten and assaying the remaining mussels to determine the amount of poison they contained. In this way he estimated that the dose was at least 20,000 MU or between 3 and 4 mg. Canadian investigators (Tennant *et al.*, 1955; Bond and Medcof 1958), however, estimated the lethal dose to be much lower, with a minimum of about 3000 MU or 0.5–1 mg. The lower dose was believed to cause death in persons that very seldom ate clams and had no tolerance for the poison. The deaths along the California coast occurred in persons that normally ate mussels regularly and most likely had consumed subclinical or sublethal doses of the poison, thereby building up a tolerance. Humans appear to be much more susceptible to oral doses of the poison than most animals. The reaction to a certain oral dose depends to a great extent on whether the mussels are eaten with other food that might delay absorption of the poison.

Another type of shellfish poisoning resulting from the poison produced by the dinoflagellate *E. mariae-lebouriae,* called oyster poisoning or asari poisoning, has occurred in the region of Hamana Bay in Japan and has resulted in much sickness and many deaths among people in this area. The disease is characterized by fatty degeneration of the liver and kidney tissue in animals. The first symptoms in humans are anorexia, abdominal pain, nausea, vomiting, constipation, and headache occurring within the first few days. These symptoms are followed by hemorrhagic spots on the skin, bleeding from the mucous membranes, and acute yellow atrophy of the liver, which results in death in many cases. The toxic principle is found in the liver or dark gland of the bivalves and has been isolated by Japanese investigators. It is quite stable to heating and causes intoxication in many animals by both the oral and intraperitoneal routes. Although the occurrence of this organism seems to be limited to one particular area of Japan, the possibility of its occurring in other areas of the world should not be overlooked by public health officials and persons in the shellfish industry.

Other dinoflagellates listed in Table I produce poisons, but their relationship to a specific type of shellfish poisoning has not been established. McFarren *et al.* (1965) reported that blooms of *G. breve* along the western coast of Florida caused oysters to become toxic; when they were consumed by humans, they produced symptoms similar to those of ciguatera poisoning. The greatest loss of animal life due to *G. breve* is caused by the poison it produces, which results in massive fish kills along the western coast of Florida.

B. Physiological Action of Dinoflagellate Poisons

The paralytic poisons produced by *G. catenella, G. tamarensis,* and *G. acatenella* are neurotoxins and produce paralysis by blocking the passage of an impulse along a nerve axon or muscle fiber. The block is due to the poison (saxitoxin and related poisons) binding to the sodium channel in the muscle or nerve cell membrane and preventing sodium ions from entering the cell. The poison acts specifically on sodium channels (Evans, 1964; Kao and Nishiyama, 1965). It does not bind to other channels, or at least in such a way that ions other than sodium are affected in their passage in or out of the cell. The major poison produced by *G. tamarensis,* 11-hydroxysaxitoxin sulfate, acts exactly like saxitoxin (Narahashi *et al.,* 1975). The action is similar to that of tetrodotoxin (an aminoperhydroquinazoline-type compound) from the puffer fish.

The specific action of the hepatotoxin from the dinoflagellate *E. mariaelebouriae* is not known, except to say that it causes atrophy of liver and kidney tissues. The specific action of the toxin produced by *G. breve* that kills fish is not known exactly, but depolarization of nerve cells is involved. The toxin produced by *G. veneficum* causes depolarization of nerve cells.

III. DETECTION OF THE POISONS PRODUCED BY DINOFLAGELLATES AND FOUND IN SHELLFISH

A. Paralytic Poisons (Saxitoxin and Related Substances)

Because these poisons cannot be detected by the appearance or the taste of the shellfish, to serve as a guide to persons collecting shellfish for food, it is important that reliable tests be available for their detection. The only dependable and the most practical method of detecting the paralytic poison in shellfish is by the mouse bioassay originally devised by Sommer and Meyer (1937). These investigators quantitatively defined a mouse unit as the minimum amount of poison required to kill a 20-gm mouse in 15 min when 1 ml of an aqueous acid extract (about pH 4) of the ground dark gland of a mussel was injected intraperitoneally. The curve relating time of death to mouse units may be constructed from the following data.

Death times of 3, 4, 5, 6, 7, 8, and 15 min are equivalent to 3.7, 2.5, 1.9, 1.6, 1.4, 1.3, and 1 MU, respectively. If the logarithm of the dose is plotted against the reciprocal of the time, a straight line is obtained. The dose may be calculated directly from the equation

$$\text{Log dose} = \frac{145}{t} - 0.2$$

where t is the time to death in seconds and death occurs between 240 and 480 sec. Although the mouse unit was originally defined in terms of the amount required to kill a 20-gm mouse in 15 min, the most consistent results are obtained when the death time is between 240 and 480 sec (Schantz et al., 1958). These times represent the portion of the death–time response curve where the dose is most accurately determined. The time of death is defined as the time from challenge to the last gasping breath of the mouse. The weight of the mouse is a factor in the quantitative assay. Usually mice weighing between 19 and 21 gm are used, where the variation due to weight is insignificant or about $\pm 3\%$. Mice weighing 17 gm die from an average dose of 0.88 MU, and those weighing 23 gm die from an average dose of 1.07 MU. Solutions to be assayed should be between pH 3 and 4.

Studies in cooperation with the U.S. Public Health Service have resulted in modifications of the assay procedure and include using the purified shellfish poison (saxitoxin) as a reference standard (Schantz et al., 1958). The mouse unit is a variable quantity depending on the species and condition of the mice and the technique used by the assayer. When the response of mice is expressed in terms of a definite weight of poison in the reference standard, the results of assays from various laboratories have been found to check very well. This procedure has been made the official method of assay for paralytic shellfish poison by the Association of Official Analytical Chemists (1975).* The assay is affected by the presence of salts and substances like ethanol. A 1% solution of sodium chloride in the assay solution increases the death time sufficiently to reduce the assay results by 50%. Salt concentrations of less than 0.1% seem to cause no trouble. The presence of ethanol also increases the death time if it is present in concentrations greater than 5%.

When shellfish are collected for assay, the contents of the shell (meat and viscera) are removed from a few mussels or clams and weighed. They are ground and extracted by heating to the boiling point in a weak solution

---

* The reference standard may be obtained free of charge by writing to the Food and Drug Administration, Division of Microbiology, 200 C Street S.W., Washington D.C. 20204.

of hydrochloric acid at about pH 2. Serial dilutions of the clarified extracts, adjusted to between pH 3 and 4, are injected intraperitoneally into white mice weighing between 19 and 21 gm. From the death times described above, the amount of poison contained in one shellfish of a certain weight or in 100 gm of the edible portion is calculated. In carrying out the assay for routine checking of shellfish, three mice are sufficient to make a reliable test for the presence or absence of poison. To comply with the official method and obtain a quantitative measure of the amount of poison present, 10 mice are used with the final dilution. The median death time of a group of mice, rather than the average, is usually taken as the value for calculations of the amount of poison in a sample, because an occasional mouse may not die. When the assay is carried out on dinoflagellates, a measured portion of the culture or seawater containing the organisms is filtered on a fast-flowing filter paper covered with about 2–3 mm of filtercell. The paper containing the cells is ground in a blender with sufficient water and hydrochloric acid to make a thick slurry at pH 2–3. This treatment lyses the cells and extracts the poison. This slurry is filtered with suction, and the filtrate is subjected to the assay exactly as described for extracts of shellfish. The number of G. *catenella* cells required to produce 1 MU of poison is about 20,000 to 30,000, and 1 MU is equivalent to 0.18 μg. The amount of poison produced by G. *tamarensis* is about the same. Because the poison produced by G. *tamarensis* is a heavier molecule, 1 MU is equivalent to about 0.23 μg.

Chemical assays have been worked out for saxitoxin (McFarren et al., 1959; Bates and Rapoport, 1975), but they are more or less complicated and are subject to inaccuracies due to impurities with structures and properties similar to those of the poison.

Another procedure used in some cases (but not for Alaska butter clams) to determine the possibility of mussels becoming toxic is the periodic examination of the water in the area around the mussel or clam beds to see if poisonous dinoflagellates are present. Organisms can be satisfactorily counted using a microscope slide for counting red blood cells or the slide used for the Howard mold count in tomato products. If only a few cells can be detected, usually harmless traces of poison will be found in the mussels. If counts of 200–500/ml or more are found in the waters surrounding the shellfish beds, the amount of poison in the shellfish might become dangerously high, and a check on the shellfish should be made. One advantage in examining the water for poisonous dinoflagellate is that one can predict at least a few days to a week in advance of the shellfish becoming dangerously poisonous, and warnings of the conditions can be posted or advertised. Proper identification of the dinoflagellate is most important in this case.

B. Tests for Other Dinoflagellate Poisons

Poison in oysters that have been feeding on *G. breve* is detected by making ether extracts of the oysters, evaporating the ether, and taking the residue up in vegetable oil. The oil solution is injected subcutaneously into mice weighing about 20 gm. A mouse unit of this poison is defined as the minimum amount necessary to kill 50% of a group of mice. Tests for this poison can also be made with goldfish by taking the residue up in water. Although the poison is not very soluble in water, enough dissolves to kill goldfish placed in the water.

The hepatotoxic shellfish poison from oysters in Japan can be detected by feeding or injecting water extracts of the oysters into mice and observing them for pathological changes in the liver and other tissues. Observations must be carried out over a period of several days to detect small quantities of the toxin.

IV. CHEMICAL NATURE OF DINOFLAGELLATE POISONS

Past developments leading up to our present knowledge of the chemical and physical nature of paralytic poisons include work dating back to 1778, when the poisonous effect was believed to be due to the accumulation of copper salts in shellfish. Over 100 years later (1885) it was discovered that the toxic effects were destroyed by ashing the shellfish, suggesting for the first time that the toxic effects were most likely due to an organic substance. Bacterial putrefaction of shellfish during the summer months also was presented as an explanation. In 1885 during a mass outbreak of poisoning in humans attributed to eating mussels near Wilhelmshaven, Germany, Brieger (1888) collected toxic mussels and isolated a poisonous substance in the form of a gold salt which he called "mylitotoxin" and believed was a quaternary ammonium base and the pure poison. His work, however, could not be substantiated by later investigators, and it was concluded that he had isolated a basic compound contaminated with the poison.

The first thorough study of the chemical nature of the poison was undertaken under the leadership of Hermann Sommer at the University of California and Byron Riegel at Northwestern University, Evanston, Illinois, and later by E. J. Schantz at the U.S. Army Biological Center, Frederick, Maryland, and the University of Wisconsin, Madison, Wisconsin. Several serious outbreaks of mussel poisoning near San Francisco in 1927 and the following years gave these investigators an opportunity to make a detailed study of the poisonous principle occurring in the California sea mussel (*Mytilus californius*). As mentioned above, Sommer and co-workers were the first to observe the relationship of the dinoflagellate *G. catenella* to the paralytic poison in mussels, which initiated interest in culturing the dinoflagellate as a source of the poison for chemical studies.

Securing the poison for chemical studies was a major undertaking for investigators in this field. California sea mussels (*M. californiaus*) that had become poisonus from the consumption of *G. catenella* could be collected only when the short toxicity period of 2–3 weeks coincided with a low tide period. Alaska butter clams (*Saxidomas giganteous*) were found to be poisonous in isolated areas of southeastern Alaska, but could be collected only at low tide periods when suitable weather conditions allowed small boats and clam diggers to reach these areas. One advantage of the butter clam is that about two-thirds of the poison found in its body is stored in the siphon and retained there for many months, in contrast to mussels that store the poison in the dark gland or hepatopancreas and excrete or destroy it within a week or two. Toxic scallops (*Pecten grandis*) from the Bay of Fundy were furnished to the laboratories at the University of Wisconsin as a gift of the Canadian Department of Health, Education and Welfare. The paralytic poison was also produced by culturing *G. catenella, G. tamarensis,* and *G. acatenella* in axenic culture at both the U.S. Army Biological Center and the University of Wisconsin. The axenic cultures of these organisms were obtained through the courtesy of Luigi Provasoli, Haskins Laboratories, New York, New York. Most of the poison for the chemical studies was isolated from butter clam siphons, some from the dark gland of sea mussels, and a small amount by laboratory culture of the dinoflagellates.

Isolation and purification of the poison from clam and mussel tissues and from the dinoflagellates were difficult, because of the extreme solubility of the poison in water and its insolubility in organic solvents immiscible with water. The poison was extracted from the tissues by grinding and mixing with Celite 545 and extracting with water at about pH 2 with hydrochloric or sulfuric acid. These extracts usually contained 2–5 MU of poison per milligram of dissolved solids which were purified by chromatography on carboxylic acid resins (Amberlite XE-64 or CG-50) followed by chromatography on acid-washed alumina in absolute ethanol (Schantz *et al.,* 1957). The best fractions had a specific toxicity of 5500 MU/mg of solids with a recovery of 60% of the poison in the original extract. Further chromatography failed to improve the specific toxicity. *Gonyaulax catenella* cells were filtered from the culture and processed in the same manner with acid (about pH 2) to lyse the cells and to liberate the poison bound in the cell. Most of the poison in the scallops and in the cultures of *G. tamarensis* cells did not absorb on the carboxylic acid resins well enough to be effectively purified in this manner, and other methods had to be employed. *Gonyaulax tamarensis* produces a mixture of poisons of which 10–15% is saxitoxin and the remainder is several neutral or slightly basic substances (Schantz, 1960; Oshima *et al* 1977). Because of the basic nature of saxitoxin, it is easily separated from the other poisons on carboxylic acid resins.

The paralytic poisons from California sea mussels and Alaska butter clams and from axenic cultures of the Pacific coast dinoflagellate *G. catenella* proved to be identical substances (Schantz *et al.*, 1966). They are extremely water soluble, to some extent in methanol and ethanol, and insoluble in most organic solvents such as ethyl and petroleum ether and chloroform. They are white solids and have a specific toxicity of 5500 MU/mg of solids (Schantz *et al.*, 1958). They contain about 26% nitrogen, have two titratable groups, and have pK_a values of 11.5 and 8.2. The molecular formula is $C_{10}H_{17}N_7O_4 \cdot 2HCl$, and the molecular weight is 372 (Mold *et al.*, 1957; Schantz *et al.*, 1961). These poisons react with the Benedict–Behre reagent (trinitrobenzoic acid) and the Jaffe reagent (trinitrophenol) to produce blue and orange-red derivatives, respectively (color reagents for creatinine), that are approximately equivalent on a molar basis to the colors produced when creatinine reacts with these reagents. Reduction of the poisons with hydrogen in the presence of a catalyst results in the uptake of 1 mole of hydrogen per mole of poison at atmospheric pressure and room temperature. This reduction destroys the toxicity and also eliminates the color reaction with the Benedict–Behre and Jaffe reagents. The poisons exist in two tautomeric forms; one is highly toxic and the other is of low toxicity, if any. When isolated separately and allowed to stand in acid solution, the tautomeric forms are reestablished. Reduction with hydrogen also eliminates the tautomeric forms. The properties of these poisons are listed in Table II.

Work on the chemical structure of the poisons was first undertaken by investigators at the University of California Berkeley, California, who published a structure, although not correct, establishing the poison as having a tetrahydropurine nucleus (Wong *et al.*, 1971). Establishment of the structure was difficult, because a suitable crystalline derivative for a crystallographic study was not available. Because of certain discrepancies between the published structure and certain chemical and physical properties of the poison, work was undertaken at the University of Wisconsin to obtain a suitable crystalline derivative. Such a crystalline derivative was obtained by reacting the poison with *p*-bromobenzenesulfonic acid and, in cooperation with Jon Clardy at Iowa State University, Ames, Iowa, the structure (1) was established as illustrated Figure 1 (Schantz

Table II Comparison of the Properties of the Paralytic Poisons Produced by Dinoflagellates from Various Sources

Property	Source of poison				
	Butter clam[a]	Sea mussel[a]	G. catenella[a]	Scallop[b]	Reduced poison[c]
Bioassay (MU/mg)[d]	5200	5300	5100	2800	0
Optical rotation	+128	+130	+128		+128
Diffusion coefficient	4.9×10^{-6}	4.9×10^{-6}	4.8×10^{-6}		
Absorption, ultraviolet and visible	None	None	None	None	None
Kjeldahl nitrogen (%)	26.8	26.1	26.3		26.2
Sakaguchi test[e]	Negative	Negative	Negative		Negative
Benedict–Behre test[e]	Positive	Positive	Positive		Negative
Jaffe test[e]	Positive	Positive	Positive		Negative
Molecular weight (dihydrochloride salt)	372	372	372	467	354
Tautomeric forms	2	2	2	2	1

[a] The poison from these sources is saxitoxin.

[b] The poison from this source is 11-hydroxysaxitoxin sulfate.

[c] Saxitoxin reduced as described by Schantz et al. (1961).

[d] Bioassay values are within experimental error of the average value 5500 ± 500 MU/mg (Schantz et al., 1958).

[e] Tests carried out as described by Mold et al. (1957).

and Clardy et al., 1975; Ghazarossian, 1977). Later the California investigators confirmed this structure with a crystalline ethyl hemiketal derivative of saxitoxin (Bordner et al., 1975).

Further studies at the University of Wisconsin showed that the carbamyl group of saxitoxin could be removed by treating with 7.5 N HCl at 100°C for 3 hr, leaving a hydroxyl group on carbon 13 in this position as illustrated by structure (2) in Figure 1 (Ghazarossian et al., 1976). This derivative, called decarbamylsaxitoxin, possesses about two-thirds of the toxicity of saxitoxin and makes possible, through reaction with this hydroxyl group, the preparation of valuable derivatives of saxitoxin such as compounds possessing radioactivity. The catalytic reduction of saxitoxin with hydrogen and with sodium borohydride takes place at carbon 10 and produces a nontoxic derivative, as illustrated by structure (3) in Figure 1.

As mentioned previously, the poisons produced by the East Coast dinoflagellate, G. tamarensis, constitute a mixture of one, accounting for about 80% of the total toxicity, that is a weak base and others that are stronger bases such as saxitoxin or compounds closely related to saxitoxin. All of these poisons adsorb on the sodium form of carboxylic

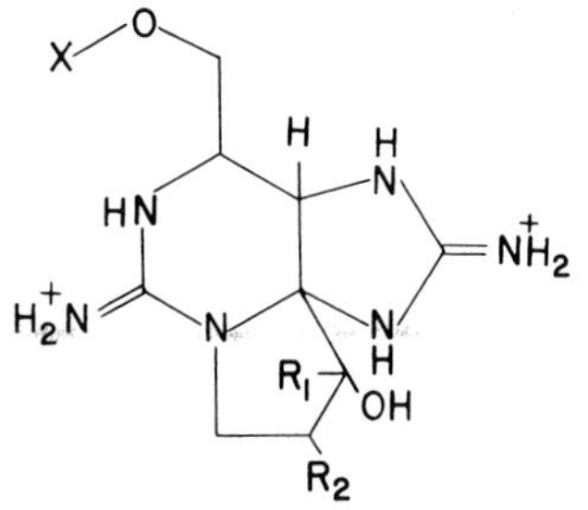

Figure 1. Basic structure of saxitoxin and some related substances. (Reprinted with permission from Schantz et al., 1975, J. Amer. Chem. Soc. **97,** 1238–1239. Copyright by the American Chemical Society.)

(1) Saxitoxin (from G. catenella); if X—$\overset{O}{\overset{\|}{C}}$—NH$_2$, R$_1$ is OH, and R$_2$ is H.

(2) Decarbamyl saxitoxin; if X is H, R$_1$ is OH, and R$_2$ is H.

(3) Reduced saxitoxin (saxitoxinol); if X is—$\overset{O}{\overset{\|}{C}}$—NH$_2$, R$_1$ is H, and R$_2$ is H.

(4) Gonyautoxin (from G. tamarensis); if X is —$\overset{O}{\overset{\|}{C}}$—NH$_2$, R$_1$ is OH, and R$_2$ is OH.

(5) 11-Hydroxysaxitoxin sulfate (from G. tamarensis); if X is

—$\overset{O}{\overset{\|}{C}}$—NH$_2$, R$_1$ is OH; and R$_2$ is OSO$_3^-$.

acid exchange resins such as Amberlite CG-50. The weakly basic poison is easily eluted at neutral or slightly acid pH and the others are eluted as the pH is lowered. Saxitoxin is eluted at about pH 2.5. The University of Wisconsin investigators have isolated the weakly basic poison from Bay of Fundy Scallops and from cultured G. tamarensis and found that the structure of this poison, as indicated by NMR spectra, chemical analyses, and chemical properties is 11-hydroxysaxitoxin sulfate as shown by structure **(5)** in Figure 1 (Boyer et al., 1978). The strongly basic character of saxitoxin and some of its derivatives is due to the guanidinium groups with pK$_a$ values at 11.5 and 8.2. These same guanidinium groups are part of the structure of the weak base from G. tamarensis but, for the most part, are neutralized by the acidic character of the sulfate group at position 11 as shown in structure **(5)** (Figure 1). Chemical analyses show that the molecule contains only one atom of sulfur and that the sulfate ester is on carbon 11, as indicated by the fact that the poison reacts with periodate only after the sulfate is hydrolyzed from the rest of the molecule. Investigators at the University of Rhode Island, Kingston, Rhode Island, have purified some poisons from the East Coast clam Mya arenaria and cultured

G. tamarensis. One of these which was found in small amounts is 11-hydroxysaxitoxin, as shown in structure (4) in Figure 1 (Shimizu *et al.,* 1976). The chemical structures of the other poisons produced by *G. tamarensis* have not been worked out.

V. PUBLIC HEALTH ASPECTS OF DINOFLAGELLATE POISONS AND THEIR CONTROL

Many marine toxins involved in cases of food poisoning are produced by dinoflagellates and reach edible marine animals consumed by humans through the food chain. These particular poisons are difficult to control from the standpoint of public safety, because of the unpredictable and sporadic occurrence of the organisms producing them. Also, they may cause any one of a number of species of fish consuming the organism to become poisonous. Another important factor regarding safety is that these poisons usually are quite stable to heat processing or cooking and are refractory to the action of the digestive enzymes of humans.

In terms of the number of cases of sickness and death resulting from poisons produced by dinoflagellates throughout the nation and the world, the public health problem is very small. However, in some local areas where considerable amounts of shellfish are used as food, the problem is very important and could become more widespread now that frozen seafoods are shipped to various parts of the world. State and federal food and drug administration authorities check the poison content of commercial fish and shellfish throughout the United States.

Many governmental agencies also carry out assays periodically from May to October on clam and mussel beds to check for poison where shellfish are collected for food. If shellfish become dangerously poisonous, warnings are posted and publicized. The most practical means of controlling shellfish poisoning is by direct sampling and assaying of shellfish in areas where they are harvested commercially and where picnickers commonly collect them for food. Education of the public regarding the danger of the sporadic occurrence of poison and its cause is very important, especially in areas where shellfish poisoning is common. Regulation of the commercial harvesting and processing of shellfish is described by the U.S. Public Health Service (1959). The U.S. Food and Drug Administration has set the maximum acceptable level for paralytic poison in fresh frozen or canned shellfish at no more than 400 MU, or about 80 μg/100 gm of edible portion. This amount or less in seafood has not been known to cause sickness.

REFERENCES

Abbott, B. C., and Ballantine, D. (1957). *J. Mar. Biol. Assoc. U.K.* **36**, 169–189.

Association Official Analytical Chemists (1975). "Methods of Analyses," 12th Ed., pp. 319–321. Washington, D.C.

Bates, H. A., and Rapoport, H. (1975). *J. Agric. Food Chem.* **23**, 237–239.

Bond, R. M., and Medcof, J. C. (1958). *Can. Med. Assoc. J.* **79**, 19–24.

Bordner, J., Thiessen, W. E., Bates, H. A., and Rapoport, H. (1975). *J. Am. Chem. Soc.* **97**, 6008–6012.

Boyer, G. L., Schantz, E. J., and Schnoes, H. K. (1978). *J. Chem. Soc. London Chem. Comm.*, pp. 889–890.

Brieger, L. (1888). *Arch. Pathol. Anat. Physiol. Klin. Med.* **112**, 549.

Burke, J. M., Marchisotto, J., McLaughlin, J. J. A., and Provasoli, L. (1960). *Ann. N.Y. Acad. Sci.* **90**, 837–842.

Connell, C. H., and Cross, J. B. (1950). *Science* **112**, 359–363.

Evans, M. H. (1964). *Brit. J. Pharmacol.* **22**, 478–485.

Gates, J. A., and Wilson, W. B. (1960). *Limnol. Oceanogr.* **5**, 171.

Ghazarossian, V. E. (1977). Ph.D. Thesis, Univ. of Wisconsin, Madison.

Ghazarossian, V. E., Schantz, E. J., Schnoes, H. K., and Strong, F. M. (1976). *Biochem. Biophys. Res. Commun.* **68**, 776–780.

Halstead, B. W. (1965). "Poisonous and Venomous Marine Animals of the World," Vol. 1, pp. 1–278. U.S. Gov. Print. Off., Washington, D.C.

Jackim, E., and Gentile, J. (1968). *Science* **162**, 915.

Kao, C. Y., and Nishiyama, A. (1965). *J. Physiol. (London)* **180**, 50–66.

Koch, H. J. (1939). *Assoc. Fr. Av. Sci., 63rd Sess., Paris* p. 654.

Konosu, S., Inone, A., Noguchi, T., and Hashimoto, Y. (1968). *Toxicon* **6**, 113–117.

Kutner, M. B., and Sassi, R. (1978). *In* "Toxic Dinoflagellate Blooms" (D. L. Taylor and H. H. Seliger, eds.), pp. 169–172. Mass. Sci. Technol. Found., Boston, Massachusetts.

Loeblich, L. A., and Loeblich, A. R., III (1975). *In* "Toxic Dinoflagellates Blooms" (V. R. LoCicero, ed.), pp. 207–224. Mass. Sci. Technol. Found., Boston, Massachusetts.

Loeblich, A. R., III, and Loeblich, L. A. (1978). *In* "Toxic Dinoflagellate Blooms" (D. L. Taylor and H. H. Seliger, eds.), pp. 83–88. Mass. Sci. Technol. Found., Boston, Massachusetts.

McFarren, E. F., Schantz, E. J., Campbell, J. E., and Lewis, K. H. (1959). *J. Assoc. Off. Anal. Chem.* **42**, 399–404.

McFarren, E. F., Tanabe, H., Silva, F. J., Wilson, W. B., Campbell, J. E., and Lewis, K. H. (1965). *Toxicon* **3**, 111–123.

Martin, D. F., and Chatterjee, A. B. (1969). *Nature (London)* **221**, 59.

Meyer, K. F. (1953). *N. Engl. J. Med.* **249**, 848.

Mold, J. D., Bowden, J. P., Stanger, D. W., Maurer, J. E., Lynch, J. M., Wyler, R. S., Schantz, E. J., and Riegel, B. (1957). *J. Am. Chem. Soc.* **79**, 5235–5238.

Nakazima, M. (1968). *Nippon Suisan Gakkaishi* **34**, 130.

Narahashi, T., Brodwick, M. S., and Schantz, E. J. (1975). *Environ. Lett.* **9**, 239–247.

Needler, A. B. (1949). *J. Fish. Res. Board Can.* **7**, 490–504.

Noguchi, T., Konosu, S., and Hashimoto, Y. (1969). *Toxicon* **7**, 325.

Okaichi, T., and Imatomi, Y. (1978). *In* "Toxic Dinoflagellate Blooms" (D. L. Taylor and H. H. Seliger, eds.), pp. 385–388. Mass. Sci. Technol. Found., Boston, Massachusetts.

Oshima, Y., Buckley, L., Alam, M., and Shimizu, Y. (1977). *Comp. Biochem. Physiol.* **57C**, 31–34.

Prakash, A. (1963). *J. Fish. Res. Board Can.* **20**, 983–996.

Prakash, A. (1967). *J. Fish. Res. Board Can.* **24**, 1589.

Prakash, A., and Taylor, F. J. R. (1966). *J. Fish. Res. Board Can.* **23**, 1265–1270.

Reyes-Vasquez, G., and Ferraz-Reyes, E. (1970). *In* "Toxic Dinoflagellate Blooms" (D. L. Taylor and H. H. Seliger, eds.), pp. 191–194. Mass. Sci. Technol. Found., Boston, Massachusetts.

Ray, S. M., and Aldrich, D. V. (1965). *Science* **148**, 1748–1749.

Ray, S. M., and Aldrich, D. V. (1967). *In* "Animal Toxins" (F. E. Russel and P. R. Saunders, eds.), pp. 75–83. Pergamon, Oxford.

Sawyer, P. J., Gentile, J. H., and Sasner, J. J., Jr. (1968). *Can. J. Microbiol.* **14**, 1199.

Schantz, E. J. (1960). *Ann. N.Y. Acad. Sci.* **90**, 843–855.

Schantz, E. J., and Magnusson, H. W. (1964). *J. Protozool.* **11**, 239–242.

Schantz, E. J., Mold, J. D., Stanger, D. W., Shavel, J., Riel, F. J., Bowden, J. P., Lynch, J. M., Wyler, R. S., Riegel, B., and Sommer, H. (1957). *J. Am. Chem. Soc.* **79**, 5230–5235.

Schantz, E. J., McFarren, E. F., Schafer, M. L., and Lewis, K. H. (1958). *J. Assoc. Off. Anal. Chem.* **41**, 160–177.

Schantz, E. J., Mold, J. D., Howard, W. H., Bowden, J. P., Stanger, D. W., Lynch, J. M., Wintersteiner, O. P., Dutcher, J. D., Walters, D. R., and Riegel, B. (1961). *Can. J. Chem.* **39**, 2117–2123.

Schantz, E. J., Lynch, J. M., Vzyvzda, G., Matsumoto, K., and Rapoport, H. (1966). *Biochemistry* **5**, 1191–1195.

Schantz, E. J., Ghazarossian, V. E., Schones, H. K., Strong, F. M., Springer, J. P., Pezzanite, J. O., and Clardy, J. (1975). *J. Am. Chem. Soc.* **97**, 1238–1239.

Schradie, J., and Bliss, C. A. (1962). *Lloydia*, **25**, 214–221.

Schurer, P. J. (1977). *Acc. Chem. Res.* **10**, 33–39.

Shimizu, Y., Buckley, L. J., Alam, M., Oshima, Y., Fallon, W. E., Kasai, H., Miura, I., Gullo, V. P., and Nakanishi, K. (1976). *J. Am. Chem. Soc.* **98**, 5414–5416.

Sommer, H., and Meyer, K. F. (1937). *AMA Arch. Pathol.* **24**, 560–598.

Sommer, H., Whedon, W. F., Kofoid, C. A., and Stohler, R. (1937). *AMA Arch. Pathol.* **24**, 537–559.

Spikes, J. J., Ray, S. M., Aldrich, D. V., and Nash, J. B. (1968). *Toxicon* **5**, 171.

Starr, T. J. (1958). *Tex. Rep. Biol. Med.* **16**, 500–507.

Tennant, A. P., Naubert, J., and Corbeil, H. E. (1955). *Can. Med. J.* **72**, 436.

U.S. Public Health Service (1959). "Manual of Recommended Practice for Sanitary Control of the Shellfish Industry." U.S. Dep. Health, Educ. Welfare, Washington, D.C.

Wong, J. L., Oesterlin, R., and Rapoport, H. (1971). *J. Am. Chem. Soc.* **93**, 7344–7345.

Yasumoto, T., Nakajima, I., Oshima, Y., and Bagnis, R. (1978). *In* "Toxic Dinoflagellate Blooms" (D. L. Taylor and H. H. Seliger, eds.), pp. 65–70. Mass. Sci. Technol. Found., Boston, Massachusetts.

Author Index

Numbers in italics refer to the pages on which the complete references are listed.

Subject Index